The International Library of Environmental, Agricultural and Food Ethics

Volume 23

Series editors

Michiel Korthals, Wageningen, The Netherlands
Paul B. Thompson, Michigan, USA

The ethics of food and agriculture is confronted with enormous challenges. Scientific developments in the food sciences promise to be dramatic; the concept of life sciences, that comprises the integral connection between the biological sciences, the medical sciences and the agricultural sciences, got a broad start with the genetic revolution. In the mean time, society, i.e., consumers, producers, farmers, policymakers, etc, raised lots of intriguing questions about the implications and presuppositions of this revolution, taking into account not only scientific developments, but societal as well. If so many things with respect to food and our food diet will change, will our food still be safe? Will it be produced under animal friendly conditions of husbandry and what will our definition of animal welfare be under these conditions? Will food production be sustainable and environmentally healthy? Will production consider the interest of the worst off and the small farmers? How will globalisation and liberalization of markets influence local and regional food production and consumption patterns? How will all these developments influence the rural areas and what values and policies are ethically sound?

All these questions raise fundamental and broad ethical issues and require enormous ethical theorizing to be approached fruitfully. Ethical reflection on criteria of animal welfare, sustainability, liveability of the rural areas, biotechnology, policies and all the interconnections is inevitable.

Library of Environmental, Agricultural and Food Ethics contributes to a sound, pluralistic and argumentative food and agricultural ethics. It brings together the most important and relevant voices in the field; by providing a platform for theoretical and practical contributors with respect to research and education on all levels.

More information about this series at http://www.springer.com/series/6215

Bernice Bovenkerk · Jozef Keulartz
Editors

Animal Ethics in the Age of Humans

Blurring Boundaries in Human-animal Relationships

Springer

Editors
Bernice Bovenkerk
University of Wageningen
Wageningen
The Netherlands

Jozef Keulartz
Radboud University Nijmegen
Nijmegen
The Netherlands

ISSN 1570-3010 ISSN 2215-1737 (electronic)
The International Library of Environmental, Agricultural and Food Ethics
ISBN 978-3-319-44205-1 ISBN 978-3-319-44206-8 (eBook)
DOI 10.1007/978-3-319-44206-8

Library of Congress Control Number: 2016947764

Printed on acid-free paper

This Springer imprint is published by Springer Nature
The registered company is Springer International Publishing AG
The registered company address is: Gewerbestrasse 11, 6330 Cham, Switzerland

Acknowledgments

This edited volume has benefitted considerably from the workshop that was held in Wageningen in November 2015, which was made possible by an innovative research grant from the Netherlands Organization for Scientific Research (OZSW), titled 'The Ethics of Domestication' (projectnr 275-20-038) and by the Wageningen School of Social Sciences (WASS). We would like to thank all the authors of this book for their thought-provoking contributions, interesting discussions, and collaborative spirit. We would also like to thank other participants and organizers of this workshop, in particular Phil MacNaghten and Inge Ruisch. We would furthermore like to express our gratitude to the editorial team of Springer, especially in the persons of Paul Thompson, Floor Oosting, and Christopher Wilby, for their flexibility and efficiency. Finally, we are grateful to our anonymous reviewers for their helpful comments.

Contents

Editors and Contributors

About the Editors

Bernice Bovenkerk is Assistant Professor of Philosophy at Wageningen University, the Netherlands. Her research and teaching deal with the issues in animal and environmental ethics, the ethics of climate change, and political philosophy. Current topics are the moral status of animals and other natural entities, with a particular focus on fish and insects, the ethics of animal domestication, animal (dis)enhancement, and deliberative democracy. She received her Ph.D. from the University of Melbourne, Australia, on a dissertation titled *The Biotechnology Debate. Democracy in the face of intractable disagreement*. She received her Master's degree from the University of Amsterdam on a thesis entitled *Pluralism in Environmental Ethics*. Her homepage is http://www.bernicebovenkerk.com.

Jozef Keulartz is Emeritus Professor of Environmental Philosophy at the Radboud University Nijmegen, and senior researcher Applied Philosophy at Wageningen University and Research Centre. He has published extensively in different areas of science and technology studies, social and political philosophy, bioethics, environmental ethics and nature policy. His books include Die verkehrte Welt des Jürgen Habermas [The Topsy-Turvy World of Jürgen Habermas, 1995] and Struggle for Nature—A Critique of Radical Ecology (1998). He is coeditor of Pragmatist Ethics for a Technological Culture (Kluwer, 2002), Legitimacy in European Nature Conservation Policy (Springer, 2008), New Visions of Nature (Springer, 2009), Environmental Aesthetics. Crossing Divides and Breaking Ground (Fordham University Press, 2014), and Old World and New World Perspectives in Environmental Philosophy (Springer, 2014).

Contributors

Stephanie Barr is a Ph.D. candidate in the School of Environment, Resources and Sustainability at the University of Waterloo, Canada. Among other degrees, she holds a Bachelor of Science in Biology from Queen's University and a Master's in Climate Change from the University of Waterloo. Stephanie has worked for both the provincial and the federal governments in climate-change policy related positions. Her doctoral work will investigate how increased exposure to climate change impacts a person's views of biodiversity conservation and whether Canadians think we should maintain ecosystems in their historic state through conventional conservation actions or use more interventionist conservation strategies.

Marc Bekoff is Professor Emeritus of Ecology and Evolutionary Biology at the University of Colorado, Boulder and co-founder with Jane Goodall of Ethologists for the Ethical Treatment of Animals. He has won many awards for his scientific research including the Exemplar Award (Animal Behavior Society) and a Guggenheim Fellowship, and has published more than 1000 essays (popular, scientific, and book chapters), 30 books, and edited three encyclopedias. His books include *The Ten Trusts* (with Jane Goodall), *The Emotional Lives of Animals*, *Ignoring Nature No More: The Case For Compassionate Conservation*, *Why Dogs Hump and Bees Get Depressed,* and *Rewilding Our Hearts: Building Pathways of Compassion and Coexistence.* In 2009 Marc was given the St. Francis of Assisi Award by the New Zealand SPCA. In 1986 Marc won his age-class at the Tour du Haut Var bicycle race (also called the Master's/age-graded Tour de France). His homepage is http://www.marcbekoff.com.

Henk van den Belt is Assistant Professor of Philosophy at Wageningen University, the Netherlands. His expertise includes science and technology studies, the ethics and politics of science and innovation, and the role of intellectual property in the life sciences. He is also interested in controversies in fields like biotechnology, synthetic biology and climate science.

Susan Boonman-Berson is a Ph.D. Candidate at Wageningen University, the Netherlands. Her research explores wildlife management practices and particularly how to deal with human–wildlife conflict situations. In her study she combines different perceptions of wildlife management from research, policy, in-the-field wildlife management and local communities and explores how to interpret human–animal relations in order to understand the current dilemmas in managing human–wild animal interactions. Therefore she studied human–wild boar management in the Netherlands and human–black bear management on the Colorado Front Range, USA. Her research interests further include topics such as human–wildlife conflict situations, interpreting human–animal relations, and the transformations occurring at each step from data collection in the field to data processing in the office, and the implications of all of these for conservation practices.

Emily Brady is Professor of Environment and Philosophy in the Institute of Geography and an Academic Associate in Philosophy at the University of Edinburgh. Her research interests span aesthetics, environmental ethics, Kant, and eighteenth-century philosophy. Within animal studies, she is interested in issues such as autonomy, respect, sentimentality, and difference in our aesthetic engagement with animals. She has written articles on the place of animals in ecological art and the role of expressive qualities in aesthetic engagement with wild animals. As an author she has published several books in aesthetics and environmental philosophy: *The Sublime in Modern Philosophy: Aesthetics, Ethics, and Nature* (2013) and *Aesthetics of the Natural Environment* (2003); and as coeditor: *Aesthetic Concepts: Essays After Sibley*; *Humans in the Land: The Ethics and Aesthetics of the Cultural Landscape* (2007); and *Human-Environment Relations: Transformative Values in Theory and Practice* (2012).

Sabrina Brando has a BSc in Psychology from California Coast University, and is currently completing an MSc in Animals Studies with Humane Society University and an MSc in Psychology with California Coast University. She has authored and coauthored peer-reviewed articles and contributed chapters to three books: The Biology of the Harbour Porpoise, (De Spil, 1997), NAMMCO Scientific Publications Volume 5 (The North Atlantic Marine Mammal Commission, 2003), and Lexikon der Mensch-Tier-Beziehungen (transcript Verlag, Bielefeld, 2015). Currently she is working on an e-book on animal welfare and training, scheduled for publication in late 2016. Sabrina owns the animal welfare consulting company AnimalConcepts, working on animal advocacy and animal well-being projects and research world-wide. Under Sabrina's leadership AnimalConcepts has organized over a 100 conferences and workshops since 2004 on a wide variety of topics.

Simon Burton originally studied philosophy as an undergraduate and postgraduate, before taking up a career as a specialist in company taxation. His current areas of focus from a philosophical perspective are philosophy of science and interpretations of transcendental idealism after Kant and Schopenhauer. Developing his early interest in the countryside, he has undertaken ornithological and botanical survey work for over 25 years, principally for the British Trust for Ornithology. He is a qualified bird ringer having ringed over 100 different species of bird in a range of habitats. His particular interests as a birder and ringer are in relation to common birds and how their life cycles interact with humans and other organisms.

Sue Donaldson is a research fellow in the Department of Philosophy, Queen's University, Canada. She is coauthor, with Will Kymlicka, of Zoopolis: A Political Theory of Animal Rights (OUP 2011), and numerous articles on interspecies justice appearing in journals such as the Oxford Journal of Legal Studies, the Journal of Political Philosophy, the Journal of Social Philosophy, LEAP, Politics and Animals, and the Canadian Journal of Political Science. She is a founding member of Queen's Animal Defence, an animal advocacy organization, and co-founder of the Animals in Philosophy, Politics, Law and Ethics (APPLE) research cluster at Queen's University.

Martin Drenthen is Associate Professor of Philosophy at the Institute for Science, Innovation & Society (ISIS) and coordinator of the ISIS Center for Nature and Society at Radboud University (Nijmegen, the Netherlands). He has published about the significance of Nietzsche's critique of morality for environmental ethics, the concept of wildness in moral debates on ecological restoration, ethics of place, and environmental hermeneutics. He is coeditor of New Visions of Nature. Complexity and Authenticity (Springer 2009), Interpreting Nature. The Emerging Field of Environmental Hermeneutics (Fordham University Press, 2013), and Environmental Aesthetics. Crossing Divides and Breaking Ground (Fordham University Press, 2014). He recently coauthored a popular science book about the wolf's return to the Netherlands (De wolf terug, Kosmos 2015, in Dutch). Currently, he leads a research project on a hermeneutic landscape ethics, which focuses on the relation between ecological restoration and rewilding landscapes, cultures of place, and moral identity.

Clemens Driessen is Assistant Professor in the Cultural Geography group at Wageningen University, the Netherlands. His research focuses on human–animal relations on farms and in the wild. Using a range of methods including ethnography, historiography, and creative design interventions he seeks new ways to explore questions in environmental philosophy and animal ethics. He has written on 'animal deliberation', on rewilding and backbreeding of wild cows, on milking robots, high rise pig farms, in vitro meat, fish welfare, and on his efforts together with game designers to create video games for bored pigs.

Anita Guerrini is Horning Professor in the Humanities and Professor of History at Oregon State University, where she teaches the history of science and medicine. She has written on the history of animals, medicine, food, and the environment. Her publications include *Experimenting with Humans and Animals: from Galen to Animal Rights* (John Hopkins, 2003) and most recently *The Courtiers' Anatomists: Animals and Humans in Louis XIV's Paris* (Chicago, 2015). Her article "Animals and Ecological Science" appeared in the *Oxford Handbook of Animal Studies* (2015). Current research projects concern skeletons as scientific and historical objects, early modern urban animals, the role of history in present-day ecological restoration, and the importance of historical data in the NSF-funded Long Term Ecological Research network in the US. She blogs at http://anitaguerrini.com/anatomia-animalia/

T.J. Kasperbauer is a postdoc in the Department of Food and Resource Economics at the University of Copenhagen. He received a Ph.D. in Philosophy and M.S. in Psychology from Texas A&M University, where he wrote a dissertation on moral psychology and animal ethics. His main areas of research are animal and environmental ethics, moral psychology, and philosophy of psychology.

Michiel Korthals studied philosophy, sociology, and German at the University of Amsterdam and the Ruprecht Karl University of Heidelberg (Germany). Korthals is interested from early on in communication and democracy, both theoretically and practically. At Wageningen University from 1993 onwards, Korthals contributed to the development of a new discipline, food and agricultural philosophy and ethics, by showing that agriculture and food are not only means to survive but also ethical and cultural factors in understanding of and giving meaning to human life and life in general. Some of her important books include the following: Filosofie en Intersubjectiviteit, Alphen a/d Rijn, 1983; Duurzaamheid en democratie (Boom, 1995); Philosophy of Development (Kluwer, 1996, with Wouter van Haaften and Thomas Wren); Between food and medicine (Utrecht, 2001); Pragmatist Ethics for a Technological Culture (with Keulartz et. al.; Kluwer, 2002); Voor het eten (Boom, 2002); Ethics for Life Sciences (Springer, 2005); Before Dinner. Philosophy and Ethics of Food (Springer, 2004); and Genomics, Obesity and the Struggle over Responsibilities (2011, Springer). Michiel Korthals retired in May 2014.

Will Kymlicka is the Canada Research Chair in Political Philosophy in the Philosophy Department at Queen's University in Kingston, Canada, where he has taught since 1998. His research interests focus on issues of democracy and diversity, and in particular on models of citizenship and social justice within multicultural and interspecies societies. He is the coauthor with Sue Donaldson of Zoopolis: A Political Theory of Animal Rights (Oxford University Press, 2011).

Brendon M.H. Larson is Associate Professor in the School of Environment, Resources and Sustainability at the University of Waterloo, Canada. His research concerns the ways that conservation is adapting to the diverse, concurrent, and human-caused ecological changes that Earth is experiencing. He has published widely, including an award-winning book, Metaphors for Environmental Sustainability: Redefining Our Relationship with Nature (Yale University Press, 2011). His current research focuses on stakeholder perceptions of assisted colonization and invasive species, and it is supported by the Social Sciences and Humanities Research Council of Canada (SSHRC) and an Early Researcher Award from the Ontario Ministry of Research and Innovation. For more information, please visit his website, http://www.brendonlarson.com.

Jeffrey Moussaieff Masson is author of some 29 books, mostly about animals, but also about Freud, and false gurus. He is now living in Sydney with his wife, a pediatrician and 14-year-old son. Masson is vegan and is writing a novel about the Holocaust called Evian, 1938. He was Professor of Sanskrit, was a Freudian analyst and is now a full-time writer.

Susan McCarthy is a science writer and a humor writer. She is coauthor of *When Elephants Weep* and the author of *Becoming a Tiger: How Baby Animals Learn to Live in the Wild*, and co-creator of *SorryWatch*, a website which analyzes apologies. Following a stay on the Western Antarctic Peninsula, she's currently working on books about Antarctic birds. She lives with her family in San Francisco, California.

Eva Meijer is currently working on a Ph.D. project in philosophy at the University of Amsterdam, titled 'Political Animal Voices', in which she develops a theory of political animal voice. She teaches the course 'Animal Ethics and Politics' at the

University of Amsterdam and is the chair of the Dutch study group for Animal Ethics, as well as a founding member of Minding Animals, the Netherlands. In addition to her academic work, Meijer works as a novelist, visual artist, and singer-songwriter.

Clare Palmer is Professor of Philosophy at Texas A&M University, USA. She is the author of Animal Ethics in Context (Columbia University Press 2010) and coauthor of Companion Animal Ethics (Wiley-Blackwell 2015). She has edited or coedited a number of collections including Linking Ecology and Ethics for a Changing World (Springer, 2014) and Killing Animals, coedited with the Animal Studies Group (Champaign-Urbana: Illinois University Press 2006) She was the founding editor of the journal Worldviews: Environment, Culture, Religion (Brill Academic Press) and has held the position of President of the International Society for Environmental Ethics from 2007 to 2010. She serves on the editorial board of a number of journals, including Environmental Values, the Journal of Agricultural and Environmental Ethics, and Environmental Humanities.

Daniel Ramp is the Director of the Centre for Compassionate Conservation at UTS and oversees its research and teaching programs across five faculties. He is a conservation biologist who has written on landscape ecology, behavioral ecology, road ecology, human–wildlife conflict, and ethics in decision-making. He is currently a Senior Lecturer in the School of Life Sciences at UTS and previously was an ARC Postdoctoral Fellow and then Senior Research Fellow in the Centre for Ecosystem Science at the University of New South Wales. The kangaroos are his greatest passion and he did his doctoral research on individual decision-making in eastern gray kangaroos at the University of Melbourne.

Aaron Simmons is Assistant Professor of Philosophy at Marywood University in Scranton, Pennsylvania, USA. He teaches courses on biomedical ethics, animal ethics, environmental ethics, and the philosophy of death and dying. His published writings include journal articles on the moral significance of empathy, animal rights theory, the harm of death, and the ethics of medical research on human embryos.

Jac. A.A. Swart is Associate Professor at the Science and Society Group (SSG) of the Energy and Sustainability Research Institute Groningen (ESRIG) of the University of Groningen. He studied biology as his major topic and geology and philosophy as minor topics and he specialized on mathematical modeling of biological processes, especially in the field of ecology. Subsequently he received his Ph.D. working on mathematical modeling of brain receptor dynamics. Thereafter he moved to the field of science and technology studies with respect to the natural and life sciences. In this field he is especially interested in ethics of life sciences and innovation. He published on ethical and social aspects of nature conservation and restoration, animal experimentation, ethics of wild animals management, and biotechnology.

Sanne van der Hout is a postdoctoral researcher in biomedical ethics at Maastricht University. She studied philosophy at Leiden University and applied ethics at Utrecht University and the Norwegian University of Science and Technology. In October 2014, Sanne defended her Ph.D. thesis at Radboud University Nijmegen, in which she critically reflects on the role of the modern life sciences in contributing to a more sustainable relationship between humans, technology and nature. Sanne is a team member of the Dutch Biomimicry foundation. She contributes to the further development and articulation of the ethos of biomimicry.

Marcel Verweij is Professor of Philosophy and was appointed as the Chair of Philosophy at Wageningen University in 2013. His research and teaching cover a broad area in applied ethics, including care for animal ethics and business ethics, but his primary focus is public health. His current projects include the ethics of zoonosis control and antimicrobial resistance, donor blood safety, health inequities, nudging, and corporate social responsibility in the food industry. Together with Angus Dawson he initiated public health ethics as a new area of studies in applied moral philosophy. They are founding editors-in-chief of *Public Health Ethics*. Verweij is member of the Health Council of the Netherlands and has been involved in more than 20 advisory reports for the Minister of Health, including reports on collective vaccination, screening, preconception counseling, special policies for high-risk groups, and expert reports on Q-Fever and Mexican Flu.

Hub Zwart studied philosophy and psychology at Radboud University Nijmegen, the Netherlands. In 2000, he was appointed as Full Professor of Philosophy of the Natural Sciences and as Chair of the Department of Philosophy at the Faculty of Science (Radboud University Nijmegen). His basic research question is how the natural sciences (from genomics and molecular biology up to environmental research) challenge and affect our understanding of nature as well as of ourselves as human beings. These issues are addressed from a continental philosophical perspective, building on research traditions such as dialectics (Hegel), phenomenology (Heidegger), and psychoanalysis (Bachelard, Lacan). He is director of the *Institute for Science, Innovation and Society* (ISIS), scientific director of the *Centre for Society and Genomics* (CSG), coordinator of the EU H2020 project PRINTEGER (on research integrity) and editor-in-chief of the journal *Life Sciences, Society & Policy* (together with Ruth Chadwick).

Changing Relationships with Non-human Animals in the Anthropocene—An Introduction

Jozef Keulartz and Bernice Bovenkerk

Abstract In this introduction, we will address the following topics. The first section will deal with the Anthropocene—What is it? When did it start? How did it develop? The second section will show how the concept works as a major bone of contention that divides the academic into those who consider the Anthropocene a planetary catastrophe and those who embrace the human domination over the Earth as a great achievement. The third section considers the biodiversity conservation options in the age of humans. The fourth and final section will provide an overview of this volume.

1 The Anthropocene—Questions of Definition and Dating

In March 2015 a young wolf entered the Netherlands via Germany, sparking a debate about the return of this top predator. Can we co-exist with this animal, especially if—as was the case—he does not respect our boundaries and trails villages rather than turn straight to the novel nature area 'the Oostvaardersplassen'? His presence led to so much unwanted attention from the media and 'wildlife tourists' that he quickly turned around and made his way back to Germany. Similarly, a so-called 'golden jackal' was spotted in the Netherlands recently, giving rise to a remarkable discussion: should we welcome this wild animal, or should we treat him as an exotic species and kill him? It was argued that only if the animal was released in our country by people he should be considered an exotic, but if he walked here by himself he was to be spared. Another animal that created a lot of media attention, and that was even named, was Johanna the humpback whale, who in 2012 washed ashore the Dutch island of Texel. Fierce discussion erupted about the question whether we should help her, euthanize her (and by what means,

J. Keulartz (✉)
Radboud University, Nijmegen, The Netherlands
e-mail: j.keulartz@science.ru.nl

B. Bovenkerk
University of Wageningen, Wageningen, The Netherlands
e-mail: bernice.bovenkerk@wur.nl

B. Bovenkerk and J. Keulartz (eds.), *Animal Ethics in the Age of Humans*,
The International Library of Environmental, Agricultural and Food Ethics 23,
DOI 10.1007/978-3-319-44206-8_1

an injection would not do the trick, but is dynamite appropriate?), or let her die in peace. Following this discussion, fierce discussion arose about the question of why we had so much discussion about this one whale in the first place, when literally millions of animals are held in poor conditions and slaughtered under intensive farming conditions. As these examples from the Netherlands show, the question of what the proper relationship between humans and animals should be is contested, and as we argue in this book needs to be redefined in light of the current level of anthropogenic influence on nature.

Reflection on the relationship between humans and other animals has in the past been characterised by a number of sharp divisions: the division between humans and animals, between individuals and collectives such as species and ecosystems, between wild and domesticated animals, and between in situ and ex situ conservation. In the Anthropocene—the age of humans—a number of developments have led to these distinctions becoming blurred. Before we explicate these distinctions, we first want to sketch the context by addressing the questions 'What is the Anthropocene, when did it start and how did it develop, and how is it evaluated by philosophers, biologists, and environmentalists?'.

The term 'Anthropocene' refers to the current time period in which human activities are so pervasive that humanity itself has emerged as a global geophysical force, at least as influential as natural forces. The term was introduced by chemist and Nobel Laureate Paul Crutzen and biologist Stroemer in 2000. Crutzen recalled how the term originated at a conference where someone said something about the Holocene, the geological epoch, which began approximately 11,500 years ago at the end of the last glacial period: "I suddenly thought this was wrong. The world has changed too much. So I said: 'No, we are in the Anthropocene'. I just made up the word on the spur of the moment. Everyone was shocked. But it seems to have stuck" (Pearce 2007, 21). The concept has indeed rapidly gained traction and became a buzz word that is included in the title of a still growing number of conferences, panels, articles, and books.

There is overwhelming evidence that atmospheric, hydrospheric, biospheric and geospheric processes are currently profoundly altered by humans. The most common indicator for the large human imprint on the global environment is the increase of greenhouse gases, together with the associated rise in global temperatures and sea levels. Other key indicators include massive global increases in soil erosion; large-scale deforestation; widespread fragmentation and destruction of natural habitats; and what is considered to be the greatest mass extinction of species since the extinction of the dinosaurs 65 million years ago.

Crutzen and Stroemer's introduction of a new geologic epoch was designed to raise awareness of the current environmental crisis created by humans. As mankind will remain a major environmental force for many millennia, so their key message goes, "a daunting task lies ahead for scientists and engineers to guide society towards environmentally sustainable management during the era of the Anthropocene" (Crutzen 2002, 23). Although the 'Anthropocene' serves as a vivid yet informal metaphor of the global environmental crisis, a proposal is currently under consideration by the Anthropocene Working Group of the Stratigraphy

Commission of the Geological Society of London to formalize it as a geological epoch at the same hierarchical level as the Pleistocene and the Holocene epochs, with a target date of 2016 (Zalasiewicz et al. 2010).[1]

The notion that humanity has come to dominate the Earth is not new but has a long history that goes back at least as far as the mid-1800s. An early precedent of the concept of the Anthropocene is the term 'anthropozoic era', coined in 1873 by the Italian geologist Antonio Stoppani. He was quoted by George Perkins Marsh in the second edition of his influential *Man and Nature: Or, Physical Geography as Modified by Human Action*, one of the first works to document humanity's impact on the environment that helped to launch the modern conservation movement. Another term that was proposed to denote this impact is the notion of the 'noösphere' (the sphere of human thought) that was jointly developed in Paris just after the Great War by three men: the Russian geochemist and naturalist Vladimir Vernadsky, the geologist and palaeontologist Pierre Teilhard de Chardin, and the mathematician Edouard Le Roy, Henri Bergson's disciple and successor at the Collège de France (Steffen et al. 2011a).

1.1 The Stages of the Anthropocene

Crutzen and Stroemer, however, offered what these predecessors had not, a starting date for the proposed new epoch. They considered the onset of the industrial era with the spread of fossil fuel-based energy systems in combination with the extensive use of steam engines as the first stage of the Anthropocene. The enormously expanding availability of fossil fuels made many new activities and technologies possible. For example, with abundant energy from fossil fuels it proved possible to produce large amounts of fertilizer by synthesizing ammonia from atmospheric nitrogen. This so-called Haber-Bosch process sharply increased crop yields worldwide, and made possible a significant growth spurt of the human population (Steffen et al. 2007, 616).

But Crutzen and Stroemer's suggestion that the beginning of the Anthropocene coincides with the onset of industrialization is certainly not uncontroversial. Some have proposed a later date—the start of the nuclear era in the 1960s—, or have contended that the Anthropocene has yet to begin; others have suggested much earlier starting dates, including "the first human use of fire to modify ecosystems (up to 1.8 million years ago), the late Pleistocene megafaunal extinctions (14,000–15,000 years ago), the arrival of modern humans on all continents except Antarctica (c. 10,000 years ago), and the rise of agriculture (c. 7000 years ago)" (Corlett 2015, 37).[2] This controversy is not yet closed, but Crutzen and Stroemer's proposal to regard the Industrial Revolution as the first stage of the Anthropocene has received by far the most support in the literature.

---

[1]http://quaternary.stratigraphy.org/workinggroups/anthropocene/.

[2]For an extensive discussion, see Braje (2015, 373–380).

At the end of the Second World War, the human transformation of the global environment is supposed to have entered a new stage: the 'Great Acceleration' (Steffen et al. 2007, 2011a). The past 50 years have been a period of dramatic and unprecedented growth in human numbers and activities. While the global population doubled to over six billion by the end of the 20th century, the global economy increased by nearly tenfold, with an acceleration of activities on all fronts: the damming of rivers, the use of water, the production of grain, the consumption of fertilizer and petroleum, the manufacturing of motor vehicles, et cetera. This burgeoning human enterprise has obviously increased the pressure on the global environment, as is evident from developments such as chemical pollution, stratospheric ozone depletion, ocean acidification, the loss of tropical rain forest, and the decline of global biodiversity.

According to Crutzen cum suis, "the Great Acceleration is reaching criticality" (Steffen et al. 2007, 614). Currently, anthropogenic environmental change takes place at such a fast speed and large scale that it definitely poses a threat to the resilience of the Earth System and could be detrimental for large parts of the human population or even for global society as a whole. Therefore a third stage is urgently needed: 'Planetary Stewardship'. We will have to develop a strategy to ensure the sustainability of Earth's life support system, otherwise "we risk driving the Earth System onto a trajectory toward more hostile states from which we cannot easily return" (Steffen et al. 2011b, 739).

1.2 *The Planetary Boundary Framework*

We cannot be content with only marginal changes to our current practices, let alone with business as usual, but have to develop more radical approaches. Crutzen cum suis envision a spectrum of such approaches, with geo-engineering at one end and a number of alternative strategies to reduce the human pressure on the Earth System at its source at the other (idem, 752). One such alternative strategy is the 'Planetary Boundaries' approach. Central to this approach is the attempt to define a 'safe operating space for humanity' by identifying biophysical thresholds which humanity should not cross in order to prevent a shift away from a Holocene-like state of the Earth System, the only one that we know for certain can sustain the conditions that enable human development and flourishing.

The Planetary Boundaries framework distinguishes nine Earth System processes with boundaries that should be respected to maintain a Holocene-like state: climate change; rate of biodiversity loss; interference with the nitrogen and phosphorus cycles; stratospheric ozone depletion; ocean acidification; global freshwater use; change in land use; chemical pollution; and atmospheric aerosol loading. The boundaries of three of these Earth System processes have already been severely transgressed: rate of biodiversity loss, climate change and human interference with the nitrogen cycle (Rockström et al. 2009, 427).[3]

[3]Recently, a fourth process—land-system change—has been added to this list (Steffen et al. 2015).

1.3 *Novel Ecosystems as Hallmark of the Anthropocene*

In ecology, the Anthropocene concept has focused attention on human-dominated habitats and novel ecosystems. The concept of the 'novel ecosystem' was first used by Chapin and Starfield (1997), but gained traction among ecologists after the publication of a paper by Hobbs et al. in 2006. They used the term for ecosystems characterized by new, non-historical species combinations that arise through land-cover change, pollution, rapid climate change, and the impacts of the deliberate and inadvertent introduction of species from other regions. Accepting the fact of novel ecosystems means that we must be ready to incorporate many alien, non-native species into management plans, rather than eradicating or drastically reducing them (Davis et al. 2011).

In fact, the conceptual framework, developed in the edited volume *Novel Ecosystems* (Hobbs et al. 2013), distinguishes three types of ecosystem: historical systems, which have remained within their historical range of variability; hybrid systems, which are composed of new species combinations and/or abiotic conditions, but can be returned to their historical states; and novel systems, which are irreversibly changed (Hallett et al. 2013). Restoration ecologists mostly work in the area of hybrid ecosystems, trying to restore them to historical ones. But more attention should be paid to novel ecosystems. It is estimated (by Perring and Ellis 2013) that already about 35 % of the world's ice-free land is currently covered by novel ecosystems, and, in light of the ongoing human impact on the environment, they are expected to become omnipresent.

In a world that is in ever-greater flux, restoration to a historic standard is becoming more and more anachronistic. Hence the suggestion that we should drop the term 'restoration ecology' with its historical focus, and replace it by the term 'intervention ecology'. This substitution of restoration by intervention signifies a shift from a 'historic', backward-looking to a 'futuristic', forward-looking approach to ecosystem management (Choi 2004; Choi et al. 2008). Rather than looking nostalgically to a past that is impossible to restore, "we should intervene with an eye to the future and toward managing for future change" (Hobbs et al. 2011, 444), which basically comes down to maintaining or repairing key ecosystem services.

2 The Anthropocene as Apple of Discord

While in ecology the Anthropocene concept has drawn attention to novel ecosystems, in conservation biology it has sparked a divisive debate on the continued relevance of the traditional biocentric aims (Corlett 2015, 36). This debate started in 2004 with an essay—*The Death of Environmentalism*—written by Michael Shellenberger and Ted Nordhaus, cofounders of the Breakthrough Institute. In this essay they express their conviction "that modern environmentalism, with all of its unexamined assumptions, outdated concepts and exhausted strategies, must die so that something new can live" (Shellenberger and Nordhaus 2004, 10). The authors present themselves as 'post-environmentalists'; the movement they initiated became known as 'ecomodernism' or 'ecopragmatism'. As Shellenberger and Nordhaus argue in their 2007 book *Break Through: From the Death of Environmentalism to the Politics of Possibility*, ecomodernist politics should shift its focus from nature preservation and environmental protection to technological innovation to achieve a new sustainable economy.

Another well-known and influential advocate of this 'new conservationism' is Peter Kareiva, senior fellow of the Breakthrough Institute and former Chief Scientist and Vice President of The Nature Conservancy, with one million members the world's largest nature organization. Kareiva and other 'neo-greens' claim that nature is not as fragile as the doom and gloom idiom of destruction, depredation and deterioration of old school conservationists suggests, but is surprisingly resilient and often recovers from even severe perturbations (Kareiva and Marvier 2012, 965).

New conservationists tend to shift the emphasis from the protection of biodiversity conservation for its own sake to the enhancement of "those natural systems that benefit the widest number of people, especially the poor" (Kareiva et al. 2012). They argue for an approach in which the centrality of humans is assumed and in which the improvement of human well-being through the management of the environment plays a key role (Kareiva and Marvier 2012, 962/3). Rather than assemblages of species, new conservationists view nature as a bundle of ecosystem services. Promoting and maintaining these services, should be conservation's primary goal, not preventing anthropogenic extinction. "Some human-caused extinctions are inevitable, and we must be realistic about what we can and cannot accomplish. We must be sure to first conserve ecosystems in places where biodiversity delivers services to people in need" (Kareiva and Marvier 2007, 56).

Not surprisingly, therefore, new conservationists distinguish themselves from old school conservationists by their willingness to work with corporations. "A small number of global corporations have a huge impact on land conversion, mining, energy extraction, and consumer choices. In essence, corporations are the 'keystone species' of global ecosystems" (Kareiva and Marvier 2012, 967). Instead of scolding capitalism, conservationists should partner with these corporations to achieve better results by dovetailing conservation aims and economic activities more effectively.

2.1 Two French Philosophers

Ecomodernists reject the doom and gloom that seems so characteristic for traditional conservationists. Instead of bemoaning the Anthropocene as a potential ecological disaster, they ask us to celebrate this new era. To them it does not represent the failure of environmentalism, but should rather be seen, more optimistically, as "the stage on which a new, more positive and forward-looking environmentalism can be built" (Marris et al. 2011), and as "the beginning of a new geological epoch ripe with human-directed opportunity" (Ellis 2012).

Ecomodernists' aversion to a pessimistic view of the 'age of man' has been expressed most forcefully in the 2011 book *The Fanaticism of the Apocalypse: Save the Earth, Punish Human Beings* by the philosophical best-selling author Pascal Bruckner, who is, not coincidental, also a senior fellow of the Breakthrough Institute. Bruckner considers today's environmentalism as a spiritual descendent of end times Christianity and Communism, too fueled by rampant guilt and self-hatred, fearful of pleasure and technological progress, and using a vocabulary of impending catastrophe to blackmail us into a dour asceticism that criminalizes our everyday behavior. Bruckner does not shy away from gross exaggerations to pillory the environmental movement, witness phrases like "If the extremists drown out the moderates, the new sobriety will have the bitter taste of concentration camps and prisons" (Bruckner 2013, 159). His book concludes as follows: "The friends of the earth have for too long been enemies of humanity; it is time for an ecology of admiration to replace an ecology of accusation" (Idem, 185).[4]

Bruno Latour, another famous French philosopher who was senior fellow of the Breakthrough Institute, has lectured the environmental movement for its alleged technofobia. He wrote an essay on Mary Shelley's novel 'Frankenstein'—*Love Your Monsters*—that was published in a volume of the same name, edited by Shellenberger and Nordhaus. According to Latour, it was not Dr. Frankenstein's crime that he invented a monster through some combination of hubris and high technology, but rather that he abandoned this creature to itself. It would have been better if he had nurtured his creature like a child to prevent it, with parental affection and supervision, from going down the wrong path.

This lesson, drawn from Shelley's novel, is what Latour wants to teach the environmental movement, that has renounced technological progress, preaching a gloomy asceticism instead, and constantly flagellating themselves for the damage they have, in their arrogance, caused to nature and the environment—*mea culpa, mea maxima culpa*! In this way, the environmental movement repeats Dr. Frankenstein's mistake. Instead of abandoning technology for fear of its destructive power, the movement should have the same type of patience and commitment to our technological creations as God the Creator himself. This comparison, Latour says, is not at all blasphemous: "We have taken the whole of Creation on our shoulders and have become coextensive with the Earth" (Latour 2012, 22).

[4]It is no surprise that Bruckner's book has been welcomed by skeptical organizations, and that he was guest of the UK's Global Warming Policy Foundation, the UK's most prominent source of climate-change denial, delivering a presentation on his book in April 2013 at the House of Lords.

2.2 Two Converts

The comparison of man to God is very popular among ecomodernists. This comparison is at the center of the book *The God Species: How the Planet can Survive the Age of Humans* by Mark Lynas, winner of the Breakthrough Paradigm Award in 2012. Lynas describes himself as a 'recovering activist'. In his student days, he was heavily involved in the anti-GM movement. But around 2010, former 'eco-warrior' Lynas became a supporter, not only of GM crops, but also of geoengineering and of nuclear energy, on which he published a book in 2014: *Nuclear 2.0: Why a Green Future Needs Nuclear Power.*

In the age of human domination over the planet, the motto applies that "Nature no longer runs the earth. We do" (Lynas 2011a, 12). Whereas environmentalists usually denounce 'playing God' as particularly dangerous, Lynas defends the opposite view: "Playing God is essential, if creation is not to be irreparably damaged or even destroyed by humanity unwittingly deploying its new-found powers in disastrous ways" (Lynas 2011b).

That is exactly the risk that greens face if they refuse to play God. Lynas calls their opposition to nuclear power "a gargantuan error, and one that will echo down the ages". He believes that the many thousands of people who marched against nuclear power in Germany following the Fukushima accident (which he qualifies as a moderate industrial accident—'nothing more') are "just as bad for the climate as textbook eco-villains like the big oil companies".[5]

If one brings out the past of prominent ecomodernists, one will encounter more radical converts. For example Steward Brand, again a senior fellow of the Breakthrough Institute. Brand played an important role in the counterculture of the 60s and 70s. He helped shape the environmental consciousness with his *Whole Earth Catalog*, a journal that appeared between 1968 and 1985, and that became the bible of the back-to-the-land movement. Around the same time as Lynas, Brand also underwent something of a Damascene conversion. In his 2009 book *Whole Earth Discipline: An Ecopragmatist Manifesto* the flamboyant ex-hippie sacrificed a lot of sacred cows of the green movement. Brand has abandoned his dream of going back to the land and became a fan of the world's new megacities which he sees as "generators of income and concentrators of efficiency and innovation". Brand pleads not only for megacities, but also, like Lynas, for genetic modification, geoengineering and nuclear energy.

2.3 Resistance and Reconciliation

Around 2013, persons and organizations who were portrayed by ecomodernists as major obstacles to successful nature management and environmental policy started

[5]http://www.theguardian.com/environment/2011/jul/02/green-movement-lost-its-way?INTCMP=SRCH.

to fight back. The counterattack was launched by Michael Soulé, the widely acclaimed doyen of conservation biology. In an editorial of Conservation Biology he denounced the new "powerful but chimeric movement" because it is to replace proven conservation tools for biodiversity protection by the provision of ecosystem services for the benefit of economic development and poverty reduction. Soulé comes to the conclusion that the new conservation, if implemented, "would hasten ecological collapse globally, eradicating thousands of kinds of plants and animals and causing inestimable harm to humankind in the long run (Soulé 2013, 896; cf. Miller et al. 2013).

In 2014, more than twenty prominent academics and activists from North and South America, Europe, and Australia joined forces in a volume entitled *Keeping the Wild; Against the Domestication of Earth*, with a contribution of Soulé, to whom the book was dedicated. The authors campaign "with rhetorical fists swinging" (as it was called on the back cover) against the techno-triumphalist glorification of the Anthropocene as the latest incarnation of human exceptionalism and human hubris. They reject the human-dominated and domesticated nature and offer bold advocacy for free nature in all its diversity.

In 2014, the year of this book publication, a petition with 240 co-signatories, led by Heather Tallis and Jane Lubchenco, appeared in the journal *Nature*. The petition called for an end to the battle between the two camps of conservationists, that is dominated by men's voices. The increasingly acrimonious debate in universities, research centers and nature organizations between those who want to protect nature for its intrinsic value (for its own sake) and those who want to save nature for its instrumental value (for ourselves) is counterproductive and severely jeopardizes the funding and progress of projects. The petition, which is now also signed by two of the main antagonists, Peter Kareiva and Michael Soulé, argues for a more inclusive approach to conservation with more openness and tolerance for a diversity of perspectives and for divers values and voices.[6]

The question is of course how far both camps of conservationists will really get along with each other. It is after all about two extremes, two antipodes: countryside versus megacities; organic farming and agroecology versus the production of genetically modified food; solar energy, wind energy and other renewable energy resources versus nuclear energy and shale gas and shale oil; preaching restraint and reducing the carbon footprint to combat climate change versus geo-engineering, et cetera. We are dealing here with what Paul Wapner in *Living Through the End of Nature* (2010) has called "the dual dreams of naturalism and mastery": on the one hand, the romantic dream of a pre-industrial paradise were humanity lives in harmony with nature and on the other hand the Promethean dream of a technological utopia in which humanity rules over nature. According to Wapner, it is time to leave both dreams behind and look for a middle way, leading

[6]As Hunter et al. (2014, 664) have pointed out regarding the battle between the two camps: "The conservation arena is large enough to accommodate many people and organizations whose diverse values lead them to different niches that can, with good will and foresight, be far more complementary than competitive".

to a "politics of ambiguity"—a winding trail full of tensions, contradictions and mixed feelings that should be embraced rather than bulldozed away, although it is questionable if this will be a passable avenue.

3 Biodiversity Conservation Options in the Anthropocene

Given the scope and scale of anthropogenic stress during the current stage of the Anthropocene era—the Great Acceleration—the animal world is under growing pressure. Traditional in situ (place-based) conservation measures such as the establishment of protected areas and the creation of connections between these areas seem no longer sufficient to save threatened species. Preserving the ecological status quo through such traditional measures increasingly resembles a Sisyphean task. As a consequence, "biodiversity conservation is entering a phase of prolific innovation" (Kueffer and Kaiser-Bunbury 2014, 131).

Today, we witness the emergence of new, more intensive forms of population and ecosystem management. Most of these forms include translocation, the intentional movement of animals from one location to another. Seddon et al. (2014) use a two-dimensional matrix diagram for the classification of translocation types. The first dimension concerns the distinction between translocations for species conservation on the one hand and translocations for restoring ecological and evolutionary processes ('rewilding') on the other. The second axis concerns the distinction between translocations within and outside the species' indigenous range.[7] These two axes yield four translocation types (see Fig. 1).

3.1 The Translocation Spectrum

The concept of **inter situ conservation** (Braverman 2014), also called pan situ conservation (Minteer and Collins 2013), refers to a new paradigm for species conservation within the species' indigenous range: the integrated and interactive management of captive and wild populations. As mentioned before, under Anthropocenic conditions many wild populations are no longer viable on their own. Mainly as a result of ongoing habitat fragmentation, an increasing number of populations are declining and are teetering on the edge of extinction. On the other hand zoo populations too are generally not viable on their own. As Robert Lacy has argued, any closed population will always lose genetic diversity in the short or long term, depending on its size. Zoos hardly, if at all, succeed in maintaining

[7]"The indigenous range of a species is the known or inferred distribution generated from historical (written or verbal) records, or physical evidence of the species' occurrence" (IUNC/SSC 2013, p. 2).

	Translocation for Species Conservation	Translocation for Rewilding
Within Indigenous Range	Inter Situ Conservation	Restoration of Keystone Species
Outside Indigenous Range	Assisted Migration	Ecological Replacement

Fig. 1 Types of translocation (adapted from Seddon et al. 2014, p. 408)

sustainable populations in their facilities, because of small population sizes, unsuccessful breeding, inappropriate founding populations, poor cooperation in managed breeding programs, et cetera. "These deficiencies are resulting in declining populations or declining gene diversity or both" (Lacy 2013, 19). Because both wild and captive populations are currently not viable on their own, it is no longer effective to manage them in isolation from one another. Practitioners of species conservation therefore increasingly use the so-called One Plan approach that was officially proposed to the IUCN World Conservation Congress in 2012. The One Plan approach promotes the interactive exchange of animals between in situ populations (in the wild) and ex situ populations (in captivity) for mutual reinforcement. With animals moving in both directions, the stability and sustainability of wild and captive populations can be greatly enhanced.

With the One Plan approach captive populations can be used for the conservation of in situ populations on the brink of extinction as a result of habitat fragmentation. But what if in situ conservation itself is being undermined by that other major environmental stressor—rapid global climate change—, that makes the species' historic indigenous ranges increasingly inhospitable? And when, moreover, populations are not able to move on their own to other areas with more suitable environmental conditions? A conservation measure that may prevent species that are unable to keep pace with rapid climate change from going extinct is **assisted migration** or assisted colonization, i.e. the intentional movement of 'climate refugees' to new habitats outside their historical range, which they otherwise could not reach.

An exclusive focus on the conservation of (endangered) species by moving them between different locations, within or outside their historic ranges, may be too narrow to meet desired nature policy targets if animals are released into severely depleted and degraded ecosystems. In such case, which is the rule rather than the exception in the Anthropocene, conservationists may resort to 'rewilding'. Rewilding is focused on the establishment of core areas, enhanced connectivity, and, above all, the **restoration of keystone species**, i.e. those species whose impact on their environment is disproportionately large relative to their numerical abundance. Keystone species are often large-bodied carnivores and herbivores that have strong impacts on landscape structure, vegetation composition and ecosystem dynamics. They are of highly functional importance for restoring and maintaining self-regulating biodiverse ecosystems; their return is vital for the restoration of the evolutionary and ecological potential that was lost with their removal (Smith et al. 2015; Svenning et al. 2015).

However, most of these megafauna species were lost after the dispersal of modern humans from Africa and Eurasia. In North America alone more than 50 species of large mammals went extinct after the arrival of the Clovis people some 13,000 years ago, including mammoths, mastodons, horses, giant ground sloths, American camels, lions, and the saber-tooth cats. This catastrophic extirpation—sometimes referred to as the 'Pleistocene overkill'—has started an ecological chain reaction that led to further extinctions and hence to severe ecosystem simplification. To correct this dramatic lose of megafauna, Josh Donlan and colleagues have launched the idea of 'Pleistocene rewilding' (Donlan et al. 2005, 2006). They propose using megafauna from Africa and Asia, such as camels, cheetahs, elephants and lions, as **ecological replacements** or proxies for extinct American species. Pleistocene rewilding is supposed to serve a dual purpose: to restore some of the evolutionary and ecological potential that was lost 13,000 years ago, and to help prevent the extinction of the world's remaining megafauna by creating new, and perhaps better protected, populations in North America. Whereas proponents of novel ecosystems have abandoned history entirely, shifting the focus from the past to the future, and replacing 'restoration ecology' by 'intervention ecology', Pleistocene rewilders try to reach back to an even deeper, more distant past, adopting what also has been termed 'resurrection ecology', thereby blurring the boundaries of time (Keulartz 2016).

The latest development in resurrection ecology is 'de-extinction': bringing extinct species back to life through new technologies, such as cloning and genetic engineering (Sherkow and Greely 2013). Currently, leading synthetic biologist George Church is working on a de-extinction project of the Long Now Foundation, to revive extinct species, starting with the iconic passenger pigeon and moving on to the woolly mammoth. Seddon calls these kind of animals "charismatic necrofauna" (Seddon 2015, 569). A more familiar de-extinction method is 'back-breeding': the selective breeding of domestic animals, in an attempt to achieve an animal breed with a phenotype that resembles an extinct wild ancestor. This technique is currently employed to breed a strain of cattle into something resembling the Aurochs, a species of wild European cattle that went extinct in 1672.

3.2 Controversies Over Translocations

The self-proclaimed 'new conservationists' and their traditional counterparts adopt different positions on both dimensions of the matrix diagram for the classification of translocation types: the dimension concerning the distinction between translocations for species conservation and translocations for rewilding, and the dimension concerning the distinction between translocations within and outside the species' indigenous range.

New conservationists are usually less interested in translocations for species conservation than old school conservationists; they have a preference for translocations to restore ecological and evolutionary processes because of their potential to provide important ecosystem services and goods. New conservationists favor a functional approach over a compositional approach, and prioritize nature's benefits to humans over its intrinsic value. "Protecting biodiversity for its own sake has not worked. Protecting nature that is dynamic and resilient, that is in our midst rather than far away, and that sustains human communities—these are the ways forward now" (Kareiva et al. 2012).

On the other hand, new conservationists are generally in favor of translocations outside the species' indigenous range, whereas traditional conservationists have a rather skeptical or even negative attitude towards this type of translocations. They fear that relocated species may behave like invasive species in the receiving environment, to the detriment of extant populations of indigenous species. Translocations such as assisted migration or ecological replacement may thus result in novel ecosystems with highly uncertain and unpredictable outcomes because they have no historic analog.

The ontological status of non-native species and no-analog ecosystems is in fact a major bone of contention between traditional and new conservationists. Traditionalists usually reject translocations outside the indigenous range as an unsound conservation practice because they consider the intended or accidental introduction of invasive species as one of the primary drivers of biodiversity loss. There is in fact a close connection between restoration ecology and invasion biology (Keulartz and Van der Weele 2008, 2009). According to Mark Davis, they emerged at about the same time as "sister disciplines" during the late 1980s. They developed an increasingly strong synergy, with the objectives of each reinforcing those of the other.[8]

New conservationists generally feel that Anthropocene conditions have rendered the presence of non-native species unavoidable. According to Davis and other scientists, "we must embrace the fact of 'novel ecosystems' and incorporate many alien species into management plans, rather than try to achieve the often impossible

[8]"Restoration ecology's emphasis on restoring environments with native species affirmed the importance of invasion ecology, and invasion ecology's emphasis on the harm caused by a small proportion of introduced species provided important justification for restoration ecology's preference for native species" (Davis 2006, 49).

goal of eradicating them or drastically reducing their abundance" (Davis et al. 2011, 153). Kareiva et al. (2012) have adopted an even more optimistic tone: "All around the world," they explain, "a mix of climate change and nonnative species has created a wealth of novel ecosystems catalyzed by human activities".

Whereas traditional conservationists consider inter situ conservation, involving the translocation for species conservation within the species' historic ranges, as the preferred option, their ecomodernist counterparts are in favor of ecological replacement, i.e. the translocation for rewilding outside the species' historic range. They have, moreover, a strong preference for de-extinction. Not surprisingly, the co-founder and president of the Long Now Foundation that firmly supports de-extinction projects using genetic engineering, is Stewart Brand, the aforementioned ex-hippie turned eco-modernist, in favor of mega-cities, genetic modification, geo-engineering and nuclear energy. De-extinction appeals to new conservationists because it promises a much cheerier story than the traditional discouraging tale of environmental destruction and loss, "a more uplifting narrative driven by sunny acts of biological creation and ecological recovery" (Minteer 2015, 14).

4 Blurring Boundaries: An Overview

Our short survey of recent conservations strategies along the translocation spectrum makes it abundantly clear why conservation biologist Kent Redford and his two colleagues speak of 'the long overdue death of the ex situ and in situ dichotomy in species conservation' (Redford et al. 2013). The distinction between in situ (on-site) and ex situ (off-site) conservation has become blurred to the point of disappearing entirely. We witness what Braverman (2015, 15) has called a shift "from bifurcation to amalgamation" of in situ and ex situ conservation: the increased development of hybrid approaches, that integrate conservation inside and outside of 'wild nature'. This development runs parallel to the blurring of some older distinctions that are reminiscent of the in situ and ex situ distinction, such as wilderness versus captivity and nature versus culture (Idem, 31f.). In this collection of essays we want to examine the boundaries of these distinctions and the (mainly moral) implications of their becoming blurred. The blurring of these distinctions is completely transforming the field of animal studies and presents us with new challenges as well as new insights and new ways of dealing with nature and non-human animals. Developments in the Anthropocene that we have described above, and the resulting blurring of old distinctions, tempt us to rethink our old conceptions of, and to devise new ways of relating to, animals and nature. For example, we may need to rewrite history to include non-human animals, as Guerrini bravely attempts to do in this volume. We may need to regard nature and animals not only as a resource, but as a mentor or teacher, as van der Hout, Korthals and Burton and Brady illuminate. We may need to examine that impacts of ubiqutous human influences on animals: With Palmer, we can ask, does climate change make animals less wild? With

Kasperbauer and Simmons, we can ask: if habitat-loss forces some animals to live in captivity, what impact does this have on their (reproductive) freedom?

4.1 Between Human and Animal

The book consists of four parts, each followed by a commentary to the chapters in that section. Every section addresses a different development and its implications for our thinking about animals. In particular, we address four old divisions in the human-animal relationship that are becoming blurred. As a matter of fact, as will become clear, the boundaries between these categories itself are blurred. Several chapters address, for example, both the distinction between human and animal and the distinction between wild and domestic. Firstly, throughout history a sharp division has been made between human beings and animals. While Darwin and later animal ethicists have called into question this human exceptionalism, in popular thinking it is still omnipresent. Recent great advances have been made in the scientific research of animal consciousness and cognition, calling into doubt the clear-cut dividing line between human self-consciousness and animal 'mere consciousness'. Similarly, recent insights in animal modes of communication have shed new light on animal capacities. These developments raise the question whether the animal perspective is really so different from the human one and whether we should rethink our relationship with other animals. Moreover, the advent of new technologies changes the boundaries between humans and animals. In the section of the book focussing on the blurred distinction between human and animal none of the authors subscribe to crude forms of human exceptionalism, but as commentator Van den Belt argues "each has different views on how far the critique of human supremacy should be pursued and where the new boundaries should be redrawn".

Guerrini pleas for a new historical discourse that de-centers human beings by also including non-human animals. Van der Hout zooms in on the field of biomimicry, the practice of mimicking nature in our technology development. The idea of this movement that biomimicry enables us to live more sustainably and in harmony with nature seems a promising response to the environmental challenges we face in the Anthropocene. But what are the hidden implications of the biomimicrists' narrative? How do we ensure that nature is not only regarded as a resource to do with as we please, but rather as a teacher, whose wisdom and creativity we can use as a source of inspiration? Korthals also wants to bring the unused teaching potential of animals to our attention. Animals are not only passive subjects of our action, but are also "active agents of symbolization, history, learning and the breaking down and reconstruction of established world perspectives". Burton and Brady think we can learn something not only about animals but also about ourselves when we try to take an animal's perspective. In their case, they imagine what it is like to be a bird and thereby they draw attention to the boundaries of our human perspectives on the world, raising the question of what attitude we should have towards knowledge. While they shy away from the behaviorist adagio that we will never know what goes on in an animal's head, they do want to develop a "stance of epistemic humility" towards our own knowledge of other animals. Meijer, on the other hand, thinks we can take our imagination of what animals experience a lot further. She blurs the distinction between human and animal considerably by focussing on the political actions of animals. Like humans, animals have culture and language, they can vote and negotiate; we just have to find out how they express these capacities. Seeing animals as political actors means that we have to rethink concepts that were in the past reserved for humans only, concepts such as 'politics', and 'democracy'. While she thinks animals are silenced politically in the Anthropocene, she also wishes to explore the potential of the blurring of old distinctions by examining the possibility to enter into an interspecies democracy. Masson and McCarthy blur the distinction between humans and animals by tracing back the psychological causes of our human exceptionalism and by throwing up the question as to why the distinction between humans and animals matters in the first place.

4.2 Between Wild and Domestic

Secondly, as ongoing urbanisation has led to the destruction of animal habitats many animals seek refuge within those urban areas, tapping into new food sources. This means that the division between wild and domesticated animals is becoming blurred. In animal ethics a recent trend has been to focus not only on animal populations in the wild or on livestock and laboratory animals and pets, but also on so-called 'liminal animals', such as foxes, mice, and hedgehogs. Relational ethicists have argued that we may have more stringent moral obligations towards animals the more influence we have on their situation. In the Anthropocene, however, we have influenced the lives of almost all animals to varying degrees and this raises the question how tenable categorisations such as 'wild', 'liminal' or 'domesticated' really are. At the same time, such categories are necessary when we are contemplating what is morally desirable in our dealings with animals. For example, should we be allowed to keep wild animals in captivity, such as in zoos, circuses or dolphinaria? And isn't the keeping of pets or production animals also a matter of captivity and if so, is this necessarily morally wrong? The latter would suggest that freedom has independent value for animals, but is this really the case? And wouldn't this suggest that we were wrong in domesticating animals in the first place? Animal rights theorists argue that domesticated animals are in a sense human artefacts that are captive in their own bodies and they point to extreme breeding practices which have adverse effects on animals' health. Others argue that there is nothing inherently wrong with domestication and that this process can often be reversed once an animal is reintroduced in the wild. This also raises the question whether dedomestication of animals is sometimes desirable, for example in order to (re)create or rehabilitate a nature area through natural grazing with large herbivores.

In society, however, objections are raised against what is regarded as the instrumental use of individual animals, who may suffer in the process of dedomestication.

As Donaldson and Kymlicka point out in their commentary to this section, the farreaching human influence on nature in the Anthropocene leads to a blurring of the old dichotomy between wild animals that are untouched by humans on the one hand and domesticated animals that are completely under human control on the other. They call for new ideas to help us understand the ethical challenges arising from this new diversity of human-animal relationships. Several alternatives to traditional animal ethical approaches are offered in this volume: Keulartz puts forward a revised version of Nussbaum's capabilities approach, Swart argues for a contextual care approach, Donaldson and Kymlicka plead for a focus on interspecies justice—which they find lacking in the other contributions to this volume—and Bovenkerk argues that traditional approaches are too inidividualistic and should focus more on the species level. All contributions raise the question of how humans and (more or less) wild animals can learn to co-exist in the Anthropocene.

Palmer examines the impact of anthropogenic climate change on the wildness-status of animals. She gives a helpful categorisation of different meanings of wildness and argues that climate change might reduce both wildness in a constitutive sense and wildness in the sense of self-willing. One feature that seems inherent in the advent of the Anthropocene is that of animal captivity on an enormous scale. Bovenkerk examines the question whether this captivity is morally justified and this inevitably leads to the further question of whether animal domestication is justified. She argues that the debate about the moral justifiability of animal justification could benefit from taking the discussion to the species-level. Donaldson and Kymlicka take domestication to be a process where animals are inherently instrumentalised and their rights violated. But if this is the case, should we dedomesticate them, as is done in the case of the Oostvaardersplassen, as discussed by Swart? While rewilding seems to make up for past rights violations, Donaldson and Kymlicka argue that it in fact "has nothing to do with respect for animals' intentional agency". But how to care for animals that are already wild or semi-wild? In Swart's view care for these animals in the anthropocene implies taking care of their threatened environment. But what happens when such care goes so far as to border on human-wild animal friendship, thereby again blurring the boundaries between humans and animals? Drenthen addresses the question of whether such friendship is even possible by discussing friendships between humans and wolves. Relationships with wild animals in his view involve a balancing act between "benevolent involvement and loving detachment". Like Boonman-Berson, he posits that certain boundaries between humans and animals are necessary to ensure peaceful co-existence. Besides such "strategies of confinement", Boonman-Berson argues that "strategies of alignment" are necessary in wildlife management if we aim to solve human-wildlife conflicts.

4.3 Between Freedom and Captivity

Thirdly, in the Anthropocene the boundary between freedom and captivity is becoming blurred. As we noted in the previous section, a variety of positions now exist between wild and domesticated animals and the old division appeared to be that domesticated animals can be kept in captivity without a problem, whereas wild animals needed freedom. However, as Simmons argues, freedom may be in the interest of all animals, whether wild or domesticated. Moreover, in the Anthropocene, due to habitat loss and habitat fragmentation, certain wild animals could not survive if it weren't for zoos, or in other words captivity. But once wild animals are kept in captivity, do they lose their wildness? If so, isn't promoting a zoo as a place to come face to face with wild animals creating an illusion? If animals have been kept in captivity, can they be successfully be reintroduced to a wild habitat? And what rights to self-determination should they be given when captive? Should they have reproductive rights, as Kasperbauer argues in the case of primates? And what can we desire of them when they are in captivity? Are we allowed to make them perform for our entertainment, as Brando critically discusses? Many of these questions show how the old division between one animal ethic for 'kept animals' such as pets and livestock on the one hand, and another animal ethic for free and wild animals is becoming problematic. With the blurring of the distinction between wild and domesticated animals, the distinction between kept and free animals is also increasingly being challenged.

It has often been argued that whereas wild animals have a right to having their freedom respected, domesticated animals, in particular farm animals, have a right to having their welfare respected. We therefore have a corresponding duty to ensure the welfare of the latter and to leave the former alone. However, Simmons argues that just like wild animals, farm animals have an interest in freedom: even when we do not make farm animals suffer, we still harm them by restricting their opportunities to "freely pursue their own enjoyments in life". As he notes, this view also has implications for other animals in captivity, such as animals in zoos and circuses. For the survival of animals that due to anthropogenic influences have lost their native habitats, it may sometimes be necessary to keep them in captivity, whether in a zoo or a sanctuary. Do these animals thereby lose all the freedoms they had in the wild? According to Kasperbauer they should not. He argues that captive primates should be granted limited reproductive rights, acknowledging that practical considerations, such as the risk of overpopulation, should be taken into account. In his chapter he raises an interesting question that also bears on the old human-animal distinction. If "regulating the reproductive behavior of human beings is generally seen as impermissible", why is this different for primates, who in many relevant respects resemble human beings?

When animals are held in captivity they no longer have to occupy themselves in the 'struggle for survival' and this can lead to boredom. Making animals work for their food, for example by entertaining an audience, could be a way to alleviate such feelings. But are entertainment goals always compatible with animal welfare? Brando gives many examples where there is a tension between entertainment and animal welfare and she argues that we always have to consider the animal's own perspective along with its species-specific needs. She furthermore raises an interesting question beyond animal welfare: does performing compromise a wild animal's dignity? Keulartz argues that all these kinds of questions that are typical for the human-animal relationship in the Anthropocene require a new way of moral thinking that can no longer be offered by traditional utilitarian and deontological approaches in animal ethics. He argues that a revised version of Nussbaum's capability approach provides a promising alternative.

4.4 Between Animal Ethics and Conservation Ethics

Fourthly, in the Anthropocene two interconnected global processes threaten animal flourishing. On the one hand globalization has broken down borders and has thereby caused dislocations of complete animal populations. International travel, transport, trade and tourism have brought along animal and plant travellers and in their wake invasive species that threaten to destroy native species and their habitats. On the other hand, global environmental changes, such as climate change, deforestation, and desertification have had a severely disrupting impact on animal life. Natural habitats have become fragmented, as nature areas are being turned into 'islands in a sea of cultivated land'. For many animal populations this means either abandoning their habitat or going extinct. This raises the question what we should do about invasive species. Should we terminate or tolerate them? In the face of habitat loss and shifting climate zones, should we assist species to migrate to areas more suitable to them? As the size and genetic diversity of the remaining wildlife populations are declining, such populations need to be managed more extensively. Such management may include translocation and (re) introduction of captive-bred animals, raising the question what the role of zoos and animal sanctuaries should be. As we have already seen in the discussion about the Anthropocene above, these developments will make the boundary separating in situ and ex situ conservation increasingly thinner.

Larson and Barr illustrate how this distinction is 'both theoretically and practically untenable' using a case study of the Monarch butterfly and touching on the discussion around assisted migration. They point out that challenges to the in situ–ex situ dichotomy echo the long-standing critique of the nature-culture divide. The interweaving of in situ and ex situ conservation poses new challenges for zoos. Keulartz raises the question of whether zoos' self-proclaimed role in conservation—captivity for conservation—is both practically and morally justified. He pleas for an intergrated approach that amongst other things calls for a better connection to in situ projects and a shift towards smaller species. Ramp and Bekoff argue that such conservation measures in the past have disregarded the welfare and intrinstic value of individual animals, because of their sole focus on species and ecosystems. Fortunately, an ethic of 'compassionate conservation' is making headway in conservation policies, where individual animals are no longer regarded as an impediment to conservation, but conservation and animal welfare are combined.

This ties in with a long-standing discussion between animal ethicists, who have traditionally focused on rights for or liberation of individual animals, and environmental ethicists, who have focussed not on individual animals, but on their habitat, biotope or ecosystem. In the debate between animal and environmental ethicists, both groups have assumed a clear dividing line between 'the wild' and 'the walled', or in other words between wild and domesticated animals. However, with the further development of both fields and the advent of new technologies, giving rise to new moral problems, the division between individual animal interests on the one hand and species or ecosystem interests on the other hand seems too stark. Many issues in animal ethics can be fruitfully analysed in terms of obligations towards individual animals, but some problems require reflection about collective dimensions of animal life in ways that individualist approaches cannot offer. For example, the question of whether it is morally permissible to genetically modify or breed animals for specific characteristics addresses the level of animal species or populations, rather than individuals. A variety of considerations exist between on the one hand the idea of animal ethics that every concern should be reducible to respect and care for individual animals, and on the other hand the ecocentric focus on care for only collectives such as ecosystems. Bovenkerk and Verweij examine the question of whether a focus on collective dimensions in animal ethics can blur this old division between atomists and holists. Finally, Zwart in his commentary raises the question of how we should recast the story of Noah's Ark in the Anthropocene. How can we make sense of the human-animal relationship when the old distinctions and categories that we used to define this relationship by have become blurred as a result of the increasing anthropogenic influence on nature?

References

Braje, T.J. 2015. Earth systems, human agency, and the anthropocene: planet Earth in the humane age. *Journal of Archaeological Research* 23(4): 369–396.

Brand, S. (1968–85). *Whole World Catalog*.

Brand, S. 2009. *Whole Earth discipline: an ecopragmatist Manifesto*. New York: Viking Penguin.

Braverman, I. 2014. Captive for life: conserving extinct in the wild species through ex situ breeding. In *The ethics of captivity*, ed. L. Gruen, 193–212. New York: Oxford University Press.

Braverman, I. 2015. *Wild life. The institution of nature*. Stanford, CA: Stanford University Press.

Bruckner, P. 2013. *The fanaticism of the apocalypse: save the Earth, punish human beings*. Cambridge, UK: Polity Press.

Chapin III, F.S., and A.M. Starfield. 1997. Time lags and novel ecosystems in response to transient climatic change in artic Alaska. *Climatic Change* 35: 449–461.

Choi, Y.D. 2004. Theories for ecological restoration in changing environment: toward "futuristic" restoration. *Ecological Research* 19: 75–81.

Choi, Y.D., V.M. Temperton, E.B. Allen, A.P. Grootjans, M. Halassy, R.J. Hobbs, M.A. Naeth, and K. Torok. 2008. Ecological restoration for future sustainability in a changing environment. *Ecoscience* 15(1): 53–64.
Corlett, R.T. 2015. The anthropocene concept in ecology and conservation. *Trends in Ecology & Evolution* 30(1): 36–41.
Crutzen, P.J. 2002. Geology of mankind. *Nature* 415(6867): 23.
Crutzen, P.J., and E.F. Stroemer. 2000. The "anthropocene". *IGBP Newsletter* 41: 17–18.
Davis, M. 2006. Invasion biology 1958–2005: the pursuit of science and conservation. In *Conceptual ecology and invasion biology*, ed. M.W. Cadotte, S.M. McMahon, and T. Fukami, 35–64. London: Springer.
Davis, M., R. Hobbs, A. Lugo, J. Ewel, G. Vermeij, J. Brown, M. Rosenzweig, M. Gardener, S. Carroll, et al. 2011. Don't judge species on their origins. *Nature* 474: 153–154.
Donlan, J., et al. 2005. Re-wilding North America. *Nature* 436: 913–914.
Donlan, J., et al. 2006. Pleistocene rewilding: an optimistic agenda for twenty-first century conservation. *The American Naturalist* 168: 160–183.
Ellis, E. 2012. The planet of no return: human resilience on an artificial Earth. In *Love your monsters*, ed. M. Shellenberger, and T. Nordhaus, 37–46. Oakland, CA: Breakthrough Institute.
Hallett, L., R.J. Standish, K.B. Hulvey, M.R. Gardener, K.N. Suding, B.M. Starzomski, S.D. Murphy, and J.A. Harris. 2013. Towards a conceptual framework for novel ecosystems. In *Novel ecosystems. Intervening in the new ecological order,* eds. R.J. Hobbs et al., 16–28. Oxford, UK: Wiley-Blackwell.
Hobbs, R., S. Arico, J. Aronson, J. Baron, P. Bridgewater, et al. 2006. Novel ecosystems: theoretical and management aspects of the new ecological world order. *Global Ecology and Biogeography* 15(1): 1–7.
Hobbs, R., L. Hallett, P. Ehrlich, and H. Mooney. 2011. Intervention ecology: applying ecological science in the twenty-first century. *BioScience* 61(6): 442–450.
Hobbs, R., E. Higgs, and C. Hall (eds.). 2013. *Novel ecosystems. Intervening in the new ecological order*. Oxford, UK: Wiley-Blackwell.
Hunter Jr., M.L., K.H. Redford, and D.B. Lindenmayer. 2014. The complementary niches of anthropocentric and biocentric conservationists. *Conservation Biology* 28(3): 641–645.
IUCN/SSC. 2013. *Guidelines for reintroductions and other conservation translocations. Version 1.0*. Gland, Switzerland: IUCN Species Survival Commission.
Kareiva, P., M. Marvier, and R. Lalasz. 2012. Conservation in the anthropocene. Beyond solitude and fragility. http://thebreakthrough.org/index.php/journal/past-issues/issue-2/conservation-in-the-anthropocene.
Kareiva, P., and M. Marvier. 2007. Conservation for the people. *Scientific American* 297: 50–57.
Kareiva, P., and M. Marvier. 2012. What is conservation science? *BioScience* 62(11): 962–969.
Keulartz, J. 2016. Future direction for conservation. *Environmental Values* 25(4): 385–407.
Keulartz, J., and C. Van der Weele. 2008. Framing and reframing in invasion biology. *Configurations, a Journal of Literature, Science and Technology* 16(1): 93–115.
Keulartz, J., and C. Van der Weele. 2009. Between nativism and cosmopolitanism. In *New visions of nature: complexity and authenticity*, ed. M. Drenthen, J. Keulartz, and J. Proctor, 237–256. Dordrecht: Springer Verlag.
Kueffer, C., and C.N. Kaiser-Bunbury. 2014. Reconciling conflicting perspectives for biodiversity conservation in the anthropocene. *Frontiers in Ecology and the Environment* 12(2): 131–137.
Lacy, R.C. 2013. Achieving true sustainability of zoo populations. *Zoo Biology* 32: 19–26.
Latour, B. 2012. Love your monsters. Why we must care for our technologies as we do our children. *The breakthrough Journal, Winter* 2012: 21–28.
Lynas, M. 2011a. *The god species: how the planet can survive the age of humans*. London: HaperCollins.
Lynas, M. 2011b. Geo-engineering, nuclear power and climate change: playing God is good for the planet. *The Telegraph*, 12 July 2011.

Lynas, M. 2014. *Nuclear 2.0: why a green future needs nuclear power.* Cambridge, UK: UIT Cambridge.

Marris, E., P. Kaveira, J. Mascaro, and E. Ellis. 2011. Hope in the age of man. *The New York Times*, December 7, 2011.

Miller, B., M.E. Soulé, and J. Terborgh. 2013. "New conservation" or surrender to development? *Animal Conservation* 17(6): 509–515.

Minteer, B. 2015. The perils of de-extinction. *Minding Nature* 8(1): 11–17.

Minteer, B.A., and J.P. Collins. 2013. Ecological ethics in captivity: balancing values and responsibilities in zoo and aquarium research under rapid global change. *ILAR* 54(1): 41–51.

Pearce, F. 2007. *With speed and violence. Why scientists fear tipping points in climate change.* Boston, MA: Bacon Press.

Perring, M. and E. Ellis. 2013. The extent of novel ecosystems: long in time and broad in space. In *Novel ecosystems. Intervening in the new ecological order,* eds. R. Hobbs et al. 66–80. Oxford, UK: Wiley-Blackwell.

Redford, K., D. Jensen, and J. Breheny. 2013. The long overdue death of the ex situ and in situ dichotomy in species conservation. *WAZA Magazine* 14: 19–22.

Rockström, J., et al. 2009. A safe operating space for humanity. *Nature* 461(7263): 472–475.

Seddon, P.J. 2015. De-extinction: reframing the possible. *Trends in Ecology & Evolution* 30(10): 569–570.

Seddon, P.J., C.J. Griffiths, P.S. Soorae, and D.P. Armstrong. 2014. Reversing defaunation: restoring species in a changing world. *Science* 345(6195): 406–412.

Shellenberger, M., and T. Nordhaus. 2004. *The death of environmentalism: global warming politics in a post-environmental world.* Oakland, CA: Breakthrough Institute.

Shellenberger, M., and T. Nordhaus. 2007. *Break through: from the death of environmentalism to the politics of possibility*. New York: Houghton Mifflin.

Sherkow, J.S., and H.T. Greely. 2013. What if extinction is not forever? *Science* 340(6128): 32–33.

Smith, F.A., C.E. Doughty, Y. Malhi, J.-C. Svenning, and J. Terborgh. 2015. Megafauna in the Earth system. *Ecography* 39: 99–108.

Soulé, M.E. 2013. The "new conservation". *Conservation Biology* 27(5): 895–897.

Steffen, W., P.J. Crutzen, and J.R. McNeill. 2007. The anthropocene: are humans now overwhelming the great forces of nature? *Ambio* 36(8): 614–621.

Steffen, W., et al. 2011a. The anthropocene: conceptual and historical perspectives. *Philosophical Transactions of the Royal Society* 369: 842–867.

Steffen, W., et al. 2011b. The anthropocene: from global change to planetary stewardship. *Ambio* 40(7): 739–761.

Steffen, W., et al. 2015. Planetary boundaries: guiding human development on a changing planet. *Science* 347(6223): 736–746.

Svenning, J.-C., P.B.M. Pedersen, C.J. Donlan, R. Ejrnæs, S. Faurby, M. Galetti, D.M. Hansen, B. Sandel, C.J. Sandom, J.W. Terborgh, and F.W.M. Vera. 2015. Science for a wilder anthropocene: synthesis and future directions for trophic rewilding research. *PNAS* 113(4): 898–906.

Tallis, H., and J. Lubchenco. 2014. Working together: a call for inclusive conservation. *Nature* 515: 27–28.

Wapner, P. 2010. *Living through the end of nature. The future of American environmentalism.* Cambridge, MA: The MIT Press.

Wuerthner, G., E. Crist, and T. Butler (eds.). 2014. *Keeping the wild. Against the domestication of Earth.* Washington, D.C.: Island Press.

Zalasiewicz, J., M. Williams, W. Steffen, and P. Crutzen. 2010. The new world of the anthropocene. *Environmental Science and Technology* 44(7): 2228–2231.

Part I
Between Human and Animal

Deep History, Evolutionary History, and Animals in the Anthropocene

Anita Guerrini

Abstract How do we write a history of animals in the Anthropocene? In the past decade, there have been several attempts to bring biological thought into historical writing. One, 'deep history' as advocated by Daniel Lord Smail, aims to push the beginnings of human history back into the Pleistocene, long before the advent of written documents. In another, environmental historian Edmund Russell advocates the study of co-evolution (of humans and other living things) to explain particular historical events. Neither of these approaches specifically comments on animals. This essay examines these approaches and others to develop a new historical discourse that de-centers humans and incorporates both human and non-human animals. Such a history will help to recapture the original moral aims of historical practice.

1 History in the Anthropocene

What is the role of history in the Anthropocene? Ecologists now speak of 'no-analogue' environments with no precedents in the past, and historians and philosophers have been predicting 'the end of history' since the 1980s from a number of different perspectives (Jackson and Hobbs 2009; Fukuyama 1992; Agamben 2004). Climate change has added new anxiety to these musings. Historian Dipesh Chakrabarty began his 2009 *Critical Inquiry* article, 'The Climate of History: Four Theses', with an arresting question: is history possible in a world without humans? Will climate change lead to human extinction and therefore to the end of history? To put it another way, has the necessary continuity from past to present to future been ruptured owing to the possible lack of a (human) future? Chakrabarty goes on to argue that anthropogenic climate change will necessarily collapse the distinction between natural history and human history, but he does not address what role animals might play in this new order. Animal studies scholars

A. Guerrini (✉)
Oregon State University, Corvallis, USA
e-mail: Anita.Guerrini@oregonstate.edu

B. Bovenkerk and J. Keulartz (eds.), *Animal Ethics in the Age of Humans*,
The International Library of Environmental, Agricultural and Food Ethics 23,
DOI 10.1007/978-3-319-44206-8_2

have made similar statements. In her well-known essay 'A Left-Handed Blow,' Erica Fudge concluded, 'No longer separate, in splendid isolation, humans must be shown to be embedded in and reliant upon the natural order' (Fudge 2002, 15). But such statements continue to assume that the subject of history is essentially and inevitably humans.

I agree with Chakrabarty that the Anthropocene merges natural history and human history. In his book *The Open,* philosopher Giorgio Agamben similarly postulates a merger of human and animal, which, he argues, will lead to the end of history. But I don't agree that history itself therefore disappears. Rather, the Anthropocene requires a new concept of history that radically de-centers humans. This might seem ironic or paradoxical in an age named for human dominance. This essay, however, proposes a different reading of the relationship between natural and human history and therefore between humans and animals. Non-human animals, I argue, also have histories, both participating in and independent of human histories. Uncovering those histories, particularly but not exclusively in evolutionary terms, will enable us to deconstruct the animal-human divide and begin to write a new history that can underpin a new ethics for the Anthropocene.

The modern Western discipline of history as an interpretive enterprise began in the late middle ages with the beginnings of humanism, the retrieval and reinterpretation of classical culture. Renaissance humanists and modern historians distinguish 'history' from 'the past': the past presumably happened, but without the interpretive enterprise of history, we would know nothing about it. That interpretive enterprise is always contingent on the kinds of evidence available (material, written, oral) and on the person who does the interpreting.[1] There is no stable entity we can simply label 'the past'. Rather, there are narratives about it that humans create. Historian of science Michael Gordin recently had this to say: 'Historians do not, *cannot*, have direct access to the past … [w]hat we all have are collections of traces—manuscripts, rocks, phonemes, fossils, genetic sequences, built structures—that we interpret today to create a narrative about what happened before' (Gordin 2014).

Early modern humanism was based largely on texts and set up two boundaries: between the human and the divine on one side, and between the human and non-human nature on the other. The space between these two ideas of what constitutes the human is what Agamben refers to as 'the open'. The question of a non-human history is therefore caught up in a larger question of the meaning and scope of the humanities. In its origins in early modern humanism, the humanities once included all of learning, and its practitioners viewed it as above all a moral and ethical enterprise. Humanism's new emphasis on the value of the human as opposed to the divine, and its revival of pre-Christian learning, included both literature and science: *scientia* referred to all of knowledge. Before 1800 at least, the humanities and the sciences were not separate but took part in a larger quest for knowledge (Grafton and Jardine 1986). Natural history and human history were components of a single story that began with the Creation. A new history for the Anthropocene

[1]For discussion of this point in relation to ecology, see Higgs et al. (2014).

returns to this larger story, with two important differences. Its essential components are the inclusion of non-human nature, particularly animals, as historical actors, and a greatly expanded time frame. As I will argue, the modern theory of evolution links these components.

2 Animal Histories

The historical study of animals crisscrosses many areas of humanistic study: literature, cultural history, the history of science, environmental history, social history. Although human histories have at times included animals in various ways, animals seldom appeared in works of mainstream academic history until the 1980s. Various histories of human mistreatment of animals appeared earlier, but their purpose was more polemical than historical.[2] As Harriet Ritvo comments in relation to her own career, 'In the early 1980s … it was considered both unusual and eccentric (which are not at all the same thing) to take animals seriously as historical subjects' (Ritvo 2010, 1). She, like many others in the field of animal studies, credits the English historian Keith Thomas with taking the history of animals out of the realm of the eccentric. Thomas's *Man and the Natural World* examined (as the subtitle to the English edition explained) changing attitudes toward the natural world in early modern England. Thomas took his topic expansively, and looked at multiple intersections between humans and early modern nature, from theology to agriculture to natural history and taxonomy to food and entertainment (Thomas 1983). Ritvo, probably the best-known American historian of animals, has followed a similar path of looking at animals in terms of a broader (human) cultural history, mainly in nineteenth-century England. Her first book, *The Animal Estate*, examined animals in Victorian England, finding that British treatment and uses of animals reflected the attitudes and anxieties of Victorian culture (Ritvo 1987).

While not precisely a historian of science, Ritvo in her second book, *The Platypus and the Mermaid*, considered scientific as well as popular ideas about the classification of humans and animals (Ritvo 1997). It touched on themes of breeding and wildness that she has continued to examine in other works, as she has moved closer to environmental history. Her third book, *The Dawn of Green* (2009), is not about animals at all, but she continues to write about them. Ritvo's main emphasis in her work is that our ideas about and treatment of animals can tell us a lot about ourselves and other humans. Through history, animals have provided a lens through which to view human society. At the same time, Ritvo clearly believes that animals merit study in themselves; the essays collected in *Noble Cows and Hybrid Zebras* range widely, from broader studies of animal use and representation to more specialized studies of classification and breeding (Ritvo 2010). Major themes underlying much of her work are the inescapable continuity between the

[2]For more on this point, see Guerrini (2003), 'Suggested Further Reading'.

past and the present, and the interweaving of animals into every aspect of human life. A moral purpose for history underlies much of Ritvo's work although she seldom takes an explicit ethical stance on human uses of animals.

In contrast, cultural studies scholar Erica Fudge has outlined an approach to the history of animals that finds its progenitor less in Keith Thomas than in Peter Singer, whose *Animal Liberation* (1975) changed the terms of moral engagement with animals and led to renewed activism on their behalf. Fudge's (2002) essay 'A Left-Handed Blow' links the history of animals to a present-day activism. In this she is not unlike the American environmental historian, Roderick Frazier Nash, who aimed to merge advocacy for nature with histories of it, particularly in his book *The Rights of Nature* (Nash 1989). To Fudge, the moral imperative of the historian, therefore, is clear-cut: 'A history of animals', she writes, 'would seem to be an obvious place where yet again the ethical nature of the historian's work should be clear' (Fudge 2002, 4). But, as she points out, animals have left no documents (written or oral), nor do they have a sense of historical time. She concludes that therefore a history of animals, as opposed to a history of human attitudes toward or relations with animals, is 'impossible' as history is currently constructed (Fudge 2002, 6).

Fudge puts aside environmental and evolutionary history to focus on text-based history, which she classifies into three types: one is intellectual (such as Thomas's work, which considers 'attitudes, not animals'); another is 'humane' history, which views animals as 'the site of [human] social change'. Fudge places Ritvo's work in a third category, 'holistic history', that looks not only at human-animal relations, but 'leads to the inevitable conclusion that the human is only ever meaningful when understood in relation to the not-human' (Idem, 8–10). Such a history, combined with a postmodern reading of documents 'against the grain', leads to Fudge's assertion of the purpose of a history of animals: 'If we identify the human as neither a given nor a transcendent truth, then intellectual attitudes that leave unquestioned the result of these assumptions—dominion—must themselves be reviewed as not true, but created' (Idem, 11). Fudge's history of animals, therefore, is above all a history of anthropocentrism, of cruelty and dominion. Only by questioning anthropocentrism can we write a legitimate history of animals.

While Fudge claims to challenge the status of the human by this approach, she nonetheless assumes the existence of a stable rhetorical entity we can call 'the human' against which is placed 'the animal'. But as we shall see, both 'human' and 'animal' are themselves highly contingent historical categories. Fudge seems to suggest this:

> We must abandon the status of the human as it is presented within humanist history; we must read against this. Instead, we need to assert and assess the ways in which 'human' is always a category of difference, not substance: the ways 'human' always relies upon 'animal' for its meaning. By refusing humanism, and implicitly, anthropocentrism we place ourselves next to the animals, rather than as the users of the animals, and this opens up a new way of imagining the past, something that has to be central to the project (Idem, 15).

But Fudge believes that this new history will be based on a reinterpretation of existing documents. She does not take the further step of looking at different kinds of evidence that is not based on texts.

3 Expanding Time

Renaissance humanism gave the discipline of history its modern form. Like the humanities, history as a discipline underwent a further shift in definition in the nineteenth century. The humanist admiration of classical models and emphasis on comparative textual analysis gave rise to the narrative form and chronology of history writing in the West: by this definition, history begins with written texts; everything before that is 'prehistory' (Smail 2008, 1–2). The Creation served as the starting point for pre-modern histories in the West, dated by Anglican Archbishop James Ussher (employing Biblical chronology) in 1655 as the nightfall preceding Sunday 23 October 4004 BC. Nascent geological studies soon suggested a vastly older age for the earth. These studies, coupled with Darwin's theory of evolution, effectively destroyed this 6000-year chronology of human existence. But nineteenth-century academic history quietly translated it to a secular chronology in which human history began with the rise of agriculture and the dawn of civilization, also approximately 6000 years ago, as we can see from any textbook in Western civilization. Historian of science Mott Greene has described how nineteenth-century scholars developed the concept of 'prehistory', defined as the era before written records. 'Prehistory' filled a gap between archaeology and history, between geological time and historical time, which scholars have until very recently been reluctant to cross. It is a 'buffer zone' (Greene 1992, 1–3).

Eliminating this 'buffer zone' of prehistory and consolidating these various chronologies are critical steps toward a redefinition of history for the Anthropocene. Over 30 years ago the iconoclastic environmental philosopher Paul Shepard suggested that evolutionary biology, archaeology and history form one unbroken story, arguing that the introduction of agriculture was disastrous for humans and nature alike. Both time and place were important; in *Nature and Madness*, Shepard wrote:

> We must stand apart from the conventions of history, even while using the record of the past, for the idea of history is itself a Western invention whose central theme is the rejection of habitat. It formulates experience outside of nature and tends to reduce place to location… History conceives the past mainly in terms of biography and nations (Shepard 1982, 47).

More recently, historian Julia Adeney Thomas argued that scale, both temporal and geographical, 'matter' to the writing of history (Thomas 2014, 1588). Rejection of the 'biography and nations' approach to history is not new, however. Beginning in the late 1920s, the *Annales* school in France and particularly Fernand Braudel brought attention to scales of history, contrasting the 'longue durée' of geological time and geographic space to the shorter and more episodic history of human events. The first section of his magnum opus, *The Mediterranean and the*

Mediterranean World in the Age of Philip II, describes the geography of the Mediterranean, what he called 'histoire profonde' (Braudel 1972). In his book on the *Annales* school, Peter Burke claims that 'Braudel has done more to change our notions of space and time than any other historian of the twentieth century' (Burke 2015, 46). In his later book *Memory and the Mediterranean*, Braudel returned to geologic time to discuss the relationships between humans and the Mediterranean environment (Braudel 2001).

But few took up Braudel's expansive approach to chronology until recently. In the past two decades, several approaches have emerged. The 'big history' approach pioneered by David Christian reaches back to the Big Bang. In *Maps of Time*, Christian aimed for a 'grand unified story', a modern 'creation myth' (Christian 2004, 4; 11). Christian and others link this expanded time frame to an expanded geographical focus beyond the West. The IHOPE (Integrated History and Future of People on Earth) project, based in Sweden, joins scientists, social scientists, and humanities scholars. The project's 2007 volume *Sustainability or Collapse?* outlined its agenda, which is organized around two big ideas: 'humans and the rest of nature' as opposed to 'humans and nature', and three timescales. The timescales are the millennial (up to 10,000 years ago), the centennial (up to 1000 years ago), and the decadal (up to 100 years ago) (Costanza et al. 2007). As others have pointed out, these scales, based closely on the ice ages in the Northern hemisphere, are not equally applicable globally (Robin and Steffen 2007).

Perhaps most influential among academic historians has been Daniel Lord Smail's 'deep history' that begins in the Paleolithic and encompasses human evolutionary history. Deeply influenced by evolutionary psychology (if not entirely convinced by it), Smail argues that one may trace the deep history of humanity by means of the human brain. Its chemistry reflects the evolutionary development of humanity, and a 'neurohistory' can both reflect and explain aspects of human culture such as patterns of dominance and submission. Smail's work tapped into ongoing studies of the history of human emotions as well as work by historians of medicine and the environment on evolution, genetics, and disease. The 'neurological turn' among historians has morphed into a more general 'biological turn' whose diversity was recently signaled in a special section of the *American Historical Review* in December 2014 entitled 'History meet Biology'. I will single out one of these approaches, evolutionary history.

Smail's work, employing the evolution of the brain as a window into the past, suggests one way to incorporate evolution into history. Edmund Russell suggested another a few years ago in the context of environmental history (Russell 2003, 2011). Evolutionary history, as Russell defines it, focuses on human impacts on the evolution of other species, as well as the co-evolution—the reciprocal impacts—on both humans and non-humans. Although Russell notes that the causal arrow points both ways, in that humans influence animals and animals influence humans, his emphasis is on anthropogenic evolution and on its impact on human history. Once again history is, by definition, about humans. Russell's focus in his book *Evolutionary History* is on domesticated species (both animals and plants) and on human diseases (Russell 2011). While evolutionary history has the potential to extend far into the human past, even into the pre-human past, Russell adopts a conventional timeline of human history by focusing on domestication and agriculture.

In his most recent work on evolutionary history, Russell returns to coevolution, particularly in the context of domestication. Although most of his examples continue to emphasize humans as drivers of evolutionary change, he points out that 'we could just as easily assume that non-human populations initiated the process', such as in the domestication of wolves into dogs (Russell 2014, 1520). His recognition of animal agency recalls the arguments of Stephen Budiansky that animals 'chose' domestication. In his 1992 book *The Covenant of the Wild*, Budiansky argued that humans were 'not… the arrogant despoilers and enslavers of the natural world, but… a part of that natural world, and the custodians of a remarkable evolutionary compact among the species' (Budiansky 1992, 24). Among the drivers of domestication, according to Budiansky, was climate change—in this case, the advent of the Ice Ages (Idem, 72). As these examples show, expanding the timeframe moves toward decentering humans but does not entirely accomplish this task.

4 History, Prehistory, and the Human

Apart from expanding the time frame for history, considering history as an evolutionary story greatly broadens the kinds of evidence that historians may consider, to include what Gordin above (echoing Smail) refers to as 'traces'. Archaeological, paleontological, geological, linguistic, biological, and even atomic evidence is therefore all fair game for the historian. But the implication that therefore non-humans might also be actors in the historical story is still not quite articulated or is even rejected by some who practice evolutionary history.

Smail justifies extending his timeframe to the Paleolithic because humanity existed that long ago. 'Humanity', he asserts, 'is the proper subject of history' (Smail 2008, 2). He approvingly cites Greene's comment that 'To abandon prehistory would be to postulate continuity between the biological descent of hominids, and the "ascent to civilization" of the abstract "mankind" of humanistic historical writing' (Smail 2008, 2; Greene 1992, 3). Prehistory, adds Greene, 'is a place where merely biological hominids turned into 'Men" (Greene 1992, 3). Smail, no doubt purposefully, avoids defining when or how hominids became humans. But this lack of definition undermines his overall argument, which excludes animals from history-as-human. Although he opines that a lack of self-consciousness does not preclude having a history, he cannot quite accept that animals might have a history on the same level of complexity as human history: animals are not, by his definition of history, historical agents. History—here apparently defined simply as events in the past—happens to both humans and animals, but only humans have the self-consciousness that allows them to make history (Smail 2008, 57). Smail admits that animals have an evolutionary history, but it is a history ultimately of aggregates, of classes, and not of individuals. As in certain kinds of social history that rely on statistical surveys of large populations, he argues that we cannot attribute agency to such aggregates (Idem, 71–72). In addition, he argues that language is a critical indicator of consciousness, so this too means that animals cannot have a history other than the 'video-recorder style of natural history' (Idem, 57).

Smail's neglect of animals nonetheless leaves openings for a broader history. An explicit acknowledgment of the unclear margins between human and non-human in the evolutionary process would constitute one step toward such a history. Looking at the reciprocal impacts of animals and humans in this process is another, as Russell has suggested. While both of these approaches are fruitful and should be pursued, neither of them gets to the heart of the issue, which (returning to the humanists and to Fudge) is the moral enterprise of history and the humanities. Smail emphasizes a long time frame, but does not add animals to his framework; Russell notes that 'Studying human evolution is not necessary (or sufficient) for evolutionary history', but does not problematize history itself except to comment that 'nearly everything historians study… would not have occurred without domestication', a conclusion 180° from Smail, who argues that such events as domestication and writing are consequences of changes in brain chemistry (Russell 2003, 205; Smail 2008). Both ultimately assume there is a category we can call 'the human' but this is itself a contingent historical category, as each implicitly acknowledges. Ideally, evolutionary history calls into question what is human, but it is not quite there yet. And neither Smail nor Russell talks about value: what is the value of animals, both in terms of human culture and in terms of themselves? How does the human evaluation of nature change over time?

5 Animals, Natural History, and Evolution

What kinds of evolutionary-historical stories might we write about animals? Aristotle was not the first to regard animals as subjects of inquiry rather than as commodities, but he was the first Western philosopher to do this systematically. His works on animals, particularly *History of Animals*, *Parts of Animals*, and *Generation of Animals*, established a science of natural history that endured until Darwin and in some ways persists today. *Historia* (Greek ἱστορία) originally meant simply an inquiry or an investigation, or an account of such an inquiry. It did not imply the passage of time, and this definition of 'history' persisted into the modern era. Aristotle's *History of Animals* offered detailed descriptions of all animals known to him. He took every opportunity to observe every animal he could: wild and domestic, native and exotic, terrestrial and aquatic. He investigated morphology, habitat, behavior, and what he called 'manner of life'; what parts were the same and what were different; how they ate and reproduced. He noted natural kinds and attempted various classifications. Broad groupings seemed obvious: birds were different than fish. Some animals had two feet, some four, others none. Some animals were 'blooded', some, like insects, were not. Aristotle believed that nature mirrored human society, and that human society mirrored nature. Hierarchy was the natural configuration of the world, and he found in animal and human generation a hierarchical system based on degrees of perfection as measured by degrees of natural heat. Thus warm-blooded viviparous animals were 'hotter' and therefore more perfect than oviparous animals, and so forth down to those animals that he believed produced larva rather than eggs. This hierarchical system, later known as the 'chain of being' or 'ladder of nature', proved to have remarkable staying power in Western thought. The chain of being was not only hierarchical but full, including every animal (and plant) that could possibly exist. It was also unchanging, so that species were fixed in time and space. And it was teleological: nature always worked toward a purpose.[3] Medieval scholastics found this system quite compatible with Christian doctrine, and they transformed Aristotle's eternal, uncreated nature into a created and temporal one.

The age of discovery in the sixteenth century led to an influx of previously unknown animals from the Americas, Africa, and Asia to Europe, and seriously disrupted the idea of the chain of being. For example, the Swiss naturalist Conrad Gessner did not quite know what to do with the armadillo in the 1550s, and he strained to fit it into a known niche on the chain of being. As translated by Edward Topsell half a century later, the 'Tatus or Guinean Beast' ('Guinean' in this era simply meant 'foreign')

> is brought for the most part out of the new-found world, and out of *Guinia*, and may therefore be safely conveyed into these parts, because it is naturally covered with a harde shell, devided and interlined like the fins of fishes, outwardly seeming buckled to the backe like coat-armor, within which, the beast draweth up his body, as a Hedghog doth within his prickled skin, and therefore I take it to be a *Brazilian* Hedghog (Topsell 1607, 705).

In the eighteenth century, Linnaeus and Buffon attempted, in very different ways, to apply human reason to the seemingly chaotic organization of the natural world. Linnaeus developed a temporally static but all-encompassing system of classification of the plant and animal worlds based on mode of reproduction. Buffon initially rejected any system of classification, claiming that identity lay in the individual, but greatly expanded the timeframe of the natural world from 6 or 7000 years to at least 75,000 years in his multiple-volume *Histoire naturelle*.[4] Both Buffon and Linnaeus later modified some of their stances: Linnaeus eventually accepted that species could change over time, while Buffon eventually accepted an idea of species. By the end of the eighteenth century, new discoveries in paleontology and new geological theories led to an even longer timeframe, in the millions of years. These discoveries also confirmed the fact of extinction. The 'traces' left by fossils became important evidence of change and contingency in the past, markers of a previously unknown animal history.

The evolutionary story developed by Darwin softened the boundary between animals and humans, but did not destroy it. Already studies of primates such as

---

[3]This paragraph and the next are adapted from a more extensive account in Guerrini (2015).

[4]Buffon secretly believed the earth could be as much as ten million years old (Roger 1997, 411).

Edward Tyson's 'ourang-outang' (actually a chimpanzee) of 1699 had thrown doubt on the rigid hierarchies of the chain of being, and Linnaeus classified both humans and certain apes such as the orang-outang under the genus *Homo*. Darwinian taxonomies separated primates into *Homo* and *Pan* as paleontological discoveries pushed the evolutionary split between humans and apes farther and farther back in time (Schwartz and Tattersall 2015). There is still fierce debate among scientists over when *Pan* and *Homo* split, or whether *Pan* and *Homo* should be separate genera at all.

The evolutionary history of humans therefore complicates the human-animal divide in one way. The evolutionary history of animals complicates it in another. On the one hand, it reveals a long and complex history of animals that often does not include humans at all. What can such a history tell us about the development of animal consciousness, social structure, migration, or interactions among animals? What can such a history tell us about present-day animals? Scientists employ the same tools to analyze animal evolution as to analyze early humans, including paleontology, genetic analysis, and neurobiology. But such analysis of animals seldom figures in historical accounts.

On the other hand, as we have seen, the tools of evolutionary science have revealed a rich history of human-animal interactions with widely-ranging implications for human and animal history. The history of domestication is only one aspect of this history; many other kinds of animal-human interactions in the past await study. New science as well as new ways of looking at history are on the verge of drastically changing our ideas of the past. In particular, the new science of inheritance based on epigenetics has implications far beyond human history. Epigenetic science studies changes in organisms caused by the modification of gene expression rather than by an alteration of the genetic code itself. Expression can be influenced by a number of external factors, particularly environment, and the changes thus induced are heritable (Brooke and Larsen 2014).

6 Evolution in Play: Rewriting History

Early modern natural historians grasped the notion of extinction with difficulty. The idea that a specific animal could simply disappear violated a number of common beliefs. Aristotle had declared that species, along with the rest of nature, were eternal. The great chain of being did not allow for spaces among its tightly packed rungs, and Christians argued that God did not make mistakes and that therefore extinction was impossible and indeed unthinkable. Yet, quite apart from the evidence of fossils, several animals had become extinct in historical times in Europe. Many commentators noted the death of the last native European ox or aurochs in 1627, a breed that had been under the protection of the king of Poland for over a century (Szafer 1968).

The discovery at the end of the eighteenth century of the bones of mammoths and of the giant sloth that Georges Cuvier named the megatherium provided convincing evidence of animals that no longer occupied the planet and had no living analogues. The fact of extinction became a key concept for Charles Darwin, who argued that species that could not adapt to changing environmental conditions would become extinct. According to evolutionary theory, other animals might occupy the ecological niches left by extinction. But extinction was forever—at least until recently.

'Rewilding' is an attempt by some ecologists to undo the effects of extinction by reintroducing animals to fill lost ecological niches. Various plans for 'Pleistocene Parks' have emerged around the world, and some have begun to be realized (Marris 2009). Jozef Keulartz's essay in this volume discusses one of these, at Oostvaardersplassen in the Netherlands, and its ethical implications. However, the recent development of new genomic technologies known as 'synthetic biology' has led to proposals for a much more radical program. Some geneticists, foremost among them George Church, are promoting the de-extinction of a number of species by genetic means. Enough genetic material remains in preserved specimens of such animals as the passenger pigeon (extinct since 1914) and the thylacine (extinct since 1936) that the prospect of revival by genomic means is a possibility (Ogden 2014). There have been attempts to bring back certain extinct species like the aurochs by 'back-breeding', a process of selective breeding for characteristics of lost species (Maas 2011). Genetic de-extinction is a high-tech version of this, an attempt not to rewind the evolutionary process (as back-breeding does) but to short-circuit it. It is very much a science for the Anthropocene, assuming human control over nature and its processes. But while bringing back certain animals, de-extinction at the same time reduces them to cells and genes, and erases their history in favor of a human-made one.

7 Conclusion

The tools, therefore, are available for a new history in the Anthropocene that de-centers humans and reconceptualizes the animal-human relationship. Such a history would be based on an evolutionary timescale and would involve many kinds of sources apart from written texts. This new history must also regain the moral ground that academic history seems largely to have lost in the past century. As the practice of history has become more specialized and less generally accessible, it has lost its ethical authority and credibility. Together with a new ethics for the Anthropocene, a new history could do much to regain the moral aims of the humanities.

References

Agamben, G. 2004. *The open: Man and animal* (trans: Kevin Attell). Stanford, CA: Stanford University Press.

Braudel, F. 1972. *The Mediterranean and the Mediterranean world in the age of Philip II* (trans: Siân Reynolds), vol. 2. New York: Harper and Row.

Braudel, F. 2001. *Memory and the Mediterranean* (trans: Siân Reynolds). New York: Knopf.

Brooke, J.L., and C.S. Larsen. 2014. The nurture of nature: Genetics, epigenetics, and environment in human biohistory. *American Historical Review* 119(5): 1500–1513.

Budiansky, S. 1992. *The covenant of the wild: Why animals chose domestication.* New Haven: Yale University Press.

Burke, P. 2015. *The French historical revolution: The Annales school, 1929–89*, 2nd ed. Stanford: Stanford University Press.

Chakrabarty, D. 2009. The climate of history: Four theses. *Critical Inquiry* 35: 197–222.

Christian, D. 2004. *Maps of time: An introduction to big history.* Berkeley: University of California Press.

Costanza, R., Graumlich, L.J., and W. Steffen, eds. 2007. *Sustainability or collapse? An integrated history and future of people on earth.* Cambridge, MA and London: MIT Press.

Fudge, E. 2002. A left-handed blow: Writing the history of animals. In *Representing animals*, ed. N. Rothfels, 3–18. Bloomington: Indiana University Press.

Fukuyama, F. 1992. *The end of history and the last man.* New York: Free Press.

Gordin, M. 2014. Evidence and the instability of biology. *American Historical Review* 119(5): 1621–1629.

Grafton, A., and L. Jardine. 1986. *From Humanism to the humanities.* Cambridge, MA: Harvard University Press.

Greene, M.T. 1992. *Natural knowledge in preclassical antiquity.* Baltimore: The Johns Hopkins University Press.

Guerrini, A. 2003. *Experimenting with humans and animals: From Galen to animal rights.* Baltimore: Johns Hopkins University Press.

Guerrini, A. 2015. Animals and ecological science. In *Oxford handbook of animal studies*, ed. L. Kalof, forthcoming. Online February 2015. doi:10.1093/oxfordhb/9780199927142.013.25.

Higgs, E., D. Falk, A. Guerrini, et al. 2014. The changing role of history in restoration ecology. *Frontiers in Ecology and the Environment* 12(9): 499–506.

Jackson, S., and R. Hobbs. 2009. Ecological restoration in the light of ecological history. *Science* 325: 567–569.

Maas, P.H.J. 2011. Recreating extinct animals by selective breeding. In *TSEW* (2016). The Sixth Extinction Website http://www.petermaas.nl/extinct. Downloaded on 3 January 2016.

Marris, E. 2009. Reflecting the past. *Nature* 462: 30–32.

Nash, R.F. 1989. *The rights of nature. A history of environmental ethics.* Madison, WI: University of Wisconsin Press.

Ogden, L.E. 2014. Extinction is forever … or is it? *BioScience* 64: 469–475.

Ritvo, H. 1987. *The animal estate. The English and other creatures in the Victorian age.* Cambridge, MA: Harvard University Press.

Ritvo, H. 1997. *The platypus and the mermaid and other figments of the classifying imagination.* Cambridge, MA: Harvard University Press.

Ritvo, H. 2009. *The dawn of green: Manchester, Thirlmere, and modern environmentalism.* Chicago, IL: The University of Chicago Press.

Ritvo, H. 2010. *Noble cows and hybrid zebras. Essays on animals & history.* Charlottesville, VA: University of Virginia Press.

Robin, L., and W. Steffen. 2007. History for the Anthropocene. *History Compass* 5(5): 1694–1719.

Roger, J. 1997. *Buffon. A life in natural history* (trans: Sarah Lucille Bonnefoi). Ithaca, NY: Cornell University Press.

Russell, E. 2003. Evolutionary history: Prospectus for a new field. *Environmental History* 8(2): 204–228.
Russell, E. 2011. *Evolutionary history. Uniting history and biology to understand life on earth.* Cambridge, New York: Cambridge University Press.
Russell, E. 2014. Coevolutionary history. *American Historical Review* 119(5): 1514–1528.
Schwartz, J., and I. Tattersall. 2015. Defining the genus homo. *Science* 349: 931–932.
Shepard, P. 1982. *Nature and madness.* San Francisco, CA: Sierra Club Books.
Singer, P. 1975. *Animal liberation: A new ethics for our treatment of animals.* New York: Random House.
Smail, D.L. 2008. *On deep history and the brain.* Berkeley, CA: University of California Press.
Szafer, W. 1968. The Ure-ox, extinct in Europe since the seventeenth century: An early attempt at conservation that failed. *Biological Conservation* 1: 45–47.
Thomas, J.A. 2014. History and biology in the Anthropocene: Problems of scale, problems of value. *American Historical Review* 119(5): 1587–1607.
Thomas, K. 1983. *Man and the natural world, changing attitudes in England (1500–1800).* Harmondsworth, Middlesex, England: Allen Lane/Penguin Books Ltd.
Topsell, E. 1607. *The historie of foure-footed beastes.* London: William Jaggard.

Organisms as Teachers? The Promise of Biomimicry

Sanne van der Hout

Abstract The learning-from-nature movement 'biomimicry' promises to guide us in living sustainably on and with planet Earth. Whereas traditional technological approaches tend to see nature as a 'resource' available for unrestricted use, the founders of biomimicry present organisms and natural systems as our 'mentors' or 'teachers'. In this chapter, I will critically reflect on the promises hidden in this 'tutorial narrative'. How did the biomimicry movement come into being and what inspired its founders to develop the tutorial narrative? What does this narrative exactly imply? Do the research activities of biomimicry practitioners match its rhetoric? To what extent do these activities reflect a new, more respectful relationship with nature in general, and animals in particular? I will conclude by arguing that the realization of a more humble relationship between humans and our fellow species entails more than the introduction of a new narrative; after all, not only the Earth's resources, but also her 'wisdom' or 'creativity' can be used in an instrumental fashion.

> Words are like empty balloons, inviting us to fill them up with associations. As they fill they begin to gain intrinsic force and at last to shape our perceptions and expectations. Worster (1994, 191).

1 The Biomimicry Revolution

How can we live sustainably on and with planet Earth? Confronted with the Earth's vulnerability to human interventions, a growing number of scientists, designers, and engineers are adopting the sustainable design principles of 'biomimicry' or 'biological design'. Co-founder Janine Benyus describes biomimicry as "a new science that studies nature's models and then imitates or takes inspiration from these

S. van der Hout (✉)
Maastricht University, Maastricht, Netherlands
e-mail: s.vanderhout@maastrichtuniversity.nl

B. Bovenkerk and J. Keulartz (eds.), *Animal Ethics in the Age of Humans*,
The International Library of Environmental, Agricultural and Food Ethics 23,
DOI 10.1007/978-3-319-44206-8_3

designs and processes to solve human problems, e.g., a solar cell inspired by a leaf" (Benyus 2002, preliminary pages). The movement is couched in promissory terms, and is even presented as a second Industrial Revolution. Whereas traditional technological approaches tend to see natural systems and organisms as resources available for unrestricted use, "the Biomimicry Revolution introduces an era based not on what we can *extract* from nature, but on what we can *learn* from her" (idem, 2; cf. Bensaude-Vincent et al. 2002, 2).

To illustrate how biomimicry distinguishes itself from classical 'resource approaches' to nature, its founders use a particular narrative: they present organisms and natural systems as our 'mentors' or 'teachers' (Baumeister et al. 2014; Benyus 2002). Using this narrative, they seek to develop a collective and shared account of what is at stake in the biomimicry movement, as well as to legitimize their research activities. Whereas closely related disciplines such as 'biomimetics' and 'bionics' continue to "rely on a conception of biological life [...] using mechanistic concepts and metaphors" (Macnaghten 2010, 31), biomimicry promises to put an end to this 'instrumentalization' of nature; the consultation of nature as our teacher will bring us into 'right relation with the rest of the natural world' (Baumeister et al. 2014, 8).

In this chapter, I will critically reflect on the promises at the heart of the 'tutorial narrative'. What does this narrative exactly imply? Do the research activities of biomimics—i.e. biomimicry practitioners—match the rhetoric of this narrative? In other words: to what extent do the actual research projects of biomimics reflect a new, more respectful relationship with nature in general, and animals in particular? Please note that, when biomimics describe nature as our mentor or teacher, they may refer not only to animals but to all life forms, i.e. from bacteria to entire ecosystems. As described in the *Biomimicry Resource Handbook*:

> Their [i.e. biomimicry practitioner's] mentors are the bacteria, fungi, plants, and animals of this planet, the organisms that clothe the landscape, cycle the nutrients, cleanse the air, sweeten the water, and create soil from rock. They are beings that can fly around the world without an engine, dive down ocean Everests without a tank, drink luxuriously from a wisp of fog, or shelter a beach dune from a hurricane gale (Baumeister et al. 2014, 10).

The structure of this chapter is as follows. I will start with explaining how the biomimicry movement came into being, and what inspired its founders to develop the tutorial narrative. Next, I will reflect on some of the questions raised by the tutorial narrative: in what ways can nature be thought of as a teacher, and what does it imply to recognize ourselves as nature's students? Finally, I will argue that, to enable a more respectful relationship between humans and the rest of nature, it is not sufficient that biomimicry practitioners present themselves as nature's students; after all, not only the Earth's resources, but also her 'wisdom' or 'creativity' can be used in an instrumental fashion. I would like to indicate that this contribution should not be read as a conclusive study; it is rather a preliminary reflection on how emerging technologies such as biomimicry no longer present human beings as situated *outside* and *above* the natural world, but as residing *in* this world, in constructive partnership with all other species.

2 The Conscious Emulation of Life's Genius

In her influential book *Biomimicry: Innovation Inspired by Nature* (2002), Benyus describes how her outlook on nature management went through a change. During her forestry studies, a reductionist, human-centered approach to nature management prevailed:

> In reductionist fashion, we studied each piece of the forest separately [...]. There were no labs in listening to the land or in emulating the ways in which natural communities grew and prospered. We practiced a human-centered approach to management, assuming that nature's way of managing had nothing of value to teach us (2002, 3).

When Benyus, started writing books on wildlife habitats, she was surprised by 'the exquisite ways that organisms are adapted to their places and to each other' (Idem, 4). She began to wonder why human beings, while facing the same physical challenges as all other living beings, sought to meet those challenges through human cleverness alone. It was then that Benyus decided to develop an alternative approach, in which organisms and natural systems are no longer regarded as resources available for unrestricted use, but as 'the ultimate teachers' (Ibidem). Benyus labelled this approach as 'biomimicry', derived from the Ancient Greek *βιος*, meaning *life*, and *μιμησις*, meaning *to imitate*. Contrary to what the term might suggest, the practice of biomimicry implies more than mere copying or slavish imitation. Actually, the term 'emulation' is more appropriate in this context, a process of *creative* imitation in which nature's lessons are translated into a technological idea. The following example illustrates how this process unfolds:

> Biomimics may study a spider to learn about sensing, fiber manufacture, adhesion, or tensegrity, but we are not actually trying to recreate the spider. What we're trying to emulate are the design principles and living lessons of the spider. How a spider meets its needs while helping to enhance its habitat is as important to a biomimic as how it spins its silk (Baumeister et al. 2014, 11).

What is so revolutionary about this "conscious emulation of life's genius" (Benyus 2002, 2)? After all, there is a long history of engineers, scientists and artificers who gained inspiration from nature. A classic example is Leonardo Da Vinci, who closely observed the anatomy of birds in the hope of enabling human flight. Although Da Vinci did not succeed in creating his own flying machine, his sketches were a source of inspiration for the Wright Brothers, who built the world's first successful airplane in 1903. Another well-known example is the Eiffel Tower, distributing its weight and stress in the same pattern of struts and spars as the human thigh bone (Ball 2001). In the 1950s, the American engineer Otto Schmitt turned these occasional applications of bio-inspired design into a more or less formal discipline. Instead of biomimicry, Schmitt used the term 'biomimetics' to describe the "transfer of ideas and analogues from biology to technology" (Vincent et al. 2006, 471). In 1960, US Air Force engineer Jack Steele coined the term 'bionics', which he defined as "the science of systems which have some function copied from nature, or which represent characteristics of natural systems or their analogues" (idem.).

As the above illustrates, the translation of biological processes into human technologies is not all that new. Even the objective to emulate nature in a less random and more systematic manner is over 50 years old. Still, there is an important difference between biomimicry and related forms of bio-inspired design. Whereas biomimetics and bionics are mainly presented as very effective approaches to solving engineering problems, biomimicry is explicitly connected with "a new way of viewing and valuing nature" (Benyus 2002, preliminary pages). This normative dimension is clearly reflected in one of the three interconnected 'seeds' underlying the biomimicry movement: the *ethos* element. The founders of biomimicry define the ethos element as "the essence of our ethics, our intentions, and our underlying philosophy for why we practice biomimicry. Ethos represents our respect for, responsibility to, and gratitude for our fellow species and our home."[1] To put it differently: the ethos element refers to a set of ideals and beliefs about *how* to consult nature, and *why*. How should nature be consulted? In a respectful manner. And why should we consult her? Because asking nature for advice will bring us "into right relation with the rest of the natural world" (Baumeister et al. 2014, 8).

3 Biomimicry as a Tutorial Practice

The ethos element forms a *leitmotif* in biomimicry discourse, and is most strongly represented by the image of nature as 'mentor' or 'teacher'. For example, in the *Biomimicry Resource Handbook*, biomimicry is put forward as a movement that regards "nature as a teacher rather than a warehouse" (Baumeister et al. 2014, 10). Moreover, Benyus argues that in a biomimetic world, humans "would consult animals and insects that have used plants for millions of years to keep themselves healthy and nourished" (Benyus 2002, 3). Biomimicry practitioners not only draw attention to the way in which *nature* appears in their research practices; they also show what this implies for us *humans*. If nature is our teacher, we are nature's 'students' or 'apprentices'. Benyus for instance suggests that the realization of a biomimetic future requires us to be "attentive to nature's lessons" (Idem, 8). Furthermore, Baumeister and colleagues describe the move towards biomimicry as a "shift in stance […] from conqueror to student" (Baumeister et al. 2014, 13).

The quotes clearly show that the tutorial narrative seeks to oppose classical 'resource approaches' to nature which "cast […] all of creation in categories of utility" (Evernden 1993, 23). According to the Australian eco-feminist Val Plumwood, a typical feature of these approaches is the strict separation between

---

[1]http://biomimicry.net/about/biomimicry/biomimicry-designlens/essential-elements/—accessed 15 December 2015.

human and non-human nature; they are based on a conception of human beings as "belonging to a superior sphere apart, a rational sphere of exclusively 'human' ethics, technology and culture dissociated from nature and ecology" (Plumwood 2002, 100). A result of separating the world in merely two realms—i.e. a human and a non-human realm—is that the complexity of the latter is severely underestimated: "Nature and animals tend to be seen as all alike in their lack of consciousness, which is assumed to be exclusive to the human, and the range and diversity of mindlike qualities found in nature and animals is ignored" (idem, 107). Plumwood argues that the variety of nature is only taken into consideration if it is expected to contribute to human prosperity: "Nature is conceived in terms of interchangeable and replaceable units, (as 'resources', or standing reserve) rather than as infinitely diverse and always in excess of knowledge and classification" (idem.).

Contrary to biomimicry practitioners, representatives of bionics and biomimetics continue to describe natural systems and organisms in instrumental or mechanistic terms, for instance when they refer to nature as a "catalogue of technological ideas from which we can select the solution of our choice" (Poelman 2015, 28).[2] This mechanization of nature is also reflected in the following quote, in which biological processes are described in terms of technical processes:

> From a technical viewpoint, organisms and their body parts can be regarded as (parts of) "machines" that fulfill all sorts of practical and recognizable functions, such as moving, observing, digging, heating, repairing, filtering, attaching, storing energy, drilling, isolating, pumping, or protecting (idem, 27).

4 Organisms as Teachers

The founders of biomimicry use the tutorial narrative to position their field in a favorable light; by presenting the relationship between nature and humans as a teacher-student relationship, they seek to demonstrate how biomimicry distinguishes itself from resource approaches to nature, and to develop a shared account of what is at stake in their field (cf. Macnaghten 2010). However, Benyus, and her colleagues seem to underestimate the complexity and multi-layered character of the tutorial narrative, notably in terms of the moral message it conveys. In the following sections, I will explore some of the questions raised by the tutorial narrative. Some explicitly concern the position of organisms, others that of humans. I will start with discussing the organism or 'teacher' side of the tutorial narrative.

In their article 'Questioning the theory and practice of biomimicry', Marshall and Lozeva (2009) argue that natural systems and organisms can be thought of both as (a) technical advisors, and (b) teachers of values and morals.

---

[2]Translated from Dutch into English by the author of this chapter.

(a) ***Nature as technical advisor***

The first form of mentorship refers to the practice of biomimicry, i.e. the actual consultation of nature's models and systems to solve complex human design challenges. The founders of biomimicry describe this 'active part' of biomimicry with the verb *emulate*. As Dayna Baumeister explains in an interview: "… emulate […] is the innovation, you know, have a challenge, find a solution, discover something cool in nature, develop a great innovation with it."[3] An illustration of how this emulation process works is the Japanese bullet train. This train could travel 300 km/h, but the sound levels exceeded environmental standards; every time it went into a narrow tunnel, it would build up an atmospheric pressure wave which created a sonic boom at the exit. Engineer Eiji Nakatsu was assigned the task to quiet this train. He asked himself the question: "How would nature solve this?" Being a birdwatcher, Nakatsu was fascinated by the fact that kingfishers move from the air, a low-resistance medium, into the water, a high-resistant medium, without a splash. As the bullet train faced the same kind of challenge, Nakatsu designed the forefront of the train inspired on the beak of the kingfisher. The more streamlined train resulting from this not only moves more quietly, but also travels 10 % faster and uses 15 % less electricity.[4]

(b) ***Nature as teacher of values and morals***

To explain what it means to regard nature as teacher of values and morals, Marshall and Lozeva refer to what Holmes Rolston calls "following nature in an imitative ethical sense" (1979, 16). With this, Rolston refers to the process of ascribing a sense of morality to nature, as if nature hides certain moral maxims just like the Bible or the Koran. Rolston nevertheless argues that morality appears in human beings alone:

> The moral dimension in human nature has no counterpart in mother nature. No being can be moral unless he is free deliberately; something must be 'up to him'; and nothing else in nature has sufficient mental competence to be moral (idem.).

Let me elucidate this with an example. Imagine yourself walking on a quiet path along the water. In the distance, a group of children are playing. Suddenly, you hear a loud scream. One of the children has fallen into the water. It is clear that without your help, the child will drown. You immediately jump into the water to rescue the child. An adult who witnesses such an accident and subsequently chooses to do nothing, would be severely addressed on his or her immoral behavior. Nobody would blame a cow in a nearby meadow for her lack of response. After all, her indifference cannot be explained in moral terms and does not result from a free choice.

Rolston argues that apart from human beings, there are no moral agents in nature: "If anyone proposes that we 'follow nature' in something like the ethical sense in which Christians 'follow Jesus', or the Buddhists, Buddha, he has very much gone astray" (idem, 17). This particular way of speaking about natural

---

[3]https://vimeo.com/25898287—accessed 22 December 2015.

[4]http://www.asknature.org/product/6273d963ef015b98f641fc2b67992a5e—accessed 10 January 2016.

systems and organisms is nevertheless present in biomimicry discourse, for instance when Benyus, maintains that "we must draw our standards from the natural world" or when she claims that "nature knows best" (Benyus 2002, 1–3). With these remarks, Benyus seems to suggest that what *is* in nature can provide guidance for our inter-human affairs. This suggestion is further strengthened by the Canon of nature's laws, a list of principles that provide guidance to those who engage in biomimicry (idem, 7):

1. Nature runs on sunlight.
2. Nature uses only the energy it needs.
3. Nature fits form to function.
4. Nature recycles everything.
5. Nature rewards cooperation.
6. Nature banks on diversity.
7. Nature demands local expertise.
8. Nature curbs excesses from within.
9. Nature taps the power of limits.

What is so problematic about using this Canon as a moral guideline? Wouldn't it make the world a much better, and especially greener place? The Canon is very inspiring indeed. Nonetheless, it is important to keep in mind that the nine laws present a romanticized and simplified version of nature. According to Marshall and Lozeva, the laws "are filtered through the values and politics of those that espouse them" (Marshall and Lozeva 2009, 5). Nature is simply too diverse to be generalizable into basic laws; she contains an infinite number of different laws that are far from exemplary. We not only follow nature's example when we, as in the 5th law, "share resources in a cooperative manner [but also] when we fight and compete for resources" (idem, 6). Nature is infinitely diverse and not a model for human conduct in every respect.

(c) ***Following nature in a tutorial sense***

By drawing a distinction between a technical and moral sense of following nature, Marshall and Lozeva seek to indicate that biomimicry entails more than the mere consultation of our fellow species for solving complex human problems. As we have seen, the authors interpret the moral dimension of biomimicry by suggesting that the 'conduct' of nature gives us guidance for our inter-human affairs. The question nevertheless arises as to whether Marshall and Lozeva's interpretation of biomimicry's moral dimension is doing justice to the message of its founders. Although the normativity hidden in particular concepts and phrases is indeed problematic, reading the work of Benyus and her colleagues does not give the impression that they have "very much gone astray" (cf. Rolston 1979, 17). In order to get a more precise view of the moral message at the heart of the tutorial narrative, it might be worthwhile to re-examine the work of Holmes Rolston. Apart from "following nature in an imitative ethical sense" (idem, 16), he distinguishes a different type of mentorship in which the technical and moral dimensions—in biomimicry jargon, the *emulate* and *ethos* elements—merge. Rolston refers to this as “following nature in a tutorial sense” (idem, 25). This form of mentorship appears to be more successful in showing us what is at stake in biomimicry. Rolston claims that, although nature gives us no guidance for inter-human ethical conduct, it has a certain leading capacity as regards “our ethic of bearing toward the natural world” (idem, 28). Or, to put it differently: nature can “teach us something of our human role, our place, and our appropriate character in the natural system as a whole” (idem, 7). If we reconsider the Canon of nature’s laws from this perspective, it suddenly appears to be far less problematic. Rather than viewing the Canon as a moral guideline per se, we should regard it as a list of principles that can guide us in developing an “appropriate form of life toward our environment, toward what the world offers us” (idem, 26).

5 Biomimicry in Action

How is the Canon—interpreted in the latter sense—reflected in the activities of biomimicry practitioners? Do these activities “help humans fit in and become a welcome species on this beautiful blue-green home” (Baumeister et al. 2014, 16)? In some biomimicry projects, we can indeed recognize the objective to live within the ecological limits of nature, for instance the Eastgate Building project in Harare, Zimbabwe. This office complex has an air conditioning system modeled on the self-cooling mounds of termites. Termites in Zimbabwe build huge mounds inside of which they farm a fungus that is their primary food source. The fungus must be kept at exactly 87 °F, while the temperatures outside vary from 35 °F at night to 104 °F during the day. The termites succeed in doing so by constantly opening and closing a series of heating and cooling vents throughout the mound over the course of the day. By emulating the designs created by termites, the Eastgate Centre uses less than 10 % of the energy required for temperature regulation as similar conventional office buildings.[5] Another example is WhalePower, a Canadian company that studies humpback fins in order to create more eco-efficient wind turbines.[6] Despite being 40–50 ft long and weighing nearly 80,000 lb, humpback whales are remarkably agile. Their dexterity can be explained by the strange bumps on the leading edge of their flippers which produce the so-called ‘tubercle effect’. By designing precisely engineered versions of those bumps, WhalePower Corporation has been successful in creating turbines that dramatically enhance wind power, even at low wind speeds.[7]

However, we can also list a number of biomimicry projects that do not help us to “fit in here over the long haul” (Baumeister et al. 2014, 8). A striking example is a

[5]http://inhabitat.com/building-modelled-on-termites-eastgate-centre-in-zimbabwe/—accessed 31 December 2015.

[6]http://biomimicry.org/biomimicry-examples/#.VknRp17ANiY—accessed 16 November 2015.

[7]http://www.whalepowercorporation.com/—accessed 31 December 2015.

research project at the University of California, focused on unraveling the offensive system of the mantis shrimp. This little creature can throw some of the fastest and most powerful punches in nature; it can deliver blow after punishing blow, breaking apart its prey without breaking apart its clubs. Engineers have unveiled the club's microscopic secrets in order to produce lighter and stronger military body armor (Patek et al. 2011). Another example is the golden-fronted woodpecker, whose plate-like spongy bone—called 'cancellous bone'—protects its brain from physical and neurological trauma during drumming. The unique structure of the woodpecker's skull is a source of inspiration for the design of heavy equipment machinery, automotive or drill pipe shock absorbers, and bullet proof jackets.[8]

In what ways do the first two projects distinguish themselves from the latter two projects? On the 'emulate' level—i.e. the 'active part' of biomimicry—the four projects appear to be quite similar: the principles, patterns, strategies and functions of our fellow species are used as sources of inspiration for tackling human design challenges. However, the difference between the four examples becomes more evident if we look at the *objectives* to which the various animals are consulted. Whereas the lessons of the termites and humpback whale are used to create more sustainable buildings and eco-efficient wind turbines, the lessons of the mantis shrimp and woodpecker are applied for military, construction, and mining purposes. It is difficult to imagine how such objectives can help us live sustainably on and with the Earth.

The above shows that in the current situation, biomimicry practitioners use the lessons of our fellow species for sustainable as well as unsustainable purposes. Therefore, in practice, it is hard to see the difference between some of the projects labeled as 'biomimicry' and those labeled as 'bionics' or 'biomimetics'; the mantis shrimp and woodpecker involved in the aforementioned projects do not appear as subjects worthy of care and respect, but as suppliers of useful information. Or, to speak with the tutorial narrative: they are transformed into *intellectual objects* (cf. Marshall and Lozeva 2009). To complicate matters, it should be noted that not only the lessons of the mantis shrimp and the woodpecker, but also those of the termites and humpback whale could be used for serving less sustainable goals. For instance, what if we emulate termite mounds to regulate the temperature in huge shopping malls that sell eco-unfriendly products?[9] And what if the wind energy created by *WhalePower* turbines is used in factories producing polluting cars? The examples show that, in order to determine whether biomimicry can guide us in, what eco-philosopher Freya Mathews calls, "resituat[ing] industrial civilization within the ecological limits of the biosphere" (2011, 380), it is not sufficient to compare the products and applications resulting from biomimicry design with its more conventional and, generally speaking, more energy-consuming counterparts. We also need to look at the broader system in which the lessons of our fellow species are

---

[8]http://www.asknature.org/strategy/a85dd1ae344aa00bb39926a4a6797cd4—accessed 24 December 2015.

[9]Currently, part of the Eastgate Centre is in fact used as a shopping mall.

implemented; are we using nature's lessons merely to meet our current consumer ends in a less harmful way, or are we also prepared "to start thinking about our ends themselves" (idem, 377)?

6 Humans as Nature's Students

The importance of reflecting on the broader cultural patterns within which nature's lessons are applied also becomes apparent if we look at the human or 'student' side of the tutorial narrative. A classical story that resonates in the presentation of human beings as nature's students is *The Sorcerer's Apprentice*. Reflecting on this story can help us to understand what is at stake in biomimicry. The tale of *The Sorcerer's Apprentice* knows many versions. Its original goes back almost 2000 years, and is attributed to the rhetorician and satirist Lucian of Samosata (2nd century AD). A famous modern version of the story is Walt Disney's 10-min animated version of *The Sorcerer's Apprentice*, which is the third segment of the film *Fantasia* (1940). The dialogue-free cartoon is introduced by a narrator, who gives a characterization of the apprentice, a role played by Mickey Mouse:

> He was a bright young lad; very anxious to learn the business. As a matter of fact, he was a little bit too bright, because he started practicing some of the boss's best magic tricks before learning how to control them (*Fantasia*).

The cartoon starts with images of the powerful sorcerer Yen Sid (Disney in reverse), who is practicing his magic. His apprentice Mickey is fetching water to fill a cauldron. After a long yawn, Yen Sid puts his hat down and retires to his chambers. Mickey is visibly happy with the opportunity to take his master's place: tired of carrying the buckets, he puts Yen Sid's hat on and tries to enchant a broom to do the work for him. The spell works, which seems to encourage Mickey in thinking that he is his master's equal. He sits down in Yen Sid's chair and falls asleep. He dreams that he is a powerful sorcerer standing on a mountaintop, commanding the stars and the sea.

Suddenly, Mickey wakes up to find the room awash with water: the cauldron is overflowing, as the broom keeps carrying water to it. Being unable to stop the broom, Mickey grabs an axe and chops it into pieces. Just when things seem to have calmed down, Mickey's attempt to break the spell turns against him: each of the pieces transforms into a whole new broom. Soon, a whole army of brooms is fetching water to the cauldron. Mickey turns over the leaves of his master's book, hoping to find the right formula to stop the brooms. Just as Mickey is about to drown, a stern and angry Yen Sid appears. With a wave of his hands, the waters recede. Mickey, looking very guilty, takes off the sorcerer's hat and returns it to its rightful owner. Then he picks up the buckets and continues with his chores.

7 Lessons from *The Sorcerer's Apprentice*

The tale of *The Sorcerer's Apprentice* confronts us in a lighthearted manner with the dangers of overestimating oneself. Mickey Mouse's hubris seems to be twofold. Firstly, he is not aware of his incompetence and assumes that he already fully understands and controls his master's powers. Having been successful in enchanting the broom, this assumption is initially confirmed. Only after things have gone wrong, he realizes that he is not yet ready to imitate Yen Sid. But there is also a more fundamental hubris at work in Mickey's attempt to use his master's magic for his own purposes. Mickey seems to be unaware of the difference between *tricks* and *magic*; he mistakenly assumes that magic is only about tricks. Mickey does not understand that there are things that cannot be learned by reading textbooks or diligent practice. This second form of hubris brings to light an(other) important difference between Mickey and his master. Yen Sid is not only smart, but also wise: he knows how to separate essentials from trivialities. Mickey, on the contrary, does not (yet) fully recognize the insignificance of some of his objectives. Unable to see the bigger picture, he uses his master's magic powers for something as trivial as fetching water. If Mickey would have used the sorcerer's powers for more serious purposes, the latter probably would not have gotten angry. Yen Sid seems especially annoyed by the fact that Mickey's decision to use his master's spells is motivated by laziness, and does not serve any serious goal (Spider-Man 2002).

What can biomimicry practitioners learn from the tale of *The Sorcerer's Apprentice*? The tale shows us that the knowledge acquired during the lessons of our fellow-organisms can bring us into trouble, especially if we do not realize that our knowledge is only fragmentary and finite. Even in the process of getting smarter and smarter, we have to remember that we are still students, and, in a certain manner, will always remain so. Benyus herself seems to be quite aware of this, for instance when she claims that we have to acknowledge that

> we are ultimately dependent on the existing natural pattern, *a pattern that we only partially understand*. Science is continuously peeling masks away, only to find another mask deeper down, one of the many worn by [...] the Great Face behind. The closer we come to that face, the greater the mystery appears (Benyus 2002, 293—emphasis in original).

However, there is also another lesson that can be learned from Mickey, one that seems especially important in the context of current developments in the biomimicry field. The tale encourages biomimicry practitioners to reflect on the question whether they keep their eyes on the bigger picture, for instance when they apply the lessons of the mantis shrimp to create stronger military body armor, or those of the woodpecker for designing drill pipe shock absorbers. I just argued that Mickey, capable of performing (at least some) spells, still needs to learn to choose his goals carefully. Or, to cite another Hollywood hero, Mickey has not yet fully grasped that "with great power comes great responsibility" (*Spider-Man*). In a similar vein, it is important that biomimicry practitioners continue to look critically at the objectives to which they apply nature's lessons. They should keep asking themselves the question: 'Why do we want what we want?' And this not only applies to the rather extreme examples of the mantis shrimp and woodpecker, but also to the more 'humble' projects involving the genius of the termites and the humpback whale. Using their lessons merely to achieve our current consumer ends "with less rather than more disruption to the life systems of planet-earth" (Mathews 2011, 373) will not help biomimics to move beyond an instrumental approach to nature.

8 Concluding Remarks

The founders of the biomimicry movement use the tutorial narrative as a more respectful alternative to classical resource approaches that reduce natural systems and organisms to 'goods' or 'products' for unrestricted use. However, they seem to underestimate the complexity and multi-layered character of the tutorial narrative, notably in terms of the moral message it conveys. In this chapter, I have shown that the realization of a more respectful relationship between humans and other species entails more than the introduction of a new narrative; after all, not only nature's resources, but also her 'wisdom' or 'creativity' can be used in an instrumental fashion. With this contribution, I hope to encourage biomimics and others to reflect on the kinds of students of nature we are, or would aspire to be: students who use the lessons of our fellow-organisms to empower ourselves, or students who apply these lessons to render our artefactual production "consistent with the ecological fabric of the greater life system" (Mathews 2011, 366)?

References

Ball, P. 2001. Life's lessons in design. *Nature* 409: 413–416.

Baumeister, D., R. Tocke, J. Dwyer, S. Ritter, and J. Benyus. 2014. *Biomimicry resource handbook: A seed bank of best practices*. Missoula, MT: Biomimicry 3.8.

Bensaude-Vincent, B., H. Arribart, Y. Bouligand, and C. Sanchez. 2002. Chemists and the school of nature. *New Journal of Chemistry* 1: 1–5.

Benyus, J.M. 2002. *Biomimicry: Innovation inspired by nature*, 2nd ed. New York: Harper Perennial.

Evernden, N. 1993. *The natural alien. Humankind and environment*. 2nd ed. Toronto: University of Toronto Press.

Fantasia. 1940. *Film. Dir. James Algar*. Burbank (CA): Walt Disney Productions.

Macnaghten, P. 2010. Researching technoscientific concerns in the making: Narrative structures, public responses, and emerging nanotechnologies. *Environment and Planning A* 42: 23–37.

Marshall, A., and S. Lozeva. 2009. Questioning the theory and practice of biomimicry. *International Journal of Design & Nature and Ecodynamics* 4(1): 1–10.

Mathews, F. 2011. Towards a deeper philosophy of biomimicry. *Organization and Environment* 24(4): 364–387.

Patek, S., D. Dudek, and M. Rosario. 2011. From bouncy legs to poisoned arrows: Elastic movements in invertebrates. *Journal of Experimental Biology* 214(12): 1973–1980.

Plumwood, V. 2002. *Environmental culture: The ecological crisis of reason*. New York: Routledge.

Poelman, Y. 2015. *De natuur als uitvinder: Miljarden jaren aan innovatie gratis beschikbaar*. Amsterdam: Carrera.
Rolston, H. 1979. Can and ought we to follow nature? *Environmental Ethics* 1: 7–30.
Spider-Man. 2002. *Film. Dir. Sam Raimi*. Culver City (CA): Columbia Pictures.
Vincent, J., O. Bogatyreva, N. Bogatyrev, A. Bowyer, and A.-K. Pahl. 2006. Biomimetics: Its practice and theory. *Journal of the Royal Society, Interface* 3: 471–482.
Worster, D. 1994. *Nature's economy: A history of ecological ideas*, 2nd ed. Cambridge, NY: Cambridge University Press.

Interspecies Democracies

Eva Meijer

Abstract Even though nonhuman animal communities make group decisions—they vote, negotiate, and sometimes even deliberate—political philosophers traditionally see democracy as human territory. This view has in recent years been challenged. Research in fields of study as biology and ethology shows that nonhuman animals have their own cultures and languages, and that differences between human and nonhuman animals are of degree and not of kind. Recent work in political animal philosophy draws on these insights and focuses on relations between groups of animals and human political communities, proposing to view nonhuman animals as political actors. This requires not only rethinking our relations with them, it also requires rethinking the concepts attached to those relations, such as 'politics' or 'democracy', non-anthropocentrically. In this chapter I focus on nonhuman animal democratic agency and investigate possibilities for rethinking democracy with other animals. I first discuss the recent political turn in animal philosophy, in which I focus in particular on the advantages of moving from seeing animals as sentient individuals to seeing them as political groups. I then turn to political animal agency in the Anthropocene and discuss how nonhuman animals are silenced politically. Building on this discussion of silencing I then contrast the liberal democratic interspecies citizenship model developed by Donaldson and Kymlicka (2011) with agonistic pluralism. In the following section I discuss the concept 'recognition' in relation to interspecies democracies. The final section investigates possibilities for rethinking democracy with other animals.

1 Introduction

Honeybees collectively decide where to build a new nest. When a new queen is born in the hive, half of the colony leaves with the old queen to find a new place to live. In order to survive, they need to find the best nest site; this can be a hollow

E. Meijer (✉)
University of Amsterdam, Amsterdam, Netherlands
e-mail: E.R.Meijer@uva.nl

B. Bovenkerk and J. Keulartz (eds.), *Animal Ethics in the Age of Humans*,
The International Library of Environmental, Agricultural and Food Ethics 23,
DOI 10.1007/978-3-319-44206-8_4

tree, or a chimneystack. The process works as follows. The transient colony swarm spends days hanging on bush or tree, and from this position scouts set out to search for promising sites. As they return, they do a waggle dance, which means they move in the shape of a figure eight. They express the location and the quality of the site they found by the intensity, direction and shape of the dance they perform. The better the location, the longer the bees dance, and so the best sites remain. Other scouts always verify the location. A unanimous decision needs to be made in a couple of days, because if they deliberate for too long, they might lose the queen, or the colony dies (Seeley 2010). Seeley and Buhrman (1999) describe the swarm's overall strategy of decision-making as a weighted additive strategy. This strategy is the most efficient but also most demanding. The bees do not only use movements and body postures in their dance, they also use sound and pheromones, and influence electric fields, which passes social information to the other bees.

Herds of red deer move when over 62 % of the adult members of the group stand up (Conradt and Roper 2005). African buffalo also make group decisions about when and where to move. Females dictate their travel preferences by standing up, staring in one direction, and then lying back down (Wilson 1997). It took scientists some time to understand this behaviour was not stretching, but voting. When the cows differ sharply in the direction of their gaze, the herd splits up and grazes in different patches. In yellow baboon communities high-ranking males and females have the final say about troop movement, but all baboons influence decisions (Norton 1986). Pigeons have complex social hierarchies in which low-ranking birds also vote on the flock's next flight. They have a flexible system of rank, in which some birds are more likely to lead and others to follow, which according to researchers makes for a particularly efficient form of decision-making (Nagy et al. 2010). African bees use pheromones to warn each other and when they attack collectively they can and do kill humans (Collins et al. 1982). Cockroaches do not have complex social structures like bees and ants, but they still seem capable of collective decision-making (Amé et al. 2006). In an experiment, a group of fifty roaches were presented with three new spaces for shelter. Rather than behaving chaotically, they split up in two groups and half filled two shelters, leaving the third empty. When large shelters where introduced, the group stayed together. In both cases they struck a balance between competition and cooperation. Many other animal species, ranging from birds, bats and fishes to different species of insects, also make collective decisions (Conradt and List 2009).

Most political philosophers see a sharp difference between human political communities and other animals' communities (Rawls 1971). This difference is usually founded on rationality, language, morality, or other characteristics that are seen as solely human. Recent research in fields of study as biology and ethology (Bekoff 2002) however shows that nonhuman animals have their own languages (see for example Gentner et al. 2006; Slobodchikoff et al. 2009) and cultures (Smuts 2001), and that differences between human and nonhuman animals with regard to these topics are, as Darwin already put it, of degree and not of kind. Human exceptionalism is also attacked from other theoretical viewpoints. Poststructuralist and posthumanist approaches point to the influence of power relations on the construction of the human as rational subject, and the violent consequences of adopting this image of the human for nonhuman and human marginalized groups (Derrida 2008; Wolfe 2003). In animal rights theory, it is argued that nonhuman animals are sentient individuals, who have a right to certain inviolable rights (Regan 1983). Recent work in political animal philosophy (Donaldson and Kymlicka 2011) shifts the focus from animal individuals to relations of animal communities to human political communities, and argues that other animals should be seen as political actors and communities. Taking other animals seriously as subjects, individuals or political actors—something the abovementioned authors convincingly argue for—requires rethinking our relations with them. It also requires rethinking the concepts attached to those relations, like 'politics' or 'democracy' non-anthropocentrically.

In this chapter I focus on nonhuman animal democratic agency, and investigate how humans can rethink, or reconfigure, democracy with other animals. I first briefly discuss the recent political turn in animal philosophy, in which I focus in particular on the advantages of moving from seeing animals as sentient individuals to seeing them as political groups. I then turn to political animal agency in the Anthropocene and discuss how nonhuman animals are silenced in a human-dominated world. Sect. 3 builds on this discussion of silencing and contrasts the liberal democratic interspecies citizenship model developed by Donaldson and Kymlicka (2011) with agonistic pluralism. In Sect. 4 I discuss the concept 'recognition' in relation to interspecies democracies. The final section investigates possibilities for rethinking democracy with other animals.

2 Political Animal Rights

In animal ethics, the focus has long been on the moral standing of sentient individuals (Singer 1975; Regan 1983). It was argued that because nonhuman animals have interests strong enough to lay a claim on others, they deserve certain inviolable rights, similar to how this works in the human case. In recent years, political philosophers have used the idea of moral animal rights as a starting point for thinking about political rights. Cochrane (2012) for instance argues for universal animal (including human) rights on grounds of sentience. Garner (2013) and Rowlands (1997) argue that nonhuman animal interests should be included in human societies and decision-making from the perspective of justice. O'Sullivan (2011) argues for including animal interests in liberal democratic laws and political institutions on the basis of equality.

Even though these authors argue for taking animal interests into account, they still assume humans should determine the larger framework (i.e. rights, citizenship, or other concepts), and should determine which animals should be granted which rights. This is problematic for normative reasons, because it reinforces a hierarchy between humans and other animals and does not take them seriously as interlocutors, and for practical reasons, because interacting with other animals can lead to new forms of understanding, which can improve relations. In their theory of political animal rights, Donaldson and Kymlicka (2011) do address this problem. They propose to view nonhuman animals as political actors, who co-shape the polis. They focus on relations between groups of animals and human political communities and argue for a group-differentiated theory of citizenship. They argue that negative rights, the rights traditionally advocated for in animal rights theory, are important but not sufficient to rethink relations. Humans and other animals share a world, and instead of advocating ending all interaction—which is factually impossible and morally problematic—we ought to concentrate on establishing just relations. In order to specify human and animal rights and duties, they propose to view domesticated animals as citizens, wild animals as sovereign communities and liminal animals, who live amongst humans but are not domesticated, as denizens.

This model provides us with a new perspective on animal communities and governments. Some wild animal communities have very little to do with human communities; they govern themselves and in their processes of decision-making they do not need to take humans into account. For domesticated animals, decision-making is usually closely intertwined with human democracy. Through processes of domestication, many of which were largely forced upon them by humans, they became not only more attuned to humans but also more dependent upon them. Between truly wild animals and truly domesticated animals we find a spectrum of human-animal relations. While some animals do not form communities with humans, human legislation might affect them; human and nonhuman animals sometimes share territories or travel through each other's land. In some circumstances negotiations between wild animal communities and human communities about land or food might resemble international relations. In many other situations the lives of humans and other animals are intertwined, because they live in the same area. If humans in these circumstances make decisions concerning nonhuman animal lives, all animals affected should have a voice in this for democratic reasons. Furthermore, when other animals are part of the demos, they should not only be given the opportunity to act within a given framework, but also be able to co-shape that framework.

Because relations between nonhuman animals and human communities, or within animal communities, are different from relations between humans, political concepts as citizenship or sovereignty, or democracy, will need to be reinterpreted in multispecies contexts. Nonhuman animals express themselves in different ways, and have many different types of relations with humans. Other animals might sometimes fit neatly into human systems, but they might also be very different. This asks us to be flexible in rethinking political concepts, and it asks for rethinking them with other animals. If we for example consider citizenship, it is not enough to impose a human fixed view of citizenship onto them, we need to take their agency into account and explore how animal citizenship might look in different circumstances. The first step in this project is to critically examine the existing situation, in which animal agency is often not acknowledged, and in which nonhuman animals are silenced as a political group.

3 Silencing Animals in the Anthropocene

The Anthropocene, understood as the epoch that began when human activity started to significantly affect ecosystems globally, is interconnected with human exceptionalism. On a descriptive level we find that the presence of humans on planet earth has a different kind of impact than the presence of other species. Scientists for example argue that the rate of extinction for nonhuman animal species in the last century was over a hundred times higher than it would have been without the impact of humans (Ceballos et al. 2015). This phenomenon also has a normative dimension. In an anthropocentric worldview, humans see themselves as fundamentally different, and hierarchically above, other animals. This leads them to act in ways that do not take into account the wellbeing of other animals, and to see other animals' interests as less important than human interests. This worldview posits the human as a moral or political standard, and presents this as neutral, while it is in fact the result of power relations (Derrida 2008).

Political animal agency is often not recognized as such in theory and practice. Nonhuman animals do not have a voice in matters that concern them, even though human political decision-making greatly influences their lives. Poststructuralist theorists argue that the view of the human, and the view of reason attached to it, on which current interpretations of citizenship and political agency are based are not universal, but the result of contingent practices and power relations (Wolfe 2003; Derrida 2008). Many current political democratic practices that are presented as equal and open to all in fact reflect the preferences of the dominant group (Wolfe 2003). Young (2000) draws attention to this mechanism in her discussion of the relation between forms of deliberation and democratic exclusion. She argues that the forms of language-use that are presented as neutral in deliberative practice and theory are in fact an expression of social inequalities and exclusionary practices. She distinguishes between two types of exclusion. External exclusion refers to forms of exclusion in which individuals or groups that ought to be included are left out of fora for discussion and decision-making. Internal exclusion occurs when groups are formally included in processes of discussion and decision-making, but whose ideas or modes of expression are ignored, dismissed, patronized or in other ways not taken seriously. In current forms of democratic deliberation, certain attitudes and attributes are favoured—such as being dispassionate, articulate, orderly, not emotional, and focused on argument. These norms of proper political communications reflect the preferences and style of the dominant group. Young shows this devalues the style of speech of people who have not learned to argue in this manner, who do not speak the language well, are less educated, more emotional and so on; these people are often members of groups that were historically excluded from participating in political discourse. This analysis is also useful for thinking about political participation and voice of nonhuman animals.

Animal exclusion currently can be seen as a form of external exclusion; other animals have languages and express themselves in many ways, but they cannot make themselves heard in the dominant political discourse because they do not speak in the language of power. To address this problem between humans, Lyotard (1988) uses the concept of 'differend'. Lyotard distinguishes between a plaintiff and a victim. A plaintiff becomes a victim if she has no means to prove the damage that is done to her. More specifically: if there is in language, because of how language is constructed, no representation possible of the wrong she has suffered. "I would like to call a differend the case where the plaintiff is divested of the means to argue and becomes for that reason a victim. If the addressor, the addressee, and the sense of the testimony are neutralized, everything takes place as if there were no damages" (1988, 10). A wrong is a damage accompanied by loss of the means to prove the damage (1988, 5). We find an example of this in situations where land was colonized and indigenous peoples could not claim it because their tribal laws were not recognized as laws. Between human and nonhuman animals, the formal political discourse is constructed so that nonhuman animals are deprived of the possibility to address damages done to them on a very fundamental level: because speaking is understood as speaking in human language, they cannot speak. The problem is not that the animals do not interact with humans or cannot draw attention to the wrongs done to them, but that a common framework in which these can be addressed is lacking.

This problem is not often acknowledged in animal philosophy. It is usually accepted that animals cannot voice their opinions or concerns, because they do not speak in human language. This is unfortunate for normative reasons, because viewing nonhuman animals as subjects means we should also think about how they can have a say in questions that concern them, which means listening and learning about their languages. Not taking these seriously leads to the internal exclusion Young, describes. It is also unfortunate for practical reasons; communicating with animals can clarify relations and build new worlds. In moving towards an interspecies model of democracy it is important to recognize that exclusion is the result of anthropocentris, and not a fixed truth. Other animals have their own voices and act politically, even if this is formally not recognized and they are often excluded from formal political practices. They resist and protest (Hribal 2003, 2007, 2010), cooperate with humans, vote with the feet, negotiate and bargain with humans (Donaldson and Kymlicka 2011). Through exercising political agency they co-shape human communities, politics, and lives (Haraway 2003, 2008; Hobson 2007). Animal acts also shape human concepts (Gaita 2002).

Donaldson and Kymlicka (2011) challenge the view that animals cannot be political actors and they argue against an overly rational interpretation of citizenship and democratic agency. They discuss three basic capacities (2011, 103), or what Rawls (1971) calls 'moral powers', that are seen as requirements for citizenship in the human case: the capacity to have a subjective good and communicate it, to comply with social norms, and to participate in the co-authoring of laws. Donaldson and Kymlicka do not dispute this list, but they do dispute the overly rationalist or intellectualist manner in which the list is often interpreted. It is for example not seen as enough to simply possess a good, it is also expected that individuals reflectively endorse a conception of the good. It is similarly not seen as enough to understand and comply with social norms, one also needs to rationally understand the reasons for them. If citizenship is interpreted in this manner, nonhuman animals, and also certain groups of humans, do seem to be incapable of citizenship, but it is not the only interpretation possible. Instead, Donaldson and Kymlicka argue for an embodied and socially embedded citizenship. They argue that nonhuman and human moral agency is grounded in pre-reflective moral sentiments and pro-social impulses. The appropriate test for animal citizenship is therefore not whether other animals engage in rational deliberation, but whether they "exhibit norm responsiveness and intersubjective recognition in actual interactions" (2013b, 4). As Young (2000) argues, a theory of democratic inclusion also requires an expanded conception of political communication, "in order to identify modes of internal inclusion and to provide an account of more inclusive possibilities of attending to one another in order to reach understanding" (2000, 56). In order to move towards multispecies citizenship, we need to learn to listen to other animals better, and to adjust democratic institutions and procedures accordingly.

4 Categorizing Animals

While some communities of animals are truly wild and do not have any interaction with humans, others have strong relations with humans. Certain groups of nonhuman animals should be seen as part of the demos. There are different reasons for this, which ones apply will differ from group to group. Human laws and regulations concern the lives of many nonhuman animals. They are the subjects of human political decision-making, and at least in a passive way part of the demos (Garner 2013; Donaldson and Kymlicka 2011); therefore their interests should be taken into account in democratic decisions. Not taking their interests into account is problematic from the point of view of equality. Nonhuman animals are sentient beings with lives that matter to them. They have interests strong enough to lay a claim on others (Cochrane 2012) and not taking these into account on the ground that they are not human is speciesist (Singer 1975; Regan 1983; Cooke 2014) (see also O'Sullivan 2011). Some animal groups are physically and psychologically influenced, or even deliberately changed, by humans, for example through processes of domestication, often involving violence, and they have become dependent on humans. They therefore should for historical moral reasons be granted membership of the community (Donaldson and Kymlicka 2011). There are however practical and theoretical difficulties in further conceptualizing the position of nonhuman animals in the demos, and their democratic participation, because they are not one homogenous group, and they do not speak with one voice.

The starting point of Donaldson and Kymlicka's political theory of animal rights is liberal democracy. To determine rights and duties, they propose to divide animals into three groups, based on their relations with human political communities. While this categorisation has advantages for determining practical guidelines, it also raises some questions. Some of these are practical; it is difficult to divide animals into three groups because they differ so much, and sometimes it is hard to draw lines between groups. Relations can be ambiguous, and there are many things we do not know about nonhuman animals. There are also normative questions related to categorizing animals. Cochrane (2013) and Cooke (2014) for example argue for cosmopolitanism instead of citizenship, because, they argue, it is unfair that some groups are treated differently than others, based on where they were born. Cochrane for example argues that wild animals should have the same right to healthcare as domesticated animals. From the perspective of animal agency, this is however problematic; one of the reasons structural healthcare for wild animals is undesirable is because it would involve trapping and taming them, or at least lead to too much human impact on their lives (see also Donaldson and Kymlicka 2013a). Additionally, Donaldson and Kymlicka (2011) mention that animals can move from group to group—domesticated animals can become liminal for example, or vice versa—and that domesticated animals should be able to choose which community they want to be part of.[1,2]

However, as we saw above, extending a human model (be it citizenship or cosmopolitanism) to include other animals runs the risk of silencing (some of) them, because the model is based on a human standard.[3] This may lead to favouring those most like 'us', obscure ways of participation beyond that model, and the model might simply not be able to facilitate some voices. Citizenship might work well for some domesticated animal species and individuals, but the human values and standards implied in the concept might not appeal to others. Furthermore, in Donaldson and Kymlicka's model, domesticated animals are represented by humans, which requires interpretation by humans, and focuses on strong, positive relations. However, not all relations are harmonious, and not all humans are good interpreters of their companions. Groups of humans often do not agree, within communities, and it is difficult to imagine how domesticated animals can structurally disagree in *Zoopolis*. Mouffe (1999) discusses these questions in relation to groups of humans. She argues that theories of justice that focus on consensus deny the political character of democracy. Democracies are in her view necessarily pluralistic; different participants come from different forms of life and it is impossible to reach consensus without exclusion. Recognizing this and fostering mechanisms that make different voices heard is a task for democratic societies.

[1]Ted Kerasote (2008) describes an interesting experiment in which he taught his dog companion Merle the rules of the town in which they lived and then installed a dog door. Merle could come and go as he pleased, and determine how he wanted to spend his days. This enlarged his freedom, made him smarter and more capable of dealing with new situation, and it improved their relationship. We could view this as letting dogs choose, step by step, how they want to shape their lives. The precise model will differ from individual to individual. Certain groups of stray dogs are good examples of independent communities who choose to co-exist with humans in different ways.

[2]Capacities for reflection, choice, and decision-making differ between species and individuals, and on whether or not they are domesticated. See for example Peterson (2012) on moral agency, Bekoff (2002) on animal emotions, and Crane (2015) on ethical animal agency.

[3]See Wolfe (2003) for a discussion of this problem.

A focus on consensus obscures exclusionary mechanisms and is ontologically problematic, because the harmony that it implies is a fiction.

Mouffe's agonistic model can shed light on differences in human-animal communities. A focus on animal struggle and conflict, for example animal acts of protest (Hribal 2010), can show us injustices that would otherwise remain hidden because by resisting animals question power hierarchies (Wadiwel 2014). Recognizing that in a democracy there is always necessarily a plurality of voices, some louder than others, might also help in challenging anthropocentrism, because human voices become one set of possible voices. Explicating disagreement and struggles on different levels, from farm animal biopolitical repressions to disciplining of domesticated animals, can help us see power relations beyond the state. For these reasons, the agonistic model might do more justice to animal agency, and the plurality of animal voices that exists, than the citizenship model put forward in *Zoopolis*. On the other hand, the citizenship model Donaldson and Kymlicka propose fares better with regard to protecting the vulnerable, and in terms of pragmatic value.

Donaldson and Kymlicka argue that rights are the best tools to come to justice because they are designed to protect the inviolable interests of individuals, and as such they are particularly important for those who cannot stand up for themselves within current formations of power.[4] Currently many animals do not, or cannot, resist in a meaningful way. The animals who do resist often cannot change their social or political position by themselves. Animal rights would not only safeguard them from human violence, they would also change their position in society. Furthermore, we live in a liberal democracy, and rights and citizenship are concepts that we know well, so they are good starting points for rethinking animal agency and participation. Existing political institutions and laws greatly influence the lives of many animals, and changing these to include other animals on grounds of justice could make a large difference. We do not need to ascribe to a universal metaphysical standard to recognize the value of this for the animal situation. Finally, Donaldson and Kymlicka also argue that humans should investigate and foster animal agency, and we cannot completely be sure what the outcome will be politically, because other animals' voices have been ignored until now. We do not know the precise scope of animal agency once we start treating them respectfully.

While categorization always carries with it the risk of silencing some groups, it can help determine rights and duties. In the animal case, the key seems to lie in developing new models of representation and interaction with them, making use of insights about their languages, cognition, and view on the world.[5] Both the liberal-democratic model of *Zoopolis* and agonistic approaches can help shed light on aspects of the animal situation. Before sketching practical ways of incorporating

---

[4]Here they explicitly move beyond the atomistic liberal individual, and emphasize the importance of relations for all beings.

[5]While dogs perceive the world in a dog manner, and fish in a fish manner, through interaction understanding is often possible, see also the discussion of Hearne (1986) below.

animal agency and voice in human democracies I will now turn to the transition from human democracy to interspecies democracy, by way of a discussion of the concept of recognition.

5 Interspecies Recognition

Currently, nonhuman animals are not seen as deserving rights, or as a political group, let alone as citizens or sovereign communities. They are regarded as objects and not as subjects, and even when they are recognized as sentient beings, they are seen as property. In order for this to change, humans need to start seeing other animals differently and they need to recognize them as beings worthy of respect. Some contemporary theorists of human social and political justice (Honneth 2014; Taylor 1994) use the concept of recognition to address dynamics between oppressed and dominant groups, for example in the context of feminism or multiculturalism. Hegel's master-slave dialectic is used to make sense of the relation between marginalized groups and groups in power, not because the former are slaves, but because Hegel argues that self-consciousness develops through relations with others. This stresses the intersubjective nature of subjectivity, and questions notions of autonomy that are foundational to liberalism. We are not atomistic individuals, we always stand in relations to others. This view has consequences for what justice entails, and shifts the focus from questions about equal distribution of goods, to questions of respect and esteem. I will not discuss recognition in detail for reasons of space, but rather explore its potential for rethinking human-animal relations. The idea of mutual recognition often comes up in arguments against animal citizenship or sovereignty. It is argued that nonhuman animals are simply not capable of the process of recognition because they lack the mental capacities or forms of self-consciousness that are needed. In order to address this problem we need to develop a non-anthropocentric interpretation of the concept, which is sensitive to different ways of being in the world.

Oliver (2015) argues that the focus on intersubjectivity and the critique of the atomistic liberal individual that theorists of recognition defend, as well as the ideal of mutual recognition, are valuable. However, even though subjectivity is in her view constituted dialogically, she sees problems with recognition, because it is distributed "according to an axis of power that is part and parcel of systems of dominance and oppression" (2015, 474). In other words, because there is a dominant group that recognizes a subordinate group, in the process of recognition the power relation is affirmed. Furthermore, when it focuses on human self-consciousness it excludes many animal others. She argues that recent attempts to link recognition to vulnerability instead of self-consciousness (Butler 2006) do open up the concept to include other animals, because we share this condition with them. However, vulnerability is according to Oliver also distributed along axes of power. Some are more vulnerable than others; who is deemed vulnerable is again the result of power structures. This is problematic, because it keeps intact the binary logic of some who have or distribute something, and others who are denied or given something. Oliver therefore concludes that we need to supplement political and ethical interpretations of recognition with a witnessing ethics that relies on the affective and imaginative dimensions of experience, in order to take radical responsibility for those who are different from us.

Oliver does not explore the political implications of extending the concept of recognition to other animals in detail, nor its relation to justice. While we do share the condition of vulnerability with other animals, only focusing on vulnerability in political relations to other animals is problematic, because it runs the risk of perpetuating stereotypes about animals as passive (contrasted with humans as active), and it ignores the dialogical character of human-animal relations, in which power relations are not given and nonhuman animals not only vulnerable. Also, there is still the question of how we should envision mutual recognition. Nonhuman animals have their own ways of perceiving the world and expressing themselves, as well as their own forms of cognition. In order to get a better understanding of interspecies recognition, 'cognition' therefore needs to be reconfigured in an interspecies context, as a step in redefining the demos. Nonhuman animals form relations with humans and each other in which they ask questions, see others, and respond to them. Respecting them means acknowledging them not just as passive receivers, but also looking for mutual recognition, taking into account their expressions and behaviors.

Animal trainer and philosopher Hearne (1986) describes how working with nonhuman animals creates common meaning. She uses Wittgenstein's concept 'language-games' to bring to light how teaching animals words creates interspecies language-games that clarify interaction and bring about understanding, even though the human and nonhuman animal perceive the world differently. Dog-human relationships are for example not the same as human-human relationships, but they have their own authenticity. In the training and working situations Hearne discusses "genuine animal partners of humans (e.g., dogs and horses) become 'kinesthetically legible' and vice versa" (Acampora 2004, 224). Both human and dog make the relationship possible, and thereby they bring into being not only a practical understanding but also a moral framework. Within the shared language-games, dog and human can be held accountable; Hearne (1986) sees this as a version of interspecies social contract theory. Another example that I will discuss in more detail below involves ethologist Barbara Smuts' interaction with a group of wild baboons (2001). In order to study and live with a group of baboons, Smuts needed to recognize them as subjects, and act so that they could recognize her as a subject. How these processes of recognition might work in an interspecies political context needs to be explored further, but these examples show that human and other animal subjects can mutually recognize each other as subject, and can build agreements on that.

6 Moving Towards Interspecies Democracies

When one discusses animals and democracy, the first remark that is usually made is that they cannot vote. As we saw in the introduction, communities of animals do make group decisions; they vote and they sometimes negotiate or even deliberate. Very little attention has been paid to the relation between animal forms of decision-making and human democracy. This is unfortunate because studying animal politics can help us imagine new forms of co-existing, both in and between communities, and working towards interspecies decision-making might improve decisions for all involved. Some theorists argue that democracy is a wise strategy for groups of human and nonhuman animals because it gives the best results. Conradt and Roper (2005) contrast animal democracy with animal despotism, the latter understood as a model in which one dominant individual makes decisions for the whole group, and argue that the former is generally more beneficial because it tends to produce less extreme decisions. Despotism is only a wise strategy if the group is small and the difference in knowledge or information large. List (2004) draws on Condorcet to argue that democracy is good at drawing information from individuals, and hence a wise strategy for nonhuman and human animal groups. Democracy is also seen as important for normative reasons, as discussed above.

In order to think and work towards social change with other animals, we need to explore how existing democratic concepts such as voting and representation can be expanded to better include animal agency, and to search for new procedures, institutions and encounters to further develop interspecies political processes and frameworks. Parallel to this we need to improve interaction with them, in order to make this a mutual project. This means learning about their languages and behaviors, and being creative in speaking back. To conclude this chapter I will explore how we can expand existing human democratic concepts, institutions and procedures to better include animal agency. The following is meant as an investigation into how we can experiment with this, not as a full list of fixed rules or principles.

6.1 Group Decisions and Material Interventions

A first step towards making democracies more inclusive would be to establish new decision-making procedures in which all who are affected are consulted. I will now discuss the example of human-seagull relations in the Netherlands to illustrate how we can experiment with interspecies decision-making. Some time ago, Dutch politician Rudmer Heerema declared war on the seagulls.[6] In The Netherlands, seagulls used to live near the coast, and twenty years ago they were hardly ever seen in other parts of the country. Due to destruction of their nesting places by human

[6] A description of his standpoint can be found on the website of the VVD: http://www.vvd.nl/nieuws/368/meeuwen-bestrijden-niet-beschermen.

activities such as farming and tourism, they moved to urban areas, where they began to nest on houses with flat roofs and changed their diet to include human foods. They can nowadays be found in most large cities in the Netherlands, especially in the west of the country, where they will eat anything humans throw away, from French fries to bread and fish. They also steal food and open garbage bags. Because the population of seagulls is declining, due to aforementioned changes in habitat, they are protected by law. Heerema wanted to change this as soon as possible, because, he argued, they open garbage bags, demolish cars, pollute neighbourhoods, and keep thousands of people awake at night; according to Heerema they also pose a danger to young children. Heerema is from Alkmaar, where they apparently steal the cheeses the city is famous for.

De Vogelbescherming, the Dutch organisation for the protection of birds, responded by strongly rejecting shooting the birds.[7] They instead proposed several material interventions to persuade the birds to move elsewhere and change their behaviour. The city of Leiden started an experiment with yellow garbage bags that are much stronger than the normal garbage bags. The first series of stronger bags were still not strong enough, so they made new ones. The birds soon found out they could not open these and moved elsewhere. Other measures that were proposed are to provide them with more natural food and create nesting islands in front of the coast. The exact locations of these islands will have to be determined together with the birds, because they will not just nest anywhere, by offering them options and following their preferences. Finally, humans need to stop offering the seagulls food. They do this because they like the birds, or because they are afraid of them because seagulls are large birds, and throw food towards them in order to get rid of them. They should learn to respectfully say no; both human and bird subjects have to change their habits in order to co-exist more peacefully.

From a political point of view, one could argue that humans should accept the presence of the gulls, since human acts made them move into towns, and since they have lived there for quite some time now.[8] Humans have no right to kill the seagulls, and killing them will also not solve the problem if the situation stays the same, because new birds will travel into the cities. However, living in cities might also pose risks to seagulls, and in some circumstances humans might have the right to keep them out. In general, seagulls have a different map of the land than humans, and sometimes humans and seagulls inhabit the same spaces. When conflicts arise, there needs to be communication about who can live where, and the recommendations from the Vogelbescherming can play a role in this. Small-scale experiments with negotiations about nesting and feeding spaces for seagulls—a form of material deliberation—could in these circumstances inform legislation.

---

[7]A summary of their response can be found here: http://www.vogelbescherming.nl/actueel/vogelberichten/q/ne_id/1484.

[8]See also Wolch (2002) about living with wildlife, and Donaldson and Kymlicka (2011) on liminal animal rights.

6.2 Animal Agency and Changing the Rules of the Game

Animal geographist Hobson (2007) argues we need to broaden our understanding of political agency and acknowledge the role that nonhuman animals play in current political constellations. She proposes to distinguish between Politics and politics. 'Politics' refers to matters of the state and international relations; 'politics' refers to practices, spaces and people which/who challenge Politics, for example social movements. It also refers to everyday politics. Taking the second type of politics into account asks for a new definition of the political subject, and in this definition nonhuman animals also can be seen as political subjects. Hobson understands political agency as relational: political subjects are not a given, but an outcome of multiple relations through which human and nonhuman animals influence one another and their worldviews. While a relational view of agency and paying attention to everyday politics are convincing, this idea is not without problems. Seeing animals as political subjects in politics and not Politics risks only conceptualizing micro animal agency, which limits their macro political position, and their scope for choice. Also, different forms of politics cannot be neatly separated into two groups, the borders are often porous and may change with time.

Animal agency is often conceptualized on the micro-level (see Donaldson and Kymlicka 2013b; Srinivasan 2015): as personal decisions within a given framework. Domesticated nonhuman animals are for example seen as capable of making certain decisions concerning food, play and behaviour in the house, but the borders are drawn by humans, and anchored in fixed institutions. Investigating questions on the level of micro-agency can help humans understand animal preferences and choices, and communicating with other animals can help extend human concepts.[9] However, if we only focus on micro-agency, the subordinate position of animals in society is not challenged. Because of human dominance, currently they cannot make decisions concerning the larger dimensions of their lives, such as where to live and whom to live with. Only focusing on micro-agency, or on politics and not Politics, reinforces the stereotype that nonhuman animals have no interest in, or are not capable of, making these decisions at all. In order to challenge this, we have to investigate how nonhuman animals (can) exercise macro-agency, and we need to re-examine the relation between formal or official forms of politics and informal forms of politics (for example practices of contestation and everyday politics).

Tully (2009) proposes to view the distinction between these forms of politics as the difference between the rules of the game of politics, and playing the game of politics. Principles, rights, goods and identities constitute in his view the framework of politics. Politics itself is the type of game in which this framework, which functions as the rules of the game, can be discussed and amended. Citizens can participate in this game in many different ways. They can do so directly, in public spheres, political parties, elections, and the like, but also through dissent, protest

[9]See for example Kerasote (2008) for dog-human freedom-experiments and Donaldson and Kymlicka (2011) for dependent agency.

and civil disobedience. They can also participate indirectly, through relations of trust with their elected representatives, courts, public servants, and media-facilitated discussions. Tully argues that political action, playing the game of politics, tends to be overlooked by modern political theorists, because of a focus on the rules of the game, or the motives of the players. These ways of theorizing fail because they disregard one of the central characteristics of the political game, namely the freedom of modifying or transforming the rules by speaking and acting differently. Drawing on the work of, Wittgenstein and Foucault, Tully locates freedom precisely there, in being able to change the rules of the game.

Shifting the focus from institutions and fixed mechanisms to acting is helpful in better understanding political animal agency, and in imagining new forms of interspecies politics with them. First of all, nonhuman animals already participate in, and challenge the borders of, different political language-games. As Hobson argues: "Animals here are not simply mapped onto the pre-existing human world or are dumb actors in diverse polities. Instead, their representations and physical presence co-creates the histories, moralities, political subjectivities and places we take as natural and/or devised through human ingenuity alone" (2007, 257). A focus on fixed institutions, laws and procedures as proper politics, or recognition-giving entities obscures this presence. Second, studying practices, instead of searching for an essence of doing politics, is a way of avoiding anthropocentrism. As discussed above, our view of politics has been modeled on a specific kind of human (male, white, able-bodied, educated, and so on). This has led to the exclusion of groups of humans and nonhumans, but also to the exclusion of sets of acts and concepts. Third, this perspective on politics allows us to see different types of animal acts in a new light. An overly rationalistic view of politics usually excludes other animals from our view by definition. If we focus on the borders of what counts as political, a whole new range of acts, challenging this border, comes into view.

Nonhuman animals challenge the rules of the game in several ways. Domesticated animals refuse to do the work they are ordered to do, and Hribal (2003) argues they thereby influenced the change from working with animals to working with machines. Wild animals in captivity try to escape or use violence, and with their acts they influence public opinion and laws (Hribal 2010). While their acts may not be the same as human political acts aimed at changing the system, they do influence it, and in some cases give animals a voice. In envisioning how these acts can influence human political systems, it is important to remember that lines between forms of politics are porous. In order to better incorporate animal agency, connections should be made stronger. Drawing on the work of Rosa Luxemburg, Maisano (2013) shows there is no necessary opposition between forms of direct democracy and institutionalized democracy. One of the challenges for social movements is to incorporate these. In the animal case we can think of small-scale human-animal decision-making organs, resembling direct democracy, that focus on topics that are important for the animals involved, and that can inform other layers of politics. The state does not have to be a model of oppression and repression but can actively foster participation.

6.3 New Forms of Politics: Greeting

In working towards democratic interaction with other animals we also need to develop new institutions and procedures. An example of this would be to establish political greeting rituals with wild and liminal animal communities. As discussed above, Young (2000) criticizes forms of political communication that are presented as universal but in fact reflect preferences of the dominant group. She aims to remedy this by discussing communicative practices that can work towards including others in the process of deliberation. One of the forms of interaction she discusses in this context is greeting. She argues that greeting is an important part of any political process, both ontologically and morally, because in greeting we acknowledge the other as individual. Levinas distinguishes a process of 'subject-to-subject recognition' from the expression of content; before we discuss thoughts and arguments, there is a moment of opening up to the other person. For Levinas, subjects respond to the physical vulnerability of the Other, without the promise of reciprocation.[10] The other, in his or her physical, material vulnerability, makes a claim on the subject that holds the subject hostage. Young argues that this moment of recognizing the other is implicit in greeting, because by greeting, we announce that we take the other seriously as individual and we express our good will to communicate. Without greeting rituals—that Young sees in non-Western and traditional societies, for example Maori societies, but also in Western political interaction (for example in the act of introducing the speaker)—the political process would not function well.

Ethologist Barbara Smuts. describes the relevance of greeting in baboon communities. The greeting rituals of male baboons serve as a means to establish, and learn about, social hierarchy (Smuts 2002). Male baboons often fight and because their teeth are sharp, they get hurt easily. They do not have many friendly encounters, such as playing or grooming and only approach each other friendly in greeting, which they often do. When a male approaches another male, the other usually avoids or threatens him. When this movement is accompanied by lip-smacking, the 'come hither' face and an exaggerated gait, it is understood as an invitation to greet, and answered by making eye contact (which is threatening under other circumstances), lip-smacking and making the come hither face in return. This is then followed by a series of gestures in which one male presents his hindquarters, allowing the other to mount him, grasp his hips, and/or touch or mouth his genitals. They sometimes nuzzle or embrace, and in rare circumstances briefly play. Their roles are mostly asymmetrical and the greeting ritual only lasts for a few seconds.

Smuts. argues that greeting involves meta-communication, which allows for behavioral asymmetries to be temporarily suspended, and for discussing the future. Patterns of greeting tend to reflect coalitional behavior: young males do not often greet, nor form coalitions; old males who have long and calm greeting sessions

[10]For Levinas this only takes place between humans, see Derrida (2008) and Wolfe (2003) for problems with this view.

most often form coalitions. Letting someone else touch their genitals poses a risk, considering the sharp teeth, and shows a willingness to cooperate. As in play (see also Smuts 2002), the safety of the greeting environment allows baboons to learn about the intentions of others, and negotiate the future, without having to fight. Greetings might change over time, as the baboons get to know the other or when their position in the hierarchy changes. The distinctive approach that announces a greeting functions as meta-communication; a baboon tells another baboon he does not want to fight but greet, and this minimizes the chance of aggression.

Greeting can play a role in interspecies contexts as well. When Smuts (2001) first came to study the baboons, she ignored them, as she was taught when she was a graduate student. Because baboons are social animals, ignoring them is not a neutral act. For the baboons it can be a sign of trust, among close friends or relatives, but with strangers it can be an indication of great tension. In order to get closer to the baboons, Smuts. needed to communicate with them; more specifically, she needed to learn to appropriately respond to their greeting signals in a given context. She usually made brief eye contact or grunted:

> It seemed that they read my signals much as they read each other's. By acknowledging a baboon's presence, I expressed respect, and by responding in ways I picked up from them, I let the baboons know that my intentions were benign and that I assumed they likewise meant me no harm. Once this was clearly communicated in both directions, we could relax in one another's company (2001, 297).

By greeting, she treated the baboons as social subjects, and expressed that she was a social subject as well.

To further investigate the political significance of greeting in interspecies situations we will need to conduct small-scale experiments. How and if greeting works will depend on the animals involved; there is no guarantee that it will improve political processes or will lead to consensus, similar to how this works between humans. As Smuts. shows, it can however lead to new ways of communicating and new forms of co-existing.

7 Future Directions

There is much we do not know yet about animal languages and politics, and about political human-animal relations. Because mutual learning and experimenting is required, in democratic interspecies interaction we need a stronger focus on context, on relations and on the situated character of interactions. We also need a stronger focus on processes, instead of single acts or moments. In addition to broadening our understanding of language, in human-animal democracies material interactions can play an important role.

To try out new forms of decision-making and co-existing, we need political human-animal experiments. An example of such an experiment can be found in Donaldson and Kymlicka (2015) proposal to view farmed animal sanctuaries as intentional communities. Currently, farm sanctuaries are usually seen as safe havens for animals who are rescued from abuse. The animals who live there are offered a life free of harm, and function as ambassadors for change. Animals who live in sanctuaries are often rescued from great abuse, which causes them to suffer from physical and mental problems, and in the current model great progress is achieved for individual animals in terms of wellbeing. While the importance of this process of rehabilitation cannot be underestimated, Donaldson and Kymlicka argue that there is a next step, which should be taken in order for animals to flourish. In the human context, rescue centers, for example refugee camps, or women's shelters, focus mainly on providing basic needs and offer a fixed framework in which individuals are offered help. As long as these centers are temporary and focused on getting the individual back into society, there is no problem with this construction. More permanent care-giving institutions (such as asylums and orphanages) however also tend to focus mostly on repairing harm, and not on fostering other capabilities and rights, which can lead to 'total institutions': institutions in which a select group of wards determines all aspects of the lives of inhabitants,[11] which leads to paternalistic control over residents. Donaldson and Kymlicka show the model currently adopted in some farmed animal sanctuaries is paternalistic in similar ways, because it "limits animals' participation in key decisions affecting their lives, (…) which results in policies and practices that may diminish animals' well-being and infringe their rights" (2015, 57). While there is scope for choice with regard to food and friends, this is often minimal. Increased freedom, space and choice might bring an increase in risks, but it also enables flourishing in new ways; a safe life is also not equal to a good life.

To challenge the existing model of animal sanctuaries, Donaldson and Kymlicka propose to view them as intentional communities in which animals should be seen as long-term residents, and in which rights of membership and participation are equally important. Animal residents should be full and equal members of the community, with the right to co-shape the community. These kinds of spaces can function as testing grounds or laboratories for developing new forms of interspecies democratic interaction, for example voting procedures and new forms of representation. Existing political concepts can function as a starting point, following which humans and other animals can together investigate different ways in which nonhuman animals can change the rules of the game. Donaldson and Kymlicka argue that there should be a connection between these communities and larger society. Insights developed in these communities can inform democratic institutions and procedures, and society at large can contribute to the communities by asking them questions related to problems that arise when animals of different species interact. Donaldson and Kymlicka's proposal is one possible way of working towards animal democracy, involving domesticated animals and Western humans; different species of animals and different cultures might ask for different models. In general, we can say that while the precise shape of interspecies democracies is

[11]Who are seen as unruly, similar to nonhuman animals.

unclear, and the role that humans will play in it as well, it is clear that there are many possible ways to work towards it, together with other animals.

References

Acampora, R. 2004. Oikos and domus: On constructive co-habitation with other creatures. *Philosophy & Geography* 7(2): 219–235.

Amé, J.M., J. Halloy, C. Rivault, C. Detrain, and J.L. Deneubourg. 2006. Collegial decision making based on social amplification leads to optimal group formation. *Proceedings of the National Academy of Sciences* 103(15): 5835–5840.

Bekoff, M. 2002. *Minding animals: Awareness, emotions, and heart*. Oxford, NY: Oxford University Press.

Butler, J. 2006. *Precarious life: The powers of mourning and violence*. New York: Verso Books.

Ceballos, G., et al. 2015. Accelerated modern human–induced species losses: Entering the sixth mass extinction. *Science Advances* 1(5): e1400253.

Cochrane, A. 2012. *Animal rights without liberation: Applied ethics and human obligations*. New York: Columbia University Press.

Cochrane, A. 2013. Cosmozoopolis: The case against group-differentiated animal rights. *Law, Ethics and Philosophy* 1(1): 127–141.

Collins, A.M., T.E. Rinderer, J.R. Harbo, and A.B. Bolten. 1982. Colony defense by Africanized and European honey bees. *Science* 218(4567): 72–74.

Conradt, L., and C. List. 2009. Group decisions in humans and animals: A survey. *Philosophical Transactions of the Royal Society of London B: Biological Sciences* 364(1518): 719–742.

Conradt, L., and T.J. Roper. 2005. Consensus decision making in animals. *Trends in Ecology & Evolution* 20(8): 449–456.

Cooke, S. 2014. Perpetual strangers: Animals and the cosmopolitan right. *Political Studies* 62(4): 930–944.

Crane, J. (ed.). 2015. *Beastly morality: Animals as ethical agents*. New York: Columbia University Press.

Derrida, J. 2008. *The animal that therefore I am*. New York: Fordham University Press.

Donaldson, S., and W. Kymlicka. 2011. *Zoopolis: A political theory of animal rights*. Oxford, NY: Oxford University Press.

Donaldson, S., and W. Kymlicka. 2013a. A defense of animal citizens and sovereigns. *Law, Ethics and Philosophy* 143.

Donaldson, S., and Kymlicka, W. 2013b. A defense of animal citizenship. Part 1: Citizen canine: Agency for domesticated animals. *Unpublished manuscript*.

Donaldson, S., and W. Kymlicka. 2015. Farmed animal sanctuaries: The heart of the movement? *Politics and Animals* 1(1): 50–74.

Gaita, R. 2002. *The philosopher's dog*. Melbourne: Text Publishing.

Garner, R. 2013. *A theory of justice for animals*. Oxford, NY: Oxford University Press.

Gentner, T.Q., K.M. Fenn, D. Margoliash, and H.C. Nusbaum. 2006. Recursive syntactic pattern learning by songbirds. *Nature* 440(7088): 1204–1207.

Haraway, D.J. 2003. *The companion species manifesto: Dogs, people, and significant otherness*. Chicago, IL: Prickly Paradigm Press.

Haraway, D.J. 2008. *When species meet*. Minneapolis, MN: University of Minnesota Press.

Hearne, V. 1986. *Adam's task: Calling animals by name*. New York: Skyhorse Publishing Inc.

Hobson, K. 2007. Political animals? On animals as subjects in an enlarged political geography. *Political Geography* 26(3): 250–267.

Honneth, A. 2014. *Freedom's right: The social foundations of democratic life*. New York: Columbia University Press.

Hribal, J. 2003. "Animals are part of the working class": A challenge to labor history. *Labor history* 44(4): 435–453.
Hribal, J. 2007. Animals, agency, and class: Writing the history of animals from below. *Human Ecology Review* 14(1): 101.
Hribal, J. 2010. *Fear of the animal planet: The hidden history of animal resistance*. New York: AK Press.
Kerasote, T. 2008. *Merle's door*. Boston: Houghton Mifflin Harcourt.
List, C. 2004. Democracy in animal groups: A political science perspective. *Trends in Ecology & Evolution* 19(4): 168–169.
Lyotard, J.F. 1988. *Le différend*. Minneapolis, MN: University of Minnesota Press.
Maisano, C. 2013. Where do we go from here? Rosa Luxemburg and the crisis of democratic capitalism. In *Rosa Luxemburg: Her life and legacy*, ed. J. Schulman, 151–166. London: Palgrave Macmillan.
Mouffe, C. 1999. Deliberative democracy or agonistic pluralism? *Social Research* 66(3): 745–758.
Nagy, M., Z. Ákos, D. Biro, and T. Vicsek. 2010. Hierarchical group dynamics in pigeon flocks. *Nature* 464(7290): 890–893.
Norton, G.W. 1986. Leadership decision processes of group movement in yellow baboons. In *Primate ecology and conservation*, ed. J.G. Else, and P.C. Lee, 145–156. Cambridge, UK: Cambridge University Press.
Oliver, K. 2015. Witnessing, recognition, and response ethics. *Philosophy and Rhetoric* 48(4): 473–493.
O'Sullivan, S. 2011. *Animals, equality and democracy*. London: Palgrave Macmillan.
Peterson, D. 2012. *The moral lives of animals*. New York: Bloomsbury Publishing USA.
Rawls, J. 1971. *A theory of justice*. Oxford, NY: Oxford University Press.
Regan, T. 1983. *The case for animal rights*. Berkeley, CA: University of California Press.
Rowlands, M. 1997. Contractarianism and animal rights. *Journal of Applied Philosophy* 14(3): 235–247.
Seeley, T.D. 2010. *Honeybee democracy*. Princeton, NJ: Princeton University Press.
Seeley, T.D., and S.C. Buhrman. 1999. Group decision making in swarms of honey bees. *Behavioral Ecology and Sociobiology* 45(1): 19–31.
Singer, P. 1975. *Animal liberation*. New York: Pimlico.
Slobodchikoff, C.N., B.S. Perla, and J. Verdolin. 2009. *Prairie dogs: Communication and community in an animal society*. Cambridge, MA: Harvard University Press.
Smuts, B. 2001. Encounters with animal minds. *Journal of Consciousness Studies* 8(5–7): 293–309.
Smuts, B. 2002. Gestural communication in olive baboons and domestic dogs. In *The cognitive animal: Empirical and theoretical perspectives on animal cognition*, ed. M. Bekoff, C. Allen, and G.M. Burghardt, 301–306. Cambridge, MA: The MIT Press.
Srinivasan, K. 2015. Towards a political animal geography? *Political Geography* 50: 76–78.
Taylor, C. 1994. The politics of recognition. In *Multiculturalism: Examining the politics of recognition*, ed. C. Taylor, A. Gutmann, and C. Taylor, 25–75. Princeton, NJ: Princeton University Press.
Tully, J. 2009. *Public philosophy in a New Key: Volume 1, Democracy and Civic Freedom (Ideas in Context)*. Cambridge, UK: Cambridge University Press.
Wadiwel, D. 2014. Do fish resist. Paper presented at human rights and animal ethics research network. University of Melbourne, December 2014.
Wilson, D.S. 1997. Altruism and organism: Disentangling the themes of multilevel selection theory. *The American Naturalist* 150(S1): S122–S134.
Wolch, J. 2002. Anima urbis. *Progress in Human Geography* 26(6): 721–742.
Wolfe, C. 2003. *Animal rites: American culture, the discourse of species, and posthumanist theory*. Chicago, IL: University of Chicago Press.
Young, I.M. 2000. *Inclusion and democracy*. Oxford, NY: Oxford University Press.

Human-Animal Interfaces from a Pragmatist Perspective

Michiel Korthals

Abstract In this chapter, I argue from a pragmatist point of view that one should focus not only on ethical principles but also, and above all, on practices that embody human-animal interactions and on their dynamic interfaces. In doing so, one can discern that animal-human interactions can have far more meanings and values than only loving, killing, or eating animals. Besides being subjects of harm and suffering, animals are also active agents of symbolization, history, learning, and the breaking down and reconstruction of established world perspectives. We admire or reject traits of our own that acquire their meaning from animals; we symbolize animals, and they inspire us to behave in a courageous or an admirable way. They can sometimes make us realize that life offers more than we imagine and that new perspectives are around that may enrich our life. They can open up hidden resources in ourselves and in our relationship with the world and therefore have world-disclosing capacities with far-reaching consequences for the way we think and behave.

1 Introduction

Sociologists like Max Weber, Anthony Giddens, and Jürgen Habermas will agree that modernity's functional differentiation of types of human–animal relationships and associations into separate practices, such as keeping companion animals, agricultural animals and laboratory animals, with their own goals of excellence, has its drawbacks. Different practices allow for the same animal to have different roles: a pig can be pet, meat producing animal or laboratory animal. Although modernization's principles such as efficiency, rational calculation, and control imply that these practices are compartmentalized and isolated from one another, there are in reality no clear boundaries between those practices; and people are continuously confronted with events that interconnect and blur the boundaries of these practices.

M. Korthals (✉)
Wageningen University, Wageningen, Netherlands
e-mail: michielkorthals@gmail.com

B. Bovenkerk and J. Keulartz (eds.), *Animal Ethics in the Age of Humans*,
The International Library of Environmental, Agricultural and Food Ethics 23,
DOI 10.1007/978-3-319-44206-8_5

Animals play an important role in these interconnecting and blurring transgressions, and their participation allows them to learn different capacities and to teach humans about them.

In some practices, animals suffer incredibly. Impressed by forms of organized animal cruelty, many ethicists focus exclusively on one or two practices and adopt abstract principles to condemn maltreatment of animals. They conceive animals as victims, as passive objects; they follow the road from pig to pork and stop there, and this often leaves people behind with negative feelings of apathy (what can one do as an individual against the big hog industry?) and not with openness to new experiences.

In this chapter, I do not focus on ethical principles but draw attention to the extraordinary learning capacities and symbolic features of animals. Animals are meaning producers per se. The symbolic values of many animals often have a practice-transgressing impact. For example, we usually see and treat farm animals in a very limited way and make restricted use of their capacities. They have so much more to offer in other types of interaction where they and humans can develop themselves much more broadly and expand their restricted use as only resources of food production or as consumers. Certain art forms, less dominant practices, and technologies can stimulate the development of competences in coping with these meaning connections, thereby bridging practices in a society with isolated practices and making those sectors more porous. They enable people to enhance and expand their capacities and those of animals in order to let all parties flourish. It is therefore worth finding out ways for newly discovered abilities of pigs and other animals to be realized in the livestock sector, so that even this sector can reach a more balanced relationship between humans and animals.

2 Animals as Objects of Moral Concern

It is striking that, in ethics, animals are seen mostly as passive victims. Many animal ethicists focus on the issue of moral considerability for animals and examine under what conditions animals can claim a treatment other than mere instrumental functioning. Animals enter the ethics discourse as objects of moral concern, because especially in agricultural practices they suffer mostly by activities driven by human interests. They are seen as moral patients in need of ethical support, because in those practices they are victims of a short-sighted human interest in cheap food for example.[1] In Peter Singer's famous book, *Animal Liberation*, the starting points of ethical deliberation are suffering, piety, and compassion. Just like the nineteenth century utilitarian Jeremy Bentham, he states: "The question is not, Can they *reason*? Nor Can they *talk*? But, Can they *suffer*?" (Singer 1975, 7).

[1]A beautiful series on BBC3 television showed how cattle becomes beef, entitled *Kill it, Cook it, Eat it* (2010).

Another ethicist with an outspoken ethics of animals is Roger Scruton. In *Animal Rights and Wrongs*, he views animals as recipients of moral concern that deserve piety and sympathy:

> There is the further and deeper question, prompted by both piety and natural sympathy, as to whether it is right to keep animals, however little they may suffer, in conditions so unnatural and so destructive of the appetite for life (Scruton 1998, 73).

Animals in ethics are exclusively seen from the perspective of piety and sympathy, and therefore count only as moral patients. Like most other ethicists, Feinberg (1974) focuses on what animals cannot do compared to humans, highlighting the animals' deficits. His list of what animals cannot do is as follows: they cannot keep promises, make contracts, obey orders and comply with duties, and in general they cannot live a moral life. They cannot claim a right or their rights in courts, and they always need spokespersons. However, these shortcomings are no reason to exclude animals from being objects of moral concern, just like babies and mentally handicapped persons. It remains striking that Feinberg considers only negative traits and compares animals with people who lack certain capacities.

In Herzog's (2011) more sociological study, *Some We Love, Some We Hate, Some We Eat: Why It's So Hard to Think Straight About Animals*, the repertoire of humans' relationship with animals is expanded a little with animals we hate and love. We try to possess them and then caress, kill, or eat them, or try to flee from them. However, other traits are not studied in this book.

This is the common thread running through most animal ethics: because animals suffer and live in unbearable circumstances, they are objects of care and deserve moral attention. Like mentally handicapped persons, animals are seen as having passive or handicapped characteristics only.[2] Other traits and capacities do not play a role in the ethicist's quest for bearers of moral worth, rights, and duties. This ethicist's attitude mirrors the way in which, in modern societies, ethical and social issues are subject to compartmentalization. They reflect only some of the practices of human–animal relationships and appeal to a very limited number of possible human-animal relationships and values. In my book, *Before Dinner* (2004), I distinguished at least six practices:

---

[2]A special case is Heidegger's (1995) radicalization of this approach. He explains: "Throughout the course of its life the animal is confined to its environmental world, immured as it were within a fixed sphere that is incapable of further expansion and contraction" (1995, 198). And further: "The animal is poor in world" (idem, 186). Because plants and animals are lodged in their respective environments but are never placed freely into the clearing of being which alone is "world", they lack language. "In its essence, language is not the utterance of an organism; nor is it the expression of a living thing. Nor can it ever be thought in an essentially correct way in terms of its symbolic character, perhaps not even in terms of the character of signification. Language is the clearing-concealing advent of being itself" (idem, 248–249). Because Heidegger makes a very sharp, quite platonic distinction between being and beings, or between ontological and ontic issues, language is also subject to this distinction. Later, I will argue that animals can have world-disclosing traits, so to deny that they have language or to state that they lack world is an indication of severe human hubris; however it is packed in onto-ontological mysticism.

1. On farms: farm animals, including beasts of burden;
2. In homes and sports facilities: pets, including animals for sport and recreational purposes;
3. In zoos: captive animals;
4. In nature parks, conservation areas: semi-wild animals and animals that are reintroduced into the wild; wild and de-domesticated animals (as in rewilding projects);
5. In laboratories: animals used for experimental purposes;
6. In industries: pharmacological animals. Examples of goods that are produced include medicine-enriched milk and meat from cattle, sheep, rabbits, fish, and chickens. Such additives (sometimes also hormones) are often introduced through genetic modification. At other times, the practice involves only the increased dosage of specific substances that can be used as medicines.

Each of these practices has its own organization, dilemmas, and standards for interactions with animals. These practices, such as keeping companion animals, agricultural animals and laboratory animals, have their own goals of excellence. There are also hybrid or transitional forms, which may even have an inconsistent character. Different practices allow for the same animal to have different roles: a pig can be pet, meat producing animal, or laboratory animal. Many humans feel uncomfortable with these practice-transgressing features; but the amazing thing is, that for animals, these transgressions do not seem to play a role in their interaction with one another. Humans, however, often protest when their pets are suddenly treated as edible and slaughtered. As I will discuss later, some urban community experiments with raising and slaughtering pigs had to deal with these protests that paradoxically are often loudest from the side of hardcore meat eaters.

This classification of animal practices on the basis of certain human-animal relationships is not as obvious as it sounds. The famous Argentinean writer, Jorge Luis Borges, refers to a Chinese classification that is totally different. Although this reference is a joke and the product of Borges' fantasy, the quotation at least relativizes our belief in the one-for-all classification. Borges (2000) states:

> Animals are divided into: (a) belonging to the emperor, (b) embalmed, (c) tame, d) sucking pigs, (e) sirens, (f) fabulous, (g) stray dogs, (h) included in the present classification, (i) frenzied, (j) innumerable, (k) drawn with a very fine camelhair brush, (l) et cetera, (m) having just broken the water pitcher, n) that from a long way off look like flies.

The differentiation into structural sectors with totally different treatment of animals is a modernization trend, finally leading to Weber's (1994, xvi) "iron cage" of differentiation. Although modernization implies that these practices are compartmentalized and isolated from one another into sectors, there are often no clear boundaries between those practices, and people are continuously confronted with phenomena that connect practices and blur their boundaries.

Prominent animal ethicists select some of these practices for ethical reflection. It is interesting to see that they thereby also select a rather limited set of moral values and intuitions and omit other values or intuitions. According to Haidt's Moral

Foundations Theory, one can in morality strike a particular balance between at least six fundamental moral values or domains of moral intuitions, although in explicit ethical treatises only a few are addressed. I have reformulated these (in bold) in terms of human-animal relationship that are most important here[3]:

1. Care and harm: **care for animals, protecting them from harm; animals care for us**
2. Fairness, Justice: **treating animals equally**.
3. Loyalty to your group, family, nation; **loyalty to (particular) animals**
4. Respect for tradition and legitimate authority; **respect for animals' experiences and cultures**.
5. Purity or Sanctity, avoiding disgusting things, foods, actions; **animals as symbols of purity and of disgust**
6. Liberty/oppression: **feelings of reactance and resentment people/animals feel toward those who dominate them and restrict their liberty**

   (Adapted from Haidt and Kesebir 2010).

It is easy to see that, in animal ethics, to date only the first and second dimensions are covered, and the others are omitted. Not all are as relevant for my aim of expanding the moral considerations of animals. In particular, the numbers 3, 4 and 5 play an important role. What would it mean to take seriously these other dimensions of moral intuition? And how can one have access to these other dimensions when the traditional animal ethical positions fail to disclose these dimensions? I will discuss in particular the fifth one and add a new one, the dimension of world disclosure.

3 Other, Often Implicit, Traits of Animals: Broadening the Meaning of Animals

Current types of explicit human-animal interaction are restricted to certain species and allow only certain traits of animals to be brought to the fore. This double restriction on interactions makes it very difficult to understand that other types of interactions are possible in which both humans and animals develop other traits.

---

[3]The six are (I quote):

1. Care for others, protecting them from harm. (Also referred to as Harm.)
2. Fairness, Justice, treating others equally.
3. Loyalty to your group, family, nation. (Also referred to Ingroup.)
4. Respect for tradition and legitimate authority. (Also referred to as Authority.)
5. Purity or Sanctity, avoiding disgusting things, foods, actions
6. Liberty/oppression: This foundation is about the feelings of reactance and resentment people feel toward those who dominate them and restrict their liberty

   (Haidt and Kesebir 2010).

These other types of interaction are often difficult to decipher, in particular for more principalist and dogmatic ethical approaches. Therefore, I recommend the pragmatist approach, which is typically contextual and bottom up, and non-foundationalist in the sense that it does not expect abstract principles to function as rock bottoms from which to derive more concrete norms and to decrease suffering. Mostly, these rock-bottom procedures tend to neglect or overlook important traits. From a pragmatist ethical point of view, one should focus not on ethical principles but on contextual and bottom-up events that express the intricacies of human-animal relationships: "Pragmatist ethics is… not only interested in the application of pregiven normative rules but in the construction of new possibilities for moral action… in the creative character of the solution of moral problems" (Joas 1993, 253).

Looking for new possibilities for moral action requires not sticking to principles, but being sensitive to the distinctness and nearness of non-human animals, in their relation with humans. As we have seen, most ethicists have the habit of considering one or two practices and construct abstract principles against suffering. They make animals victims only, passive collaborators. Although we do not deny that, in some practices, animals suffer badly, this exclusive focus limits the range of possible human-animal relationships and therefore also that of the potential solutions to this suffering.

Firstly, animals play an important role in the traffic streams between practices, and their participation in these crossovers allows them to acquire different capacities and to teach humans. These connection-bridging practices in a society with isolated sectors make those sectors more porous and probably also more livable for humans. Often, we are so immersed in one or two practices that we simply do not have access to sufficient metaphorical and novel resources to do the necessary crossovers.

Secondly and consequently, we can learn from these animalistic crossovers how to do it ourselves, or they can inspire us to try to bridge the different practices and to deal with our own irresistible urge to classify and compartmentalize and, simultaneously, with our discontent or even grief about that urge.

In particular, the symbolic values of many animals often have a practice-transgressing impact, even when the distance between these sectors is huge. Animals have symbolic value, negative or positive, standing for objects of fear, hate or of love, tenderness, courage, and strength. In particular in the past, before the industrial revolution, all kinds of symbolic values, attributed to animals, were invented, and, through the clever interpretations and associations of theologians and artists, they could tell us something about ourselves. However, since the industrial revolution, animals are more and more placed at a distance from our daily life, with the exception of companion animals.

So pigs are perceived both as clever and curious and as dirty and lazy and thereby exemplify important ethical values people appreciate or hate in humans, although they are also perceived and used as mere material for pork. Pigs in use are not only instruments—their nakedness tells us something about procreation and lactation; their fear and courage affect how we can lead our life. In striving for the good life, humans are in need of these kinds of meanings. The gap between practices of food production and consumption, however, entails consumers being confronted with impoverished flows of meaning regarding animals, in particular due to the lack of differentiated personal experience.

Animals as symbols, as icons, and as providing meaning strategies appear in most religions. In Christianity, as can be read in the bible, lots of comparisons, meaning elucidations, and illuminations are made by referring to certain types of animals. Believers are often compared to sheep, as in Isaiah 53: 6, "We all, like sheep, have gone astray, each of us has". It is well known how the relation between predators and prey, such as lions or wolves and sheep and eagles and small birds, are often used to legitimize social inequality and power differences.

Some of these crossovers are quite limited in discovering new animal traits or capacities, although in certain cases they can result in unexpected findings. Examples include experiments with animals to discover their sensory and cognitive possibilities in brain machine interfaces. Monkeys in particular are used for these types of experiments. Other examples include communication studies developing languages with crows, dolphins, or monkeys.

A very old way of trying to decipher new traits in the human-animal relationship is (day-)dreaming about animals or symbolizing them in paintings, music, and sculpture. Indeed, the oldest mural paintings by humans depict animals. These artistic achievements, however, by no means exhaust the possibilities for symbolizing animals and their traits.

Berger gives some more profound examples of chimps that show feelings of boredom, or play or just simulate play, or deal with a troublesome body, and thereby directly confront us with similar human behavior: "They suffer from a surplus of curiosity. They can momentarily forget their needs, or any single unchanging role… And then too, starved of events, they suffer boredom" (2009, 44). "Apes don't live entirely within the needs and impulses of their own bodies—like the cats do. (…) They have gratuitous curiosity." There is indeed a big difference between man and animals, but: "Just because of this distinctness, however, an animal's life, never to be confused with a man's, can be seen to run parallel to his" (idem, 15). Maybe "parallel" is not the right concept here; probably "concomitance" is better, because both animal and man deal with their worlds via suffering, curiosity, play, work, and more.

There is, however, an even more radical way animals can inspire us, and that is when we suddenly see in a different light what counts for us as a world, as if a switch has been turned on and things and events that earlier were in the dark are now attracting our attention. For this direct apprehension of new phenomena, Dewey uses the word "experience". In *Experience and Nature* he writes: "We need a cautionary and directive word, like experience, to remind us that the world which is lived, suffered and enjoyed as well as logically thought of, has the last word in all human inquiries and surmises" (1951, 372). Experience reminds humans of the structure of the world as it is given to them in dealing with it. Learning—opening oneself to new, hitherto unidentified, objects, events, traits, and other phenomena—is directly connected with this type of experience.

Animals are meaning producers per se both for themselves and for humans; this has been and is taken up by art and religion as symbol-renewing and inventing activities. The world-disclosing, symbolic values of many animals often have a practice-transgressing impact. It is interesting to note that some environmental ethicists do indeed recognize animals, and, more broadly, life, as a source of value. Rolston III (1994) for example emphasizes "the value of life arising as a creative process on Earth…", and that process is according to him the real *raison d'ệtre* for environmental ethics. However, it is a pity that he never gives more details about this process of value development and the various values that result from it.

Animals have so much more to offer; in our interactions with them, they and we can develop ourselves much more broadly when we expand our restricted use of them as functioning only as food. Certain art forms, technologies, and design devices can increase our competences in coping with these meaning connection-bridging practices in a society with isolated sectors and make those sectors more porous. Art, design, and technologies enable us to enhance our capacities and those of animals in order to let both parties thrive (Driessen and Korthals 2012).

It is interesting to see that these symbolic and broader cultural values of animals are totally neglected in many recent books on animals (for example, Sandøe and Christiansen 2008; Arluke and Sanders 1996). For example, Hal Herzog's previously mentioned study on the different ways we humans use animals, the symbolic values do not come to the fore. Indeed, we kill animals, eat them, or care for them, but we also are afraid of some and honor others, play with them, admire and venerate them as brave, curious, intelligent, or very well adapted to life and see them as role models, and this in one way or another is simultaneously enacted by parallel or concomitant attitudes of the animals involved.

4 Learning from Animals in Biomimicry[4]

A new branch of applied biology, which is actually a mix of biology, engineering, and physics, is biomimicry. This discipline aims to study animals and plants and to learn from the technologies they apply. For example, termites maintain virtually constant temperature and humidity in their termite mounds: how do they achieve this without investing a large amount of energy? Bats can orient themselves perfectly in dark places. A kind of echolocation enables them to find their targets. Butterflies do a lot of color modeling on their wings to attract, distract, or repel others. Spiders make incredibly strong fibers that can hold more than twice their weight. In her 1997 book, *Biomimicry: Innovation Inspired by Nature*, Benyus

[4]Another interesting branch is Bradshaw's (2009) trans-species psychology, which argues that all mammals have the common capacity to think, feel, and experience themselves and their lives, including different ways of thinking, communicating, and experiencing (see also Safina 2015).

defined biomimicry or biomimetics "a new science that studies nature's models and then imitates or takes inspiration from these designs and processes to solve human problems".

Biomimicry starts with the assumption that animals and plants have traits and capacities that sometimes widely surpass human ones, or that humans even lack. The negative things no longer prevail—the fact that they do not use language to communicate or do not have future expectations; rather, what is important is their amazing competences and learning capacities to lead their own life. Their technologies are seen as resources that engineers can make use of in inventing and designing devices that make human life more sustainable or comfortable. Biomimicry in itself provides an argument for saving as many species of plants and animals as possible, because one never knows what type of animal or plant capacity will be useful for humans.

5 Animals Have Their Own Life, Meaning, and Connections with Humans

Animals are more than flesh and models for design engineers. They identify and struggle with problems that are in many instances the same as ours. For that reason, they can teach us a lot about the life tasks with which we have to deal, like coping with misfortune and fortune, an uncooperative body, an ailing body, or with some extraordinary talent, with power and frustration.

They teach us not only tricks but how to deal with other species, with themselves, and with us. In coping with reality in a fruitful way, with one's body and/or with one's surroundings, it is not always necessary to talk or to deliberate. Dreaming, playing, just searching, and trial and error attitudes are strategies to make life livable and to endure illnesses, loss of a loved friend, aging, or hunger and thirst.

Philosophers often think that these strategies to make life livable require deliberation or at least talk, but these life tasks can also be performed in other ways. Instead of considering the animal ethicists' question, Can they suffer? We should therefore also ask: Can they play? Can they dream? Can they come up with alternatives? Examples of even better questions include: How do they endure stress? What do they find relaxing? Can they stir our imagination? How do they deal with aging in general and with their bodies in particular?

It is not only the range of life tasks that is structurally equivalent to that of humans, the reciprocity of activities in animals is also just as striking as it is in humans. When humans interact with animals and correspondingly change their behavior, so animals learn in these interactions with humans and with other animals in their surroundings. Therefore, Donna Haraway speaks of "responsive interaction" in her 2008 book *When Species Meet*. So, all parties are responding to one another in their own way. Sheepdogs learn how to deal with a herd and specific sheep, and sheep learn how to deal with sheepdogs (and supervising humans).

However, as Clare Palmer has argued with respect to some practices of domestication, in interaction with behaviorally flexible animals (such as those that are under domestication), "we are not just responding to (...) what animals are like" in terms of their capacities; we are actually in part creating "what animals are like" (2010, 47). However, in other practices of human-animal interaction, this "in part creating" covers only a rather limited aspect. First, these practices create humans as well (humans fit in with animals), and, secondly, the development of their capacities is often more triggered by new circumstances than straightforwardly produced by human beings. Nevertheless, domestication sometimes brings new animal capacities to the fore: animal-human interaction can enrich not only humans, but also animals because both can learn from their new environment. Normally, apes do not speak human languages, but they can be taught to learn one.

6 World-Destroying and World-Disclosing Values of Human-Animal Interaction

Now, to take stock, I have argued that animals do a lot with us, and vice versa. A component of the previously discussed practices, and of other animal-human interactions, is a constant process of responding from both sides. In entertaining us, in making us compassionate, in feeding us, both culturally and mentally, in being our companions, in playing with us (see Driessen 2014), in giving us meaning, they teach us something about ourselves, contribute in an essential way to our identity, and help us to deal with our emotions of fear, threats, hope, desires, and shortcomings. Animal symbols bestow meaning on sometimes otherwise not very intriguing phenomena and give us an identity. Being a snake, rat, lion, eagle; not being a cow, lamb is for many an essential trait (Myers 2002). These perspectives on animals belong to Haidt's categories of respect for animals' experiences and cultures and of animals as symbols of purity and of disgust.

In socializing us with our worlds, in educating us, animals still live their own lives. They learn mostly not only tricks from us (if they do), but how to deal with us (eventually in learning tricks from us), and with other species. And in this process of dealing and connecting, they develop meaning. According to the pragmatist animal ethics here developed, one should not forcefully imprint animals with one's favorite features. So, it is not always necessary to identify talking or deliberation to have fruitful human-animal relations. We can also ask ourselves: Can they play with us or other animals? Can they stir our imagination?

Besides these meaning processes, there is a special one that, for its enactment, requires a certain feeling of uneasiness that suddenly makes someone sensitive to a new experience. This process starts, for example, with experiencing animals in an unexpected place. In one of the largest industrialized areas in the world—in the

Rotterdam island of Rozenburg—one can discern amidst oil pipes, huge freighters, and noisy traffic, wild Konik horses and Galloways. The giant ships move deftly along, and the animals scarcely pay attention to them. This encounter can stir in visitors a new feeling of an intimate relationship between animals and heavy and high-tech technology. Animals living in an unexpected place like a desolate piece of nature in a highly industrialized area act as the source of a world-disclosing experience. They teach us that technologies, even those that look incredibly distanced from human proportions, do not impress animals, with the result that they continue living thriving lives and do not mourn the loss of rural areas because they feel estranged. They even contribute strongly to a kind of mental pulling down of the giant metallic structures to more livable shapes (Donaldson and Kymlicka 2011).

But there is more. As one nears the area, one sees cattle and horses grazing and running around, and assumes that they are not fenced. Until one dismounts from one's bicycle and sees the quite ingeniously constructed fence: it is a kind of two meter ditch that makes it impossible for the horses or the cattle to jump over it. Here are technologies and nature combined to give those animals a place. From DNA analysis to a new type of fencing that does not give the impression of a deer camp, everything is used to provide the animals a place that is for them and simultaneously allows humans to come into interaction with them (the park is accessible to everyone).

What happens in these two examples is a kind of reconstruction of our experience, our interaction with animals and with their environment. This type of reconstruction, which can be very far-reaching, exemplifies a world-disclosing experience. The realization that Galloways quietly, royally, graze, while simultaneously surrounded by these machineries and technological dynamics, actively shows that in such surroundings humans also could maybe act in a quiet, detached, but still interested way. We discover a new world, a different way of dealing with technology, when we are struck by this encounter. One of the conditions of this reinterpretation is that we are aware of some recurrent and unsolvable problems in our current interpretation of the (technological) world. In this, we feel already that something is wrong by making sharp and rigid distinctions between nature and technology, between a place for animals and a place for machinery. Only when we feel that something is wrong with our current world are we socially and mentally able to look for something new that can replace our destroyed world.

In the same way, we can study types of communication between dolphins, bees, and even plants; they teach us about hitherto unexplored modalities of communication and associations other than language. Frans de Waal—who has researched the social strategies of primates to negotiate their role in a group and to solve sometimes quite heated conflicts about hierarchy—had shown that at least two different strategies can be successful, depending on the context. In doing so, he lays bare neglected traits that humans can also source. As practiced by bonobos (pygmy chimpanzees), love and caring reconciliation strategies are sometimes more fruitful than chimps' social hierarchical strategy. With these types of moral learning, apes explore new social responsibilities, attribute moral qualifications in a new way, and define anew moral agency (de Waal 2006). Finally, it has a relativizing effect for humans, because it shows that humans are not the only living beings that deal with these problems and that have invented, as it now turns out, only a small number of strategies. Enhanced self-knowledge is the result, concomitant with ways of enriching human-animal interaction. The ethical value at stake here is enhancing living beings' quality of life.

7 Is It Possible to Integrate These Less Prominent Traits in Pig Farming?

Are these, often implicit, features of human-animal interaction only fringe phenomena that amuse bored citizens and other non-farmers looking for new diversions, but who will move on to the next trend when their boredom resurfaces?

Or can these features—better, the encouragement of these features—also be built into the banal meat production system? Given the vast and expanding system, covering more or less the whole world, it is difficult to give a plausible answer. Nevertheless, attempts have been made to give these features a chance in—or better, alongside—this intensive and complex system, possibly making cracks in the system. Pig farming with happy pigs is not a mere illusion nowadays in the Netherlands, where more than 15 % of consumers are concerned about this intensive system, and the number of local food production initiatives is growing considerably (with a market share of more than 5 % of total food production). Some farmers are experimenting with pig comfort barns and quite considerable barn enrichments.

More radical are the experiments with pig farming in cities and villages, like in Annerveenschekanaal, Rotterdam, and Tilburg. In raising pigs with the neighbors,[5] citizens and farmers take care of two or more pigs and upskill themselves to give the pigs a good life before they are slaughtered. Experiencing the animals, their own reactions, and those of the neighborhood worked like a continuous awareness project. The discussions and emotions about the role of pigs in cities, ritually slaughtering them, implied a disclosing process of a totally new world in two distinct ways. In the first place, the plan to slaughter the animal had the effect of destroying the idea that meat is peacefully produced somewhere in a factory. It

[5]"The Pig House is a project that connects theatre and the arts, both visually and tastefully, with agriculture and cattle breeding. The experience of living together with pigs will be put down into words, images and plates to be shared with the public. The Bentheim Black Pieds or Bunte Bentheimer Landschweine is a breed of pig that genetically most closely resembles the now extinct Dutch Landvarken. The pigs are being raised and cared for to be, in the end, killed by a professional butcher and prepared as meat. In 2011 the pigs were living on the grounds of the PeerGroup in Donderen." (pighouse.org).

brought to mind that a living being had to be killed. This destruction of a now former world interpretation was followed by a more sophisticated world disclosure of a vulnerable living being that belongs to the edible and meaningful world. Moreover, for the first time, neighbors had intimate interactions with one another and with the animals, they learned to deal with the soft and strong sides of their own mortality and morality, and they had intense emotions in eating the meat. This is a third meaning of the world-disclosing effect of this pig house (see MEATing.nu). The event illustrates the necessity of a kind of preparation before any world-disclosing effect can be achieved: First, people have to feel that something is wrong with their current way of making sense of their world; they have to be aware that their current world of eating and edible things is somehow "rotten". Next, the social experience of being together in experiencing these new ways of dealing with the world of living beings that become meat prompted the people involved to reinvent their world, with new types of responsibilities and actions.

This type of urban pig farming is an example of new food networks, with very close interactions between citizens and farmers, and can give occasion to the identification of new characteristics of animals and of human relations with them. Renting et al. (2008) attribute to these networks a more than local significance because it resonates with consumers concerns in the Netherlands.

I discern the same kind of phenomena with the new trend among designers and citizens to focus on food and everything that is connected with food, be it spoons[6] and tables, or ways of sharing food or methods of raising pigs, designing new beehives that elicit new traits in bees (I am referring here to pollinators' paradise[7]) and other animals. They have the potential to trigger new types of human-animal relations and to bring animals back into the human-animal world, where they indeed belong. The same story can be told of the pig or mouse as laboratory animals.[8]

[6]See www.jjhyun.com.

[7]http://www.benjaminspoth.com/pollinators-paradise; this type of design brings to consciousness (or: creates awareness of) the important role of insects; see also studioroosegaarde.net, which is inspired by mimicry.

[8]Another example is the laboratory mouse and pig in modern life sciences: *The Laboratory Mouse and Pig as a Hero*, the title of the dissertation of the Dutch philosopher Ellen ter Gast (2007). Laboratory mice are the main actors in the life sciences: they perform the acts that scientists do not dare to perform themselves; they get enhanced muscular mass and become incredibly strong (and can climb a ladder with a big water bottle tied to their tail) or they become prematurely old. Nevertheless, even in those very restricted contexts, mice have something of a symbolic value, often they are seen by many as monsters, for example that can escape and take over world domination. Another example is animals that play in movies, like the Jack Russell Uggie in relation to which Spielberg suggests us "give that barking actor an award", NRC 18 January 2012; see also the movie *Babe*.

8 Conclusion

In this chapter, I have argued from a pragmatist point of view that one should focus not on ethical principles but on practices that embody human-animal interactions and on their dynamical interfaces. In doing so, one can discern that animal-human interactions can have a lot more meanings and values than only loving, killing, or eating them. Besides being subjects of harm and suffering, animals are also active subjects of symbolization, learning, history and the breaking down and reconstruction of established world perspectives. We admire or reject traits in us that get their meaning from animals, we symbolize animals, and they inspire us to behave in a courageous or an admirable way. They can sometimes make us realize that life is more than we expect and that we are surrounded by new perspectives that enrich our life. They can open up hidden resources in ourselves or in our relationship with the world and therefore have a world disclosing-capacity that can have far reaching consequences for the way we think and behave. Even their relative lightness, whereby they are not hindered by a fixed system of classification of leisure, work, and household, can show us that frontiers between sometimes rigidly classified sectors can be permeated because of other, more important, considerations. The values at stake are more than just harm and injustice.

Facing the problem of intensive factory farming with enormous suffering, on a very large scale, one can plead for abolition of the whole system and even for non-animal food production systems. However, even though the reduction in scale and suffering is necessary, it seems to me that we have to face the impossibility of getting rid of animal production immediately. Even if we decide to stop with production now or in the near future, we still have to come to terms with billions of domesticated animals. Experience (be it directly or be it through mass media or movies like *Food Inc* or *Our Daily Bread*) of the suffering of intensively farmed animals prepares the ground by casting doubt on beliefs about current animal farming and by creating a readiness to be open to new worlds. This world-destroying effect can be replaced or followed up by a world-disclosing effect on the basis of new experiences of human-animal relations. Moreover, taking these experiences of animal desires, interests, and capacities seriously can trigger substantial improvements in animal farming. Just as with the humans working in those practices, so with the animals, it proves to be a big win when the interaction between humans and animals is enriched by animals' hitherto hidden and unused or even suppressed other capacities. By learning the unknown capacities of animals and by considering animals in a new way, we learn to understand ourselves better; and we can get a deeper understanding of ourselves and our relations to other humans by reminding ourselves of the non-productive functions and meanings animals have for us. So, I am pleading not for an abolitionist program, but for a radical reformist program that integrates animals' non-profitable or to-be-discovered traits and human-animal interactions. This discovery process, this enrichment of our interactions, can happen not only for farmers but for consumers and producers alike. I have discussed several examples of artful imagery and technological devices that have meaning-enhancing features. They are only a few compared to the many more possibilities that unrestricted human-animal interaction can provide. Intentional design with the aim of enhancing our and animals' capacities to discover meanings and to develop their responsive interaction can also have a place in food farming.

References

Arluke, A., and C. Sanders. 1996. *Regarding animals*. Philadelphia, PA: Temple University Press.

Berger, J. 2009. *Why look at animals*. London: Penguin.

Benyus, J. 1997. *Biomimicry: Innovation inspired by nature*. New York: William Morrow & Company.

Borges, J.L. 2000. 'John Wilkins' analytical language. In id., *Selected non-fictions*, ed. Eliot Weinberger, 229–32. New York: Penguin Books.

Bradshaw, G. 2009. *Elephants on the edge: What animals teach us about humanity*. New Haven: Yale University Press.

de Waal, F. 2006. *Primates and philosophers*. Princeton: Princeton University Press.

Dewey, J. 1951. Experience and nature. In J. Dewey, *The later works*, 1925–1953, ed. Jo Ann Boydston, Vol I. Carbondale: Southern Illinois University Press.

Donaldson, S., and W. Kymlicka. 2011. *Zoopolis: A political theory of animal rights*. Oxford: Oxford University Press.

Driessen, C. 2014. *Animal deliberation. WUR dissertation*. Wageningen: The Netherlands.

Driessen, C., and M. Korthals. 2012. Pig towers and in-vitro meat: disclosing moral worlds by design. *Social Studies of Science* 42: 797–820.

Feinberg, J. 1974. The rights of animals and unborn generations. In *Philosophy and the environmental crisis*, ed. W. Blackstone, 43–68. Athens: University of Georgia Press.

Haidt, J., and S. Kesebir. 2010. Morality. In *Handbook of social psychology*, 5th ed, ed. S. Fiske, D. Gilbert, and G. Lindzey, 797–832. Hobeken, NJ: Wiley.

Haraway, D. 2008. *When species meet*. Minneapolis, MN: University of Minnesota Press.

Heidegger, M. (1995). *The fundamental concepts of metaphysics: world, finitude, solitude* (trans: McNeill, W. and Walker, N.). Bloomington and Indianapolis: Indiana University press.

Herzog, H. 2011. *Some we love, some we hate, some we eat: Why it's so hard to think straight about animals*. New York: HarperCollins.

Joas, H. 1993. *Pragmatism and social theory*. Chicago: University of Chicago Press.

Korthals, M. 2004. *Before dinner. Philosophy and ethics of food*. Dordrecht: Springer.

Myers, C. 2002. Symbolic animals and the developing self. *Anthrozoös* 15(1): 19–36.

Palmer, C. 2010. *Animal ethics in context*. New York: Columbia.

Renting, H., M. Schermer, and D. Rossi. 2008. Building food democracy: Exploring civic food networks and newly emerging forms of food citizenship. *International Journal of Sociology of Agriculture and Food* 19(3): 289–307.

Rolston, H. 1994. *Conserving natural value*. New York: Columbia University Press.

Safina, C. 2015. *Beyond words; What animals think and feel*. London: Holt.

Sandøe, P., and S. Christiansen. 2008. *Ethics of animal use*. London: Blackwell.

Scruton, R. 1998. *Animal rights and wrongs*. London: Demos.

Singer, P. 1975. *Animal liberation*. London: Pimlico.

ter Gast, E. (2007). *Biotech pioneers, a philosophical inquiry concerning the genetically engineered mouse*. Dissertation. RTODTO Nijmegen/Amsterdam.

Weber, M. 1994. *Political writings* (Cambridge texts in the history of political thought), ed. Peter Lassman. Cambridge: Cambridge University Press.

What Is It Like to Be a Bird? Epistemic Humility and Human-Animal Relations

Simon Burton and Emily Brady

Abstract Birds are everywhere. One of the reasons for this ubiquity is the power of flight, allowing the exploitation of a wide range of habitats which might be otherwise inaccessible. That they participate in so many domains and do remain relatively abundant, allied with at times breathtaking beauty, has meant that they have provided a rich source of aesthetic, cultural and scientific reflection. These deliberations can provide an opportunity for us to reflect on the very boundaries of our own human perspectives on the world. This diverse group of organisms may provide a heuristic device to think of ourselves as if from nowhere, freed from the entanglements of being human. In this chapter, we consider some of the ontological, epistemological and, ultimately, ethical issues thrown up by an attempt to become placed outside of ourselves, imagining the terms of other beings with very different lives to our own, lives largely indifferent to our own. We argue that the 'difference' of these winged creatures might help us, in this potential age of the Anthropocene, to develop a stance of 'epistemic humility'. Such humility recognizes the limits of our knowledge in a way that enables us to become receptive to listening to nature's story.

1 What Is It Like to Be a Bird?

Birds are everywhere. It was ever thus: Shakespeare's Banquo, observing with approval the House Martins breeding on Macbeth's castle, shares this with Duncan:

Banquo This guest of summer,
The temple-haunting martlet, does approve,
By his loved mansionry that the heaven's breath
Smells wooingly here: no jutty, frieze,
Buttress, nor coign of vantage, but this bird

S. Burton · E. Brady (✉)
University of Edinburgh, Edinburgh, UK
e-mail: emily.brady@ed.ac.uk

B. Bovenkerk and J. Keulartz (eds.), *Animal Ethics in the Age of Humans*,
The International Library of Environmental, Agricultural and Food Ethics 23,
DOI 10.1007/978-3-319-44206-8_6

> Hath made his pendent bed, and procreant cradle:
> Where they most breed and haunt, I have observed
> The air is delicate.
>
> *Shakespeare, Macbeth Act I Sc. VI*

Of course the air proves anything but 'delicate' for Duncan, but our observation of the lives of birds, and our readiness to read across into our own lives has a rich history. When Lady Macbeth had earlier remarked that "The raven himself is hoarse, That croaks the fatal entrance of Duncan…", she perhaps ascribes to the bird insight into Duncan's fate. We might say that this is a commonplace way of talking about birds.

Nagel's (1974) paper, 'What is it like to be a *bat*?', in asking us to *look* at the world through another's eyes, critically examines this activity. Attempting to understand the subjective character of experience of any creature brings into relief both epistemological and ethical questions about the limits of our knowledge and how humans act based on that knowledge. Even characterizing in brief the import of Nagel's paper, of course, gives the game away—using a visual metaphor here confirms how strongly the senses we possess shape our world.

We are going to explore here what it is like to be a bird, or more particularly, certain species of bird, and what that might tell us about being human in relation to other species. This enterprise is, we believe, interesting in its own right. But we will also argue that the exercise can serve us in other ways. We argue, in particular, that it can allow us to engage with and in some way further develop an understanding of 'epistemic humility', as it emerges out of Schopenhauer's synthesis of Kant's philosophy in *The World as Will and Representation.* Building on this, we explore how this idea might explain our engagement with other species 'on their own terms', as it were. Our approach in the chapter will be to interweave empirical observations of bird lives with philosophical reflections about how we might understand them.

Our understanding of birds has been greatly widened and deepened by the explosion of interest in birds, birdwatching and ornithology in the UK and elsewhere in the late 19th and 20th century. Even in the time that has elapsed since Nagel's paper was first published in 1974, our appreciation of birds, or the class *Aves*, has grown enormously.

Birds are thankfully familiar to all of us. But that very familiarity prevents us from appreciating some striking features of their lives. We might use the air to travel from place to place, and birds do too. But many species of bird *inhabit* the air in a way that is hard to appreciate from simple observation. Common Swifts, for example, feed, copulate and sleep on the wing. The Swift you see arriving in Scotland in early May has been on the wing since it left its nest the previous July. Apart from incubating eggs, or feeding young in the nest, it does not use any solid substrate at all. The air *is* its habitat.

We need to appreciate that this means that the air, the atmosphere, provides its food. Aerial plankton, insects, spiders too inconspicuous to see without the aid of scientific instruments, provide its nutritional needs. Perhaps we should consider air quality in a different way; should we ask how rich in biodiversity a given cubic meter of atmosphere might be, as a way of assessing the quality of the Common Swift's habitat?

This is a habitat that is shared with other species of bird. In Britain, Barn Swallows, House Martins and Sand Martins all exploit the air in this way, spending their lives to a greater or lesser extent on the wing as well. If we try to reflect on what this might be like we have a range of difficulties to overcome at the very least. These species' dazzling flight speeds, allied with almost miraculous ability to maneuver in confined spaces, exhausts the imagination. And for the Common Swift and House Martin roosting—sleeping—on the wing, we struggle to understand how they never stop moving.

Here we are imagining ourselves with our recognizably human point of view, entering into that world. Is that what is needed for 'seeing as' another? We will return to that but it does illustrate in a simple way our understanding of 'difference', and isolating and illuminating that difference is a key part of what we need to do here.

Let's take a more challenging example, that of migration. Many species of bird migrate. In the northern hemisphere, for example, a common bird like the Willow Warbler breeds in temperate zones and migrates to spend the winter in Africa. Ringing recoveries show that the Willow Warbler, a bird with a wing length of around 65 mm, weighing no more than 10 g may travel around 4,380 km from its breeding site in Kippo in Scotland to Mauritania where it overwinters. How does it do this? How does it find its way?

The latest research indicates that certain species of bird are able to recognize and respond to variations in the Earth's magnetic field. The Willow Warbler, migrating under cover of darkness senses and responds to what is to us an invisible force field, with astonishing accuracy. It is quite commonplace, for example, to catch and ring a long-distance migrant like a Willow Warbler in wet scrub and for the same bird to be re-caught a year later in the same thicket, having travelled to east Central Africa and back in the meantime. So we no longer need to use a bat as an example of an animal that has senses we do not possess. Thanks to that recent explosion of interest, we now know that other vertebrate families make use of capacities we do not share and process data that we are unable to sense.

Even when we are on apparently familiar ground, our ability to think this through from the inside is brought up short. Consider another trans-Saharan migrant, the Blackcap. Slightly larger than a Willow Warbler, this UK breeder also winters in sub-tropical and tropical Africa. While breeding it may weigh about 17 g, but when fuelling for migration, fat deposits are accumulated and migrant birds may weigh as much as 25 g. This process is facilitated by a switch in diet from invertebrates to fruits, such as Elderberries, as any ringer will attest. Imagine preparing for your trip south by eating voraciously and gaining half as much weight again, yet still being able to perform physically at your peak. While we share physiological capacities and abilities in broad terms, when we look at this in fine grain can we really be confident that we do share that bird's eye view?

Migration is a really interesting example because much recent work is causing us to reevaluate apparently well-supported assumptions. The migration of Bar-tailed

Godwit is a startling instance. This wader, breeding over much of the northern hemisphere, can be classified into a number of different sub-species. The population that breeds in Alaska over-winters in Australia. Until the recent advent of satellite tracking technology, there was uncertainty as to its migration route and of course the duration of the journey time. Remarkably, it now seems that the birds travel non-stop over the Pacific Ocean, traveling approximately 11,000 km in a journey of around 7 days, making it probably the longest non-stop journey of any bird. Besides simple astonishment, this example again illustrates how a process that we feel we do understand—endurance exercise, perhaps—takes place in terms which seem to be well beyond what might be possible for us.

This should at the very least make us pause and reflect on what we are doing when we consider the world from a bird's point of view. How confident can we be that the process we engage in here is at all appropriate? Our use of analogy here dominates when we undertake this activity. In simple terms, we take a familiar activity as seen from the inside by us as subject, and posit ourselves as the bird—object as subject, subject as object. These examples from the world of birds should at least suggest that our grounds for being confident here that we are getting this right, that when we engage in the activity of trying to see as another, and we feel that we are successful in that enterprise, is not as robust as we might at first think.

We need to take care here to identify when we are talking of error and when we are talking of differences in kind. Error must of necessity involve us in having a correct position in principle. An example would be a situation where we feel we have some insight into the life of the bird and further investigation or observation shows our view to be incorrect. The Robin singing in the garden in autumn, becomes 'our' Robin. We identify with him as our bird, and readily recall his presence from last year. After all, he was in the habit of taking mealworms from the hand and once again, as the days cool, he reappears and readily takes the proffered meal.

Ringing studies, particularly color ringing studies, have enriched our observational data here immeasurably. We now know that both male and female Robins will sing in the autumn. We know that the turnover in garden birds is high, and birds we think of as 'ours' are actually different individuals of the same species using the same habitat, the garden. Robins are both resident and migrants in the UK, some wintering as far south as north Africa, while many individuals reach our shores each autumn from Scandinavia, for example, and occupy territories alongside the resident breeders. We now know that small bird mortality in the first year is very high and while some birds are long lived—the oldest ringed Robin was nearly 8 years 5 months—a typical lifespan is not more than 2 years.

So we reevaluate based on our updated knowledge and think afresh about the life of a familiar species. But the improvement or, more neutrally, the enhancement of our knowledge of the bird is of course only the latest point on a process of continuous discovery. We know a lot more than we once did, but we are not prepared to say in principle that we now know all there is to know. We can readily accept that in principle here the gathering and revising that we engage in can and will continue as long as we are prepared to take an interest.

Because this process is continuous and because we have discovered very readily that some apparently incontrovertible beliefs are ill-founded, we must be cautious in our attempts to see as another. So there is also a difference in kind here, since it does not prove possible for us in principle to identify what would be right. We might say, provisionally, that this is what the exercise might look like but we can never collate the 'facts' in a definitive way and then undertake our exercise of looking *through* the eyes of the bird.

In both of these ways then, our knowledge is provisional. The argument here can be taken further. There are sources of error but we recognize that the very possibility of further error means we are in *principle* unable to look at the world in the way that a bird does. When we look at differences of degree between ourselves and a bird, we are tempted into over-confidence and all too readily *seem* to look at the world through the eyes of the bird. But when we reflect on more fundamental differences, such as migratory navigation using magnetic fields, or mate selection in Blue Tits by using the ultra violet part of the electromagnetic spectrum, we might be inclined to accept for the time being that there is an insurmountable obstacle to the effort. This is not to deny the value of this endeavor. The very activity has, after all, helped to shape our enquiry into other animals and has yielded extraordinary riches in this respect. Instead, the argument is that we should accept more limited aims and that acceptance must shape our relationships with other animals.

2 Schopenhauer's Transcendental Idealism[1]

Schopenhauer's reworking of Kant's transcendental idealism informs all of *The World as Will and Representation.* Although distinctively different from his predecessor, Schopenhauer was absolutely clear that his work would not have been possible without Kant: "…evidently my line of thought, different as its content is from the Kantian, is completely under its influence, and necessarily presupposes and starts from it…" (Schopenhauer 1969a, pp. 416–417). He was also explicit that the work that had been done by Kant that allowed his brand of transcendental idealism to emerge was Kant's greatest achievement:

> *Kant's greatest merit is the distinction of the phenomenon from the thing-in-itself,* based on the proof that between things and us there always stands the *intellect*, and that on this account they cannot be known according to what they may be in themselves (Ibidem).

The world is representation and the building blocks that make up that world—space, time and causality—while providing tools if you will for building an extraordinarily rich and diverse world for us, at the same time set radical limits on what can, as a matter of principle, be an object of perception. We can no more

[1]We should note that we are drawing from transcendental idealism a tool for thinking about human-animal relations, rather than intending to provide a detailed account of this idea in Kant and Schopenhauer.

perceive 'outside' this framework as we could fly unaided, from here to Cameroon to overwinter. But we can be stronger than this, and by now this should not be surprising. On this view, it is simply misconceived to talk of perception at all outside of the space, time and causality framework.

> ...the whole of the material world with its bodies in space, extended and, by means of time, having causal relations with one another, and everything attached to this – all this is not something existing *independently* of our mind, but something that has its fundamental presuppositions in our brain-functions, *by means of* which and *in* which alone is *such* an objective order of things possible...he who has clearly grasped this soon reaches the conviction that the assumption that things exist as such, even outside and independently of our consciousness, is really absurd (Schopenhauer 1969a, pp. 8–9).

Because the world of perception is determined by our perceptual apparatus, and in a more profound way, by space, time and causality, this cannot be all that there is, as Bryan Magee puts it:

> Our perceptions and conceptions cannot be all that there is, but cannot be 'like' what exists in addition to them, so what else there is cannot consist of an independently existing world which corresponds to them; however, since they constitute the limits of what we can envisage, we cannot form any notion of what there is besides (1997, p. 74).

This leaves room to posit what is literally an endless array of 'things', what there is outside of this world, safe in the knowledge that we can no more strike them down than we can properly describe what they might be. The exercise is literally pointless. Amongst other things, this should lead us towards a proper humility. We see that we operate with tools, limited by 'design', as it were, so we must exercise caution when we contemplate or examine a metaphysical account.

The Dotterel is a plover, a wading bird, wintering in the Mediterranean and breeding in small numbers in Scotland. In breeding plumage, the plumage we see here in the summer months, the bird is a riot of rich reds, gold and buff. It breeds on high montane plateaus, the typical whale back hills of the Cairngorm massif. Scientific study of such a bird in this type of remote habitat is hard and arduous, but dedicated workers have rewarded us with insight into their lives. The female, the more brightly colored of the pair, lays the eggs and then leaves the male to incubate and look after the chicks. Ringing studies have shown, quite astonishingly, that some females then travel further north, to Norway, where they breed with another male, leaving that second male to travel south to its wintering grounds.

Simple observation alone, while providing us with rich and memorable experiences, does not allow us to begin to grasp what it is this species actually gets up to—the simplest level at which we reflect on other species' lives. Detailed observation armed with rigor and disciplines honed by testing elsewhere reveal to us aspects of their lives that really challenge our ability to think or see as them. We might argue, as we have here, that this is the first stage to our recognizing that we have an in kind obstacle to that activity. This is interesting in itself but it is also beneficial to see these examples as tools to help us to try to grasp what transcendental idealism might mean, and to assimilate what this might mean to us in practice.

3 Epistemic Humility

The Kantian account to which Schopenhauer responds reveals the limits of the human perspective, of human knowledge, however it also indicates the very *receptivity* of human knowledge, that "we are receptive creatures, who must be affected by things of which we have come to have knowledge" (Langton 1998, p. 2). In her study of Kant's first *Critique*, Rae Langton describes this stance as 'epistemic humility'. Receptivity for Kant is a kind of passive sensibility, a very minimal empiricism in the sense that any kind of knowledge begins with experience, with our being affected by objects (Langton 1998, pp. 44–45). It does not mean that knowledge comes from experience or through a range of empirical properties.

Why does this matter? It provides some way towards the more positive reading of Kant's transcendental idealism, that although we may not know things in themselves, the limits of our knowledge become indications of what we can know and what we cannot know. Kant's intelligible/sensible distinction is intended to set up a contrast: "it is to emphasize that they [things in themselves] are not sensible, that their intrinsic nature is epistemically inaccessible to receptive creatures. That is not dogmatism but Humility" (Langton 1998, p. 41). Epistemic humility is a form of epistemic finitude, but not one that is oriented towards skepticism. Rather, it recognizes the limits of our knowledge by resisting "traditional claims that there is nothing significant at all beyond what we can determine, at least indistinctly, through our own representations" (Ameriks 2003, p. 139). Langton argues that on Kant's view, receptivity actually *follows* from epistemic humility.

A significant moment in the genealogy of epistemic humility can be found in Plato's *Apology*. At his trial, as part of his defense and in the face of the accusation that he showed disrespect for the gods and corruption of youth, Socrates provides an account of wisdom. Here, he recounts his investigation of how he could be claimed by the god, Delphi, to be the wisest of men:

> I went to interview a man with a high reputation for wisdom, because I felt that here if anywhere I should succeed in disproving the oracle and pointing out to my divine authority. 'You said that I was the wisest of men, but here is a man who is wiser than I am.' Well, I gave a thorough examination to this person…and in conversation with him I formed the impression that although in many people's opinion, and especially in his own, he appeared to be wise, in fact he was not. Then when I began to try to show him that he only thought he was wise and was not really so, my efforts were resented both by him and by many of the other people present. However, I reflected as I walked away, 'Well, I am certainly wiser than this man. It is only too likely that neither of us has any knowledge to boast of; but he thinks that he knows something which he does not know, whereas I am quite conscious of my ignorance. At any rate it seems that I am wiser than he is to this small extent, that I do not think that I know what I do not know (Plato 1969, p. 50).

Because Socrates has been accused of having great wisdom, his defense attempts to persuade his accusers of what wisdom consists in. We see here the strong exercise of epistemic humility, or at least a plea for epistemic accuracy. As Sharon Ryan puts it, "Wise people, one might argue, possess epistemic self-confidence, yet lack epistemic arrogance. Wise people tend to acknowledge their fallibility, and wise people are reflective, introspective, and tolerant of uncertainty. Any acceptable theory of wisdom ought to be compatible with such traits" (Ryan 2014).

On this account of wisdom, a key ingredient is recognizing the limits of our knowledge, the possibility that we may err. It is especially interesting for our purposes here because wisdom is situated within an approach that *values* the quest for accurate knowledge. Where Kant and Schopenhauer provide metaphysical insight into epistemic humility, we might say that Plato moves us toward a virtue-oriented approach of what we can or cannot know. In other words, epistemic humility not only indicates the limits of our knowledge, but it can also serve as a guide to action, to how we might conduct ourselves in terms of trying to understand and be receptive to the world.

4 Humility and Receptivity

This epistemic perspective potentially yields moral benefits where human-animal relations are concerned. Recent work in environmental virtue ethics identifies humility in a more general sense, and there is a close link between humility and respect, as well (Hill 1983). Humility originates in the Latin, *humilis,* which means 'low' (OED 2015). This meaning points to the moral development of the concept, suggesting not pride but a willingness to serve or bow to others. The feeling of insignificance associated with experience of the sublime in nature is commonly linked to being humiliated—countering the self-aggrandizement of presumed human superiority over nature (Brady 2013). As the sublime humbles the self, an awareness is opened out to other things in the world, or universe, that are greater than humanity in terms of material size, power or as beyond what we can fully know. In Kant's terms this is the sublime as the supersensible. For Schopenhauer (1969a, p. 206), the sublime becomes: "an exaltation beyond our own individuality". He also writes:

> [I]f we lose ourselves in contemplation of the infinite greatness of the universe in space and time, meditate on the past millennia and on those to come; or if the heavens at night actually bring innumerable worlds before our eyes, and so impress on our consciousness the immensity of the universe, we feel ourselves reduced to nothing… (Schopenhauer 1969a, p. 205).

The sublime is offered here to present a case of experiences which force receptivity to things other than ourselves. We do not mean to suggest that the kinds of knowledge or interactions with birds we have are experiences of the sublime, though some will be. Experiencing at close hand thousands and thousands of Northern Gannets on or around the Bass Rock in Scotland is likely a case of the mathematical sublime, the seemingly infinite number of things which make the single, sole human individual feel outnumbered and insignificant. Or, the very *idea* of the Bar-tailed Godwit's migration evokes the mathematical sublime, given the challenge to imagination of this creature flying such a distance.

Many of our interactions with birds are likely to be more reminiscent of wonder. Both the sublime and wonder provide limits through perceptual, immediate experiences of a range of things. For sublimity, it will be great things that belittle us. For wonder, it can be something large but also small, such as a delicate spider's web, and even smaller, the tiny, incredible form of a snowflake. Wonder relates to things that are marveling and enables us to grasp what is different from our human selves —fascinating and wonderful at least because of that.

We also find that a sort of sustained attention can bring us out of ourselves and into a more receptive state (James 2015, p. 89). In a well-known passage from the *Sovereignty of the Good*, Iris Murdoch describes her encounter with a kestrel in order to illustrate the idea of 'unselfing', which we might construe as an act of epistemic humility in which we become receptive to the world:

> I am looking out of my window in an anxious and resentful state of mind, oblivious of my surroundings, brooding perhaps on some damage done to my prestige. Then suddenly, I observe a hovering kestrel. In a moment everything is altered. The brooding self with its hurt vanity has disappeared. There is nothing left but kestrel. And when I return to thinking of the other matter it seems less important (1991, p. 84).

While 'unselfing' presents a much more receptive and less egotistical approach to the kestrel, a worry arises that forms of unselfing can leave the experiencing subject *without* a self, moving too far in the direction of humility, where perhaps *nothing* of ourselves is left to engage. Something like this perspective is taken by Godlovitch (1994, p. 26) in adopting the idea of 'mystery' to characterize his 'acentric theory' of aesthetic appreciation of nature, which commentators have dubbed the 'aloofness' or the 'mystery model.' The acentric perspective places the subject in a position of radical de-subjectivity where all cultural, and even scientific knowledge, is removed. In a position of being acutely aware of nature's independence from us, and lying beyond our knowledge, he argues that our only appropriate aesthetic response is a sense of mystery.

The 'mystery' approach is not what we are after here. We suggest that it is important to avoid a position where we become overly sensitive to the question of what we do and do not know, a sort of epistemological skepticism driven by the quest for epistemic humility and respecting other forms of life. Such skepticism would endanger our very relations with other forms of life, leaving little room for a critical anthropomorphism, where the latter stance at least brings us into range with other beings.

The critical anthropomorphism we have in mind objects to scientific accounts which so often argue that we are liable to make mistaken and distorting attributions of mental states, and that we cannot have knowledge of other minds. Midgley (1995), for instance, argues that in so far as we think our attributions of various emotions to humans are apt and at least in the right ballpark, we should have a fair degree of confidence when it comes to mammals. Also, importantly, why should we be so concerned about mistaken readings of animals and not of humans? We do make mistakes in reading humans, and will also make mistakes with animals, but Midgley denies that the gap between the two forms of experiencing emotions in others are so vastly different. Her chief criterion for ensuring appropriate judgments is close, careful attention, especially in cases of animals with whom we are less familiar. That this also holds for humans supports a more robust account of anthropomorphism, e.g., people with less experience of children will have less honed skills in reading behavior in them. In Midgley's account, attributing human qualities to animals is actually more one of being carefully aware of similarities and being watchful of projection and misattributions.

These thoughts about critical anthropomorphism suggest something else we can take from Nagel's article. That we cannot hope to know everything it is like to be a bat, given our own experience we can at least grant there is something that it is like to be a bat, and we would do well to make the effort to use imagination and epistemic tools to try to understand as best we can what that experience is like. Our remarks here may be susceptible to the objection that critical anthropomorphism conflicts with epistemic humility because the former is interested in credible similarities and commonalities while the latter seeks a more respectful distance by recognizing the limits of our knowledge (though not one which removes the human perspective altogether).

While this chapter has focused on difference and what we cannot know, we do not mean to deny the similarities, or indeed the continuum, that exists between humans and other animals, including birds. Our focus has been on birds in the wild rather than the (potentially) more common worlds that may be formed between humans and birds in captivity, or in the liminal spaces where human and wild birds meet, such as the garden, backyard or home (see, for example, Len Howard's work, 1952, 1956). In these spaces we will find cases of closer interaction, bonds of trust and, potentially, shared understanding. The stance of epistemic humility seeks a balance between what we can and cannot know; it is not a scientific perspective which merely observes animals in a detached fashion, nor is it a humanizing anthropomorphism.

Where does this leave us in finding a role for humility and receptivity? Epistemic humility involves an attempt to grasp nature on its own terms. Yuriko Saito provides some guidance here, that it will mean adopting "…an attitude which would involve listening to nature's own story….recognizing and respecting nature as having its own reality apart from our presence" (Saito 2004, p. 142). There is respect for nature at work here, putting our own story in the background in order to be open to the story of a Bar-tailed Godwit or a Common Swift. We will need to be careful not to assume there are terms we could ever know, just as we cannot know the supersensible or noumenal. We may not be successful for this reason, and because of other limitations in getting to know the full story, yet in our effort towards such a grasp, we exercise humility and receptivity.

5 A Caspian Tern

We conclude with a recent personal encounter with an individual bird, a Caspian Tern, which we hope illustrates how we might listen to the stories of other creatures unlike ourselves. The Caspian Tern is the world's largest tern, a "Huge, gull-like tern, with deep, dagger-shaped bill on heavy head, noticeably round body, large, rather blunt wings…and relatively short tail. Flight recalls that of large …gull, with slow and heavy wing-beats" (Cramp 1978–1996, p. 17). It is a widespread breeding bird of the northern hemisphere and North American populations over winter around Atlantic and Pacific coasts, breeding principally in the Great Lakes area and Arctic Canada.

They feed, like all terns, by plunge-diving for fish and although capable of swimming it is unusual to encounter them on the water. In early spring 2015, walking by the Potomac River in the city of Alexandria, Virginia, we noticed a Caspian Tern gliding into a small bay area close by. This was unusual in itself, so attracted our attention. As it landed, the tern seemed uncertain, and not fully in control. Once on the water, it appeared in distress; repeated attempts to take to the air again failed, the bird appearing weak and now clearly unable to fly. The tern's efforts took it closer to some driftwood corralled in the corner of the bay, where it remained, occasionally repeating attempts to propel itself into the air.

With no obvious means of help available, we moved on our way, knowing that we would return to the site later in our walk. On our return, the bird had died. It floated, freshly dead, amidst the logjam of driftwood where we had last seen it alive. Navigating the floating timbers allowed a close approach, and we collected the tern, intending to examine the bird for rings. This bird had in fact been ringed, and having recorded the unique combination of numbers, we examined its physical condition. Ringers routinely assess a bird's physical condition by recording the condition of the pectoral muscle, the bird's principal flight muscles. The muscle mass is scored using a standardized scale, with muscle scores on the scale of 0–3, and this is used as a standard proxy for the bird's overall physical condition. An active, healthy bird would typically score 2 or 3.

This Caspian Tern, which the ring recovery showed had been ringed as a chick on Little Galloo Island, Lake Ontario in 2008, had a muscle score of between 0 and 1. It seemed likely that it had starved and died at the relatively young age of 7; Great Lakes Caspian Terns live to 12 years on average. The ring recovery data told us a story about where it had hatched, the breeding grounds of its parents. This in itself was remarkable, and quite moving as well, for it gave us insight into the life of an individual bird which we could know little about without such data. Of course, we could not *know* very much more about this particular tern's life, apart from where it likely spent time in terms of over wintering and breeding grounds. But in trying to 'listen' in an informed way to the story of the 7 years of this tern's life, we could try to imagine its activities and interactions, based on our still limited knowledge of the species.

And we had witnessed its last struggle. We were not alone. As we walked on the first time before the tern had died, we looked back and noticed a passerby, coffee cup in hand, looking inquisitively at the bird. He would have been drawn to seeing a tern so close in, adjacent to the path—a close-up view. Maybe he also recognized the bird's distress. Perhaps he empathized, and felt compassion. For us, while we cannot say we had a common world with this tern, there was shared suffering.

That very compassion is for Schopenhauer the only genuine moral incentive. In his metaphysics, it stems from the recognition of the essential identity of all living things:

> Hence, if plurality and separateness belong only to *appearance*, and if it is one and the same essence that presents itself in everything living, then that apprehension which suspends the distinction between I and Not-I is not in error. Rather it must be the opposite that is in error... This view, then, would be the metaphysical basis of ethics, and it would consist in the fact that one individual immediately recognizes in another himself, his own, true essence (Schopenhauer 2010, p. 267).

It is clear that Schopenhauer does not intend recognition in 'another' to mean 'another human'. For example, he writes, "One must be truly blind...not to recognize that the essential and primary thing is the same in the animal and in the human..." (Idem, 241). This is a reminder of what is shared with the tern. From procreation, nest building, the search for food to shared suffering, we acknowledge what we have in common. Adopting a stance of epistemic humility, again, is not to deny what we have in common with the tern.

We had witnessed this tern's last flight, and its death. We brought to our observations much assimilated knowledge and were able to establish much more after the event. But we were reminded of the limits of our knowledge and also how such limits, rather than constituting dead-ends, become openings for developing a less arrogant epistemic perspective. If we are in fact in the age of the Anthropocene, such limits remind us too that the effects of humans on the planet do not, by necessity, translate into human domination. While such effects may be widespread, epistemic humility has the moral benefit of resisting a conception of the planet as human-centered.

References

Ameriks, K. 2003. *Interpreting Kant's critiques*. Oxford, NY: Clarendon Press.

Brady, E. 2013. *The sublime in modern philosophy: Aesthetics, ethics, and nature*. New York: Cambridge University Press.

Cramp, S. 1978–1996. *Birds of the Western Palearctic*, Vol. IV. Oxford: Oxford University Press.

Godlovitch, S. 1994. Icebreakers: Environmentalism and natural aesthetics. *Journal of Applied Philosophy* 11(1): 15–30.

Howard, L. 1952. *Birds as individuals*. London: Collins.

Howard, L. 1956. *Living with birds*. London: Collins.

Hill Jr., T. 1983. Ideals of human excellence and preserving natural environments. *Environmental Ethics* 5: 211–224.

James, S. 2015. *Introduction to environmental philosophy*. Cambridge, UK: Polity.
Langton, R. 1998. *Epistemic humility*. Oxford: Oxford University Press.
Magee, B. 1997. *The philosophy of Schopenhauer*. Oxford: Oxford University Press.
Midgley, M. 1995. *Beast and man*. London and New York: Routledge.
Murdoch, I. 1991. *The sovereignty of the good*. London: Routledge.
Nagel, T. 1974. What is it like to be a bat? *The Philosophical Review* 83(4): 435–450.
OED, *Oxford English Dictionary* (oed.com; Accessed 2 Nov 2015).
Plato. 1969. *Apology*. In *The last days of Socrates* (trans: Tredennick, H.). London: Penguin.
Ryan, S. 2014. Wisdom. In *The Stanford Encyclopedia of Philosophy*, ed. E.N. Zalta. http://plato.stanford.edu/archives/win2014/entries/wisdom/. Accessed 15 Oct 2015.
Saito, Y. 2004. Appreciating nature on its own terms. In *The aesthetics of natural environments*, ed. A. Berleant, and A. Carlson, 141–155. Peterborough, Canada: Broadview Press.
Schopenhauer, A. 1969a. *The world as will and representation*, Vol. I (trans: Payne, E.F.J.). New York: Dover.
Schopenhauer, A. 1969b. *The world as will and representation*, Vol. II (trans: Payne, E.F.J.). New York: Dover.
Schopenhauer, A. 2010. *On the basis of morals* (trans: Cartwight, D. and E. Erdmann). Oxford: World Classics.
Shakespeare, W. 1984. *Macbeth Act I Sc. VI*. London: Arden.

Unfeeling Brutes

Jeffrey Moussaieff Masson and Susan McCarthy

Abstract In this reprint from the thought-provoking book *When Elephants Weep*, Jeffrey Masson and Susan McCarthy call into question human exceptionalism. Why have humans throughout history sought to distinguish themselves from other animals? Why is such a distinction even important? Traditionally, two lines of reasoning have been put forward: either humans are unique in their failings and need to take animal virtues as their example, or—more predominantly—human advantages have been emphasized. It has been claimed that humans are the only beings who are intelligent, have culture, have a sense of humor, have knowledge of death, can make tools, and understand morality. In seeking an explanation for the alleged importance of the human-beast distinction, the authors draw a parallel with the way dominant human groups have claimed their superiority over the humans they were subordinating. Pointing out distinguishing characteristics 'between man and beast serves to keep man on top'. In a similar vein, pointing out distinguishing characteristics between white and black people or men and women has served to keep dominant groups on top. Another answer can be found in convenience: 'animals' presumed lack of feeling has provided a major excuse for treating them badly', providing a rationale to hunt and eat them and to use their labor. Moreover, in science, the desire to avoid anthropomorphism has been a major obstacle to investigating the emotions of animals.

Throughout human history, there has been much concern with differentiating humans from beasts. *We* speak; *we* reason; *we* imagine; *we* anticipate; *we* worship; *we* laugh. They do not. The historical insistence on an unbridgeable gap suggests that it serves some need or function. Why do we humans define ourselves by

This is a reprint from Jeffrey Moussaieff Masson and Susan McCarthy (1995), *When Elephants Weep. The Emotional Lives of Animals* (pp. 24–44). New York: Dell Publishing.

J.M. Masson (✉) · S. McCarthy
University of Toronto, Toronto, ON, Canada
e-mail: jeffreymmasson@gmail.com

B. Bovenkerk and J. Keulartz (eds.), *Animal Ethics in the Age of Humans*,
The International Library of Environmental, Agricultural and Food Ethics 23,
DOI 10.1007/978-3-319-44206-8_7

distinction from animals? Why should the distinction between man and beast matter?

Any dictionary of quotations reveals many attempts to make this distinction, which fall largely into two categories. Many cite human failings as unique, chief among which is fighting among ourselves. Usually the writer is attempting to inspire his readers with moral resolutions. In the first century AD, Pliny the Elder in his *Natural History* admonishes: "Lions do not fight with one another; serpents do not attack serpents, nor do the wild monsters of the deep rage against their like. But most of the calamities of man are caused by his fellow men." When, in 1532, Lodovico Ariosto in *Orlando Furioso* says that "Man is the only animal who injures his mate", this too is meant as an admonition. James Froude in his *Oceana* of 1866 claimed that "Wild animals never kill for sport. Man is the only one for whom the torture and death of his fellow creatures is amusing in itself." And even William James, in this century, wrote that man "is simply the most formidable of all the beasts of prey, and, indeed, the only one that preys systematically on his own species." Samuel Butler in the nineteenth century made a slightly different but equally admonishing distinction concerning man's relations to other animal species: "Man is the only animal that can remain on friendly terms with the victims he intends to eat until he eats them." In these examples, animals are not so much being observed as men are being exhorted to cease killing (usually) other men. They are intended to shame men into recognising that they behave worse than animals. This is perhaps what Carolus Linnaeus had in mind in a letter he wrote in 1747: "What is the difference between man and ape, based on natural history? Most definitely I see no difference. I wish some one could show me even one distinction."

The other—and by far the larger—category of man/beast contrasts cites human advantages: our intelligence, our culture, our sense of humour, our knowledge of death. In the nineteenth century William Hazlitt maintained that "man is the only animal that laughs and weeps; for he is the only animal that is struck with the difference between what things are and what they ought lo be." And in our century, the philosopher William Ernest Hocking claimed that "man is the only animal that contemplates death, and also the only animal that shows any sign of doubt of its finality." Uniqueness is claimed for the human sense of humour, the ability to understand virtue, the ability to make and use tools. Again the authors seem more interested in making a didactic point for humans than in observing or understanding animals.

Contemporary renditions of this contrast have been scarcely more grounded in animal reality and have not shed much more light on animals—or humans. Recently Nicolas Humphrey wrote that "human beings have evolved to be the most highly social creatures the world has ever seen. Their social relationships have a depth, a complexity and a biological importance to them, that no other animals' relationships come near" (Humphrey 1976). Considering how little is known about "other animals' relationships", this seems unwarranted. Until very recently it was held as a canon of animal behaviour that only the human female experienced an orgasm: non-human females did not. As recently as 1979, anthropologist Donald Symons pronounced that the "female orgasm is a characteristic essentially restricted to our own species" (Symons 1979, 78–9). When the question was actually investigated in the stump-tailed macaque, using the same physiological criteria as used for humans, it was found that the female macaques did appear to experience orgasm (Goldfoot et al. 1980; cited in de Waal 1989, 151–3). Primatologist Frans de Waal notes that, from behavioural evidence, this is also true of the female bonobo (pygmy chimpanzee) (de Waal 1989, 198–206). The truth is that, like many questions specifically involving the human female, not many scientists had ever considered the question systematically, let alone done the necessary field observation studies to find an answer. Perhaps it pleased most male scientists to imagine that while animal females sought sex only during an oestrus cycle, and hence had sex only for reproduction, human females, due to their unique orgasmic capacity, wanted sex all the time.

People have always exalted certain 'higher' feelings which we claim single us out from animals. Only humans, it has been claimed, feel the noble emotions, such as compassion, true love, altruism, pity, mercy, reverence, honour and modesty. On the other hand, people have often attributed so-called negative or 'low' emotions to animals: cruelty, pride, greed, rage, vanity and hatred. At play here appears to be the seemingly unbearable injury to our sense of uniqueness, to our entitlement to the nobility of our emotional life. Thus not only whether animals can feel; but what they feel, is used to strengthen the species barrier. What lies behind this 'us/them' mentality, the urge to define ourselves by proving we are not only different, but utterly different, including emotionally? Why should this distinction between man and beast be so important to humans? Examining distinctions humans draw among themselves may provide a partial answer. Dominant human groups have long defined themselves as superior by distinguishing themselves from groups they are subordinating. Thus whites define blacks in part by differing melanin content of the skin; men are distinguished from women by primary and secondary sex characteristics. These empirical distinctions are then used to make it appear that it is the distinctions themselves, not their social consequences, that are responsible for the social dominance of one group over the other.

On this model, one could say that the distinction between man and beast serves to keep man on top. People define themselves either as distinct from animals, or similar when convenient or entertaining, in order to keep themselves dominant over them. Human, beings presumably want to treat animals the way they do—hunting them, jailing them, exploiting their labour, living on their flesh, gaping at them and even owning them as signs of social cachet. Any human being who has a choke does not want to be treated like this. In this analysis, the distinction between man and beast serves to direct various forms of human aggression away from human beings and towards animals as such.

A blatant example of many of these prejudices, with a suggestion of some of their social consequences, can be found in the article on 'Animals' in the *Encyclopaedia of Religion and Ethics,* written in 1908:

> Civilization, or perhaps rather education, has brought with it a sense of the great gulf that exists between man and the lower animals…in the lower stages of culture, whether they be found in races which are, as a whole, below the European level, or in the uncultured portion of civilized communities, the distinction between men and animals is not adequately, if at all, recognized…The savage… attributes to the animal a vastly more complex set of

> thoughts and feelings, and a much greater range of knowledge and power, than it actually possesses… It is therefore small wonder that his attitude towards the animal creation is one of reverence rather than superiority.[1]

Only a lower man, one close to animals, would value them. The ingenuity of human rationalisations of this gap is analysed in an elegant book on hunting by Matt Cartmill:

> In policing the animal-human boundary, scientists have shown considerable ingenuity in redefining supposedly unique human traits to keep them from being claimed for other animals. Consider our supposedly big brains. Human beings are supposed to be smarter than other animals, and therefore *we* ought to have larger brains. But in fact, elephants, whales, and dolphins have bigger brains than ours; and small rodents and monkeys have relatively bigger brains (their brains make up a larger percentage of body weight than ours do). Scientists who study these things have accordingly labored to redefine brain size, dividing brain weight by basal metabolic rate or some other exponential function of body weight to furnish a standard by which these animals' brains can thus be deemed smaller than ours. The unique bigness of the human brain thus turns out to be a matter of definition (Cartmill 1993, 222).

Animals' presumed lack of feeling has provided a major excuse for treating them badly. If they have no feelings, humans need not respect those feelings or them. This has been so extreme that animals were long regarded as unable to feel pain, physical or emotional. When an animal is hurt in a way that would hurt a person, it generally reacts much as a person would. It cries out, it gets away, then examines or favours the affected part, and withdraws and rests. Few veterinarians can be found who doubt that it is pain that animals feel when wounded. The only criterion that an animal fails to meet for feeling physical pain as humans understand it is the ability to express it in words. Yet the fish on the hook is said not to be thrashing in pain (or fear) but in a reflex action. A lobster in boiling water or puppies whose tails are being docked are said to feel nothing. To rebut such rationalisations a recent German book on animal consciousness argues that "The fact that we so immediately understand these signals is just a further sign that we share with other animals the grand construction of our pain apparatus" (Artz and Birmelin 1993, 154). When the subject is actually researched, the findings are in line with common sense. The pain of the fish twisting on the hook is real. Even insects are found to be sentient and experience pain.[2]

---

[1]This splendid example of benightedness is also quoted in Midgley (1992, 223). The article is a very long and learned one written by Thomas (1908). The article actually begins (p. 483) by citing the "great gulf that exists between man… and the elephant and the anthropid ape".

[2]"Another assessment of pain in fish comes from a team of researchers under the direction of Professor John Verheijen at the University of Utrecht in the Netherlands (in 1988). They concluded that fish do feel pain and experience fear" (Orleans 1993, 148). Can an ant feel pain? For years, the received wisdom was that insects did not feel pain, hence one could do anything one liked to them. But Wigglesworth (1980) in an article in the prestigious journal *Antenna* argued they do. An even more recent article by Lockwood (1987) states that "existing evidence indicates that insects qualify as sentient and their lives ought to be included in moral deliberations" (p. 70). For another positive response to the question see Eisemann et al. (1984).

In considering the role of the denial of sensate experience in the way animals have been made into a lesser form of life by many, it is important to observe that rationalising away the pain of others is practised on people too, Krutch writes, "Man has never denied anything of the animals without coming shortly thereafter to deny it of himself also" (1950, 95). It is usually, in fact, a dominant group denying feelings in a subordinate group. And surely the feelings of animals have value whether or not their denial ultimately hurts people.

It has always been comforting to the dominant to assume that those in subservient positions do not suffer or feel pain as keenly or at all, so they can be abused or exploited without guilt and with impunity. That lower classes and other races ore relatively insensitive has been a frequent assertion in the history of prejudice.[3] Similarly, until the 1980s, it was routine for surgery on human infants to be performed with paralytic agents but without anaesthesia, in the long-held belief that babies are incapable of feeling pain. It was believed, without evidence, that their nervous systems were immature.[4] The notion that babies do not feel pain is directly counter to their screams and can only be classified as scientific myth. Yet it has been a tenet of human medicine, only recently acknowledged to be false in the wake of studies showing that infants who do not get pain medication take longer to recover from surgery.

A similar bigotry has extended to the presence of emotions in the poor, the foreign, those raised in impoverished or unenlightened cultures, and in children, who supposedly have not yet learned to feel in fully human ways. It is often asserted that when an infant smiles, it is a physical response to gas in the intestines. The baby is said not to be smiling in response to other people, or out of happiness, but in response to digestive events. Despite the fact that adults do not smile as a result of discomfort in the stomach, this notion is widely repeated—though often not believed by the infant's parents. Studies showing that infant smiles are not correlated with burps, regurgitation and flatulence have made little impact on this idea (Emde and Koenig 1969; cited in Izard 1977). Many people are gratified to drink of infants as having diminished or no feelings.

If it is so easy to deny the emotional lives of other people, how much easier it is to deny the emotional lives of animals. In science, the greatest obstacle to investigating the emotions of other animals has been an extreme desire to avoid anthropomorphism.

---

[3]E.S. Turner recently commented about his 1964 book *All Heaven in a Rage*, one of the first books to challenge attitudes toward animals: "In my original introduction I commented that in our attitude to animals we are hopelessly, perversely inconsistent. Reviewing this book in the Observer, Philip Toynbee followed up the point, remarking that the rage of English foxhunters knew no bounds when they learned that the Russians had shot a dog into space. He thought that a certain pattern could be traced in these bewildering inconsistencies. 'We abominate the cruelties which we are not tempted to perform, and we abominate them all the more when they are practiced by people who do not belong to our own group'. He could have added 'or when they are practiced by people of another nation'." (Turner 1992, 323–4).

[4]This incredible practice is well attested in medical sources. See Anand and McGrath (1993), Schechter et al. (1993), Bagnall et all. (1987), Gauntlett et al. (1987), Hall (1992).

Anthropomorphism means the ascribing of human characteristics—thought, feeling, consciousness and motivation—to non-human things. When people say that their car is out to get them, or that a tree is their mend, they are anthropomorphising. Few believe their vehicle is plotting against them, but anthropomorphic ideas about animals are held with greater conviction. Outside scientific circles, it is common to speak of the thoughts and feelings of pets and of wild and captive animals. Yet many scientists regard even the notion that animals feel pain as the grossest sort of anthropomorphic error.

Science considers anthropomorphism towards animals a grave mistake, even a sin. It is common in science to speak of 'committing' anthropomorphism. The term originally was religious, referring to the assigning of human characteristics to God—the hierarchical error of acting as though the merely human could be divine—literally a sin. In science, the sin against hierarchy is to assign human characteristics to animals. Just as humans could not be like God, now animals cannot be like humans (note who has taken God's place).

Young scientists are indoctrinated with the gravity of this error. As animal behaviourist David McFarland explains, "They often have to be specially trained to resist the temptation to interpret the behaviour of other species in terms of their normal behaviour-recognition mechanisms" (McFarland 1987, 17). It is still treated as axiomatic that "scientists must keep a constant vigil against anthropomorphic thinking and interpretation when performing animal research" (Davies and Balfour 1992, 23). In his recent book, *The New Anthropomorphism*, behaviourist John S. Kennedy laments that "the scientific study of animal behaviour was inevitably marked from birth by its anthropomorphic parentage and to a significant extent it still is. It has had to struggle to free itself from this incubus and the struggle is not over. Anthropomorphism remains much more of a problem than most of today's neobehaviourists believed…If the study of animal behaviour is to mature as a science, the process of liberation from the delusions of anthropomorphism must go on" (Kennedy 1992, 3–5). He ends his book hoping that "anthropomorphism will be brought under control, even if it cannot be cured completely. Although it is probably programmed into us genetically as well as being inoculated culturally that does not mean the disease is untreatable" (*Id.*, 167).

The philosopher John Andrew Fisher has noted that "The use of the term 'anthropomorphism' by scientists and philosophers is often so casual as to almost suggest that it is a term of ideological abuse, rather like political or religious terms ('communist' or 'counter-revolutionary') that need no explication or defense when used in criticism" (Fisher 1991, 49).

Women have been deemed especially prone to anthropomorphic empathy in a science dominated by men. Women, long considered inferior to men precisely on the grounds that they feel too much, were thought to over-identify with the animals they studied. This is one reason why male scientists did not encourage female field biologists for so long. They were too emotional;[5] they allowed emotions to sway judgments and observations. Women, it was felt, were more likely than men to attribute emotional attitudes to animals, to project their own feelings on to them. Thus did gender bias and species bias converge in a supposedly objective environment.

To accuse a scientist of anthropomorphism is to make a severe criticism. It is regarded u a species-confusion, an unprofessional merging, a forgetting of the line between who one is and what one is observing, between subject and object, womanish. To assign thoughts or feelings to a creature known to be incapable of them would indeed be a problem, but to ascribe to an animal emotions such as joy or sorrow only makes the error anthropomorphic if one knows that animals cannot feel such emotions. Many scientists have made this decision, but not on the basis of evidence.

The situation is not so much that emotion is denied but that it is regarded as so dangerous an area that it should not be part of the scientific agenda—such a minefield of subjectivity that no investigation of it should take place. As a result, none but the most prominent scientists risk their reputations and credibility by venturing into this area. Thus many scientists may believe that animals have emotions but are unwilling to say that they believe it, and unwilling to study it or encourage their students to investigate it, They may also attack other scientists who try to use the language of the emotions.

From the belief that anthropomorphism is a desperate error, a sin or a disease, flow further research taboos, including the appropriate use of language. A monkey is not angry; it exhibits aggression. A crane does not feel affection; it displays courtship or parental behaviour. A cheetah is not frightened by a lion; it shows flight behaviour. In keeping with this, de Waal's use of the word 'reconciliation' in reference to chimpanzees has been criticised: wouldn't it be more objective to say 'first post-conflict contact'? (De Waal 1982, 41–2). What is really being sought here may be that part of objectivity which distance and disidentification provide.

The biologist Julian Huxley has argued, against this scientific orthodoxy, that to imagine oneself into the life of another animal is both scientifically justifiable and productive for knowledge. In 1961 Huxley introduced one of the most extraordinary accounts of a very deep and mysterious emotional tie between a human being and a free-living lioness, Joy Adamson's *Living Free*, as follows:

> When people like Mrs Adamson (or Darwin for that matter) interpret an animal's gestures and postures with the aid of psychological terms - anger or curiosity, affection or jealousy - the strict Behaviourist accuses them of anthropomorphism, of seeing a human mind at work

---

[5]"Male/female differences in attitudes and knowledge of animals were substantial and implied the need for better understanding and appreciation of female activities toward and interests in wildlife. Particularly provocative were variations in basic feelings and ethical concerns for animals. The most outstanding result was the much greater humanistic concern for animals among females" (Kellert and Berry 1980, 59). Kellert and Berry's publication gives the results of a U.S. Fish and Wildlife Service-funded study of 'American Attitudes, Knowledge and Behaviours Toward Wildlife and Natural Habitats'.

> within the animal's skin. This is not necessarily so. The true ethologist must be evolution-minded. After all, he is a mammal. To give the fullest possible interpretation of behaviour he must have recourse to a language that will apply to his fellow-mammals as well as to his fellowman. And such a language must employ subjective as well as objective terminology - *fear* as well as *impulse to flee, curiosity* as well as *exploratory urge, maternal solicitude* in all its modulations in welcome addition to goodness knows what complication of behaviourist terminology (Adamson 1961, xi).

Huxley's argument ran counter to mainstream scientific thinking at the time, and still does so. A contemporary example is provided by Alex, the African grey parrot, who was being trained or tested by experimenters who varied the requests they made of him for several reasons, one being to avoid cueing, another to prevent Alex from becoming bored. When reviewers of a paper the researcher Pepperberg submitted to a scientific journal vetoed her use of the term 'boredom', she responded:

> I had a referee go ballistic on me. And yet, you've watched the bird, he looks at you, he says 'I'm gonna go away.' And he walks! The referee said that was an anthropomorphic term that had no business being in a scientific journal…I can talk in as many stimulus-response type terms as you want. It turns out, though, that a lot of his behaviours are very difficult to describe in ways that are not anthropomorphic.[6]

What is wrong with exploring the idea, based on many such observations in a research setting, that parrots and humans may have a shared capacity for boredom? The preconceived belief, the dogma, the unresearchable, has gone from the uncritical projection of human feeling-slates on to animals to the refusal to recognise animal feeling-states even when ample evidence is produced over and over again.

A related taboo in the study of animal behaviour has been that the scientist should not name the animals. To separate individuals, they might be called Adult Male 36, or Juvenile Green. Most field workers over the generations have resisted this precept, naming the animals they spent their days watching, at least for their own use, Spot-Nose and Splotch-Tail, Flo and Figan, or Cleo, Freddy and Mia. In their published work, some reverted to more remote forms of identification; others continued to use the names. Sy Montgomery reports that, in 1981, anthropologist Colin Turnbull declined to provide a supporting statement for Dian Fossey's book of observations on mountain gorillas because she assigned names to the gorillas (Montgomery 1991, 143). It is even more common not to name animals in laboratories, perhaps for the same reason that farmers often avoid naming animals they expect to slaughter: it is harder to kill a friend, and proper names humanise.

Rebutting the view that naming animals only causes one to assign them human traits, elephant researcher Cynthia Moss notes that the opposite happens to her: people remind her of elephants. "When I am introduced to a person named Amy or Amelia or Alison, across my mind's eye flashes the head and ears of that elephant" (Moss 1988, 37). The no-name norm has gradually changed, particularly among primatologists, perhaps because of the outstanding work of researchers who

[6]This theme is also expressed in Barber (1993).

named—and admitted that they named—the subjects they studied. Yet as recently as 1987, researchers studying elephants in Namibia (then South West Africa) were instructed by park authorities to assign the animals numbers because names were too sentimental (Thomas 1990, 99). Granted that a number is more dehumanised than a name, does that make it more scientific? Assigning names to them—referring to a chimpanzee as Flo or Figan—can be called anthropomorphic, but so is assigning numbers. Chimpanzees are no more likely to think of themselves as F2 or JF3 than as Flo or Figan.

We do not know if animals name themselves or each other. We do know that they have and recognise individuality. Animals recognise other animals as individuals and distinguish between them; names are the way humans label such distinctions. Names make the same distinctions animals do. There is speculation that bottlenose dolphins may identify and imitate one another's signature whistles, something very close to a name (Tyack 1989). Some animals clearly respond emotionally to being given a name. Mike Tomkies in *Last Wild Years* writes that "only the ignorant pour scorn on this habit of mine of giving names to the creatures that, over the years, have shared my home. And also others that have not. So long as it is not a harsh sound, it matters little what the name is, but there can be no doubt whatever that an animal or bird will respond differently, become more trusting, once it is given a name" (Tomkies 1992, 172).

If naming the animals one studies promotes empathy towards them, this may help rather than occlude insight. The essential fact glossed over in the attack on anthropomorphism is that humans are animals. Man's relation to animals is not a literary exercise in creating charming metaphors. As Midgley puts it: "The fact that some people are silly about animals cannot stop the topic being a serious one. Animals, are not just one of the things with which people amuse themselves, like chewing-gum and water-skis, *they are the group lo which people belong.* We are not just—rather like animals; we *are* animals" (Midgley 1973, 114). To act as if humans are a completely different order of beings from other animals ignores reality.

Even fierce opponents of anthropomorphism concede that it often works when trying to predict animal behaviour. By considering what an animal feels or thinks, we may improve our ability to project how it will act. Such guesses have a high success rate. It has frequently been pointed out that a successful prediction does not prove that the animal actually felt or thought what was imagined. But successful prediction is a standard test of scientific theories. J.S. Kennedy, the animal behaviourist who views anthropomorphism as a disease, concedes nevertheless that, it is a useful way to predict behaviour. In fact, he advocates using 'mock' anthropomorphism for this purpose, although successful predictions are not to be construed as indicating that the basis for prediction had any value. Kennedy argues that anthropomorphism works because animals have evolved to act *as if* they thought and felt: "it is natural selection and not the animal that ensures that what it does mostly 'makes sense', as we are wont to say" (Kennedy 1992, 87).

Even though Kennedy disavows what he calls the "assumptions that they have feelings and intentions", he does acknowledge that empathy can be useful for generating questions and malting predictions. Thus one might predict that a cheetah, fearing for the lives of her cubs, may run close to a lion to lure it away from them. Under Kennedy's formulation, if the cheetah does run close to the lion, it does not mean she fears for the lives of her cubs. It only means that she has evolved to act for survival *as if* she fears for their lives. To speculate that leaving more offspring is the ultimate cause of her behaviour is permitted. Not permitted is to speculate that fear for their lives is its proximate cause, far less about how she may feel seeing the lion grabbing them. Why is it so impossible to know what animals feel, no matter how much or what kind of evidence there is? How is knowing about their feelings different, in truth, from the assumptions made routinely about the feelings of other people?

Short of being another person, there is no way to know with certainty what another person is feeling, although few people, even philosophers, carry their solipsism this far. In learning of the feelings of others, people are not always led by words alone, but watch behaviour—gestures, the face, the eyes—and its patterns and consistency over time, Conclusions are based on this, and ground everyday life decisions. We love certain people, hate others, trust some, fear others, and act on this basis. Although belief in the emotions of others is indispensable to life in human society, some will argue that ultimately it cannot be proved. Nicolas Humphrey writes that "for all l know no man other than myself has ever experienced a feeling corresponding to my feeling of hunger; the fact remains that the concept of hunger, derived from my own experience, helps me to understand other men's eating behaviour."[7]

On the question of humans knowing animals' pain, Mary Midgley has said: "If a torturer excused her activities by claiming ignorance of pain on the grounds that nobody knows anything about the subjective sensation of others, she would not convince any human audience. An audience of scientists need not aim at providing an exception to this rule" (Midgley 1978, 41; cf. Midgley 1983). She locates the basis of human assumptions of natural superiority underlying the position of the solipsist when she quotes an astonishing passage from the *Ethics* of the seventeenth-century Dutch philosopher Benedict de Spinoza:

---

[7]A fuller exposition is given as follows: "If consciousness has evolved as a biological adaptation for doing introspective psychology, then the presence or absence of consciousness in animals of different species will depend on whether or not they need to be able to understand the behaviour of other animals in a social group. Wolves and chimpanzees and elephants, which all go in for complex social interactions, are probably all conscious; frogs and snails and codfish are probably not…The advantage to an animal of being conscious lies in the purely private use it makes of conscious experience as a means of developing an ideology which helps it to model another animal's behaviour. It need make no difference at all whether the other animal is actually experiencing the feelings with which it is being credited; all that matters is that its behaviour should be understandable on the assumption that such feelings provide the reasons for its actions. Thus for all I know no man other than myself has ever experienced a feeling corresponding to my own feeling of hunger; the fact remains that the concept of hunger, derived from my own experience, helps me to understand other men's eating behaviour" (Humphrey 1980, 68/9).

> It is plain that the law against the slaughtering of animals is founded rather on vain superstition and womanish pity than on sound reason. The rational quest of what is useful to us further teaches us the necessity of associating ourselves with our fellow-men, but not with beasts, or things, whose nature is different from our own; we have the same rights in respect to them as they have in respect to us. Nay, as everyone's right is defined by his virtue, or power, men have far greater rights over beasts than beasts have over men. Still I do not deny that beasts feel; what I deny is that we may not consult our own advantage and use them as we please, treating them in the way which best suits us; for their nature is not like ours, and their emotions are naturally different from human emotions.[8]

Spinoza refrains from discussing how he knows that animal emotions are different from human ones, or from explaining how this justifies the human exploitation, plunder and murder of them. He simply says we have more power than they do. Might makes right. J. Ortega y Gasset's defence of hunting comes to the same conclusion, insisting that the victim is always asking for it:

> [Hunting] is a relationship that certain animals impose on man, to the point where not trying to hunt them demands the intervention of our deliberate will…Before any particular hunter pursues them they feel themselves to be possible prey, and they model their whole existence in terms of this condition. Thus they automatically convert any normal man who comes upon them into a hunter. *The only adequate response to a being that lives obsessed with avoiding capture is to try to catch it* [Gasset's italics] (Ortega y Gasset 1972, 136–8; cited from Cartmill 1993, 240).

Such delusional anthropomorphism is based on deep and hidden structural assumptions and interests. Ortega y Gasset's buried premise—that hunted beings seek their own demise—closely resembles rationales about rape. A common excuse of rapists is that women ask for rape, thus seeking and causing their own violation, especially when trying to avoid it. A similar exoneration of hunters is sought here by justifying the capture of animals by calling animal flight from capture an 'obsession'—meaning they most desire what they most strenuously flee.

More simple forms of anthropomorphism can also interfere with observation and distort understanding. Carolus Linnaeus, the eighteenth-century Swedish naturalist who developed the classification system of living things in his *Systema Naturae* (1758), wrote of the frog: "These foul and loathsome animals are… abhorrent because of their cold bodies, pale colour, cartilaginous skeletons, filthy skin, fierce aspect, calculating eye, offensive smell, harsh voice, squalid habitation and terrible venom." The words are all very emotive, referring to emotions Linnaeus felt when *he* saw a frog. They are pure projection. 'Calculating' is not a scientific term to describe a frog's eye. This passage describes little in the physical world, but powerfully conveys the scientist's subjective state. It is art.

Both cats and dogs are prime targets of anthropomorphism, for good and bad. It is common to ascribe unlikely thoughts and feelings to pets: "She understands every word you say"; "He sings his little heart out to show how grateful he is". Many people seem to enjoy believing that cats are selfish, unfeeling creatures who heartlessly use their deluded owners, compared with loving, loyal and naive dogs.

---

[8]Note I to Proposition XXXVII, Part IV of the *Ethics*.

More often, however, people have what seem to be quite realistic views about their pets' abilities and attributes. The experience of living with an animal often provides a strong sense of its abilities and limitations—although even here, as for people living intimately with people, preconceptions can be more persuasive than lived evidence, and can create their own reality. Perhaps the richest source of actual anthropomorphic error is our thinking about wild animals. Since people live with domestic animals, erroneous theories about their behaviour are likely to be disproved in the course of events. But since most people's contact with wild animals is so limited, theories about them may never run up against actual facts, and we remain free to imagine ravelling wolves, saintly dolphins or crows who follow parliamentary procedure.

Another problem with anthropomorphism has been that human views of gender—often as wrong as human views of animals—have been attributed to animals. A recent television nature programme featured a family of cheetahs in Tanzania's Serengeti National Park. The male cub was called Tahu and the female Tamu—Swahili for Trouble and Sweetness. One expects different things from a Sweetness than a Trouble. Surely the sentence "Trouble is prowling around my tent" is more threatening than "Sweetness is prowling around my tent". Sociobiology has tended to encourage prejudices men have about women by insisting that they are 'natural', by which is meant they can be found among members of the animal kingdom. As already noted, one can prove almost anything by careful choice of species. It does not seem accidental that human society has for so long been compared to baboon society, despite the facts that baboons are far more sexually dimorphic than humans and that baboons do not form mated pairs. The idea seems to be to impose greater gender inequality on human females by enforcing a supposedly natural template.

One way to avoid such errors is to pretend that animals are neuter in gender. Writing about creative behaviour in dolphins for a scientific journal, researcher Karen Pryor was told to call the rough-toothed porpoise Hou 'it' rather than 'she', on the grounds that 'she' should be reserved for referring to humans (Pryor 1975, 240).[9] Not that being called 'she' has humanised women with any security. Refusal to discuss observable facts (Hou was unquestionably female) is hardly scientific. Such language use springs from a desire to treat animals as things rather than beings. At the same time, by treating animals as creatures like ourselves· we risk projecting unexamined assumptions about human gender on to observations and analysis.

A serious problem with careless and uncritical human-animal comparisons is the inadequacy of our present knowledge, especially of crucial matters like the role of culture in animal learning in the wild. Elephants, for example, learn from their elders which humans to fear based on the history of the herd with humans. Mike Tomkies describes watching an eaglet in the wild being taught to fly so as to hunt and kill by repeated demonstrations on the part of its parent, who was dearly showing the youngster what to do rather than engaging itself in search of prey.

[9]Reference is to Pryor et al. (1969).

Evidently the eaglet is not born knowing this (Tomkies 1987, 136/7). It is transmitted by learning—that is, by culture. That a skill has to be learned does not make it unnatural. The distinction between innate and natural, on the one hand, and cultural and learned on the other, loses much of its force in the light of more recent observations on what animals teach each other. To use the word natural to describe how the eaglet kills simply means that an animal was observed doing it.

The real problem often targeted by many of the criticisms of anthropomorphism is actually *anthropocentrism.* Placing humans at the centre of all interpretation, observation and concern, and dominant men at the centre of that, has led to some of the worst errors in science, whether in astronomy, polities or animal behaviour. Anthropocentrism treats animals as inferior forms of people and denies what they really are. It expresses a passionate wish to differentiate ourselves from animals, to make animals 'other', presumably in order to maintain humans at the top of the evolutionary hierarchy and the food chain. The notion that animals are wholly other from humans, despite our common ancestry, is more irrational than the notion that they are like us. But even if they were not like us at all, that is no reason to avoid studying them for their own sakes. The point has been made by J.E.R. Staddon that "Psychology as a basic science should be about intelligent and adaptive behavior, wherever it is to be found, so that animals can be studied in their own right, for what they can teach us about the nature and evolution of intelligence, and not as surrogate people or tools for the solution of human problems" (Staddon 1989, 123). The knowledge obtained from this study, whether or not it contributes to the solution of human problems, is still knowledge.

Although not nearly as frequent as the denigration and demonisation of animals, idealising animals is another kind of anthropocentrism, the belief that animals have all the virtues to which humans aspire and none of our faults. It is anthropocentric because its core is an obsession with the ways of humans, negative in this case, which animals are used to highlight. In this view the natural world is a place without war, murder, rape and addiction, Animals, in this sentimental formulation, never lie, cheat or steal. This view is embarrassed by reality. Deception has been observed in animals from elephants to Arctic foxes (Rüppel 1986). Ants take slaves. Chimpanzees may attack other bands of chimpanzees, unprovoked and with deadly intent. Groups of dwarf mongooses battle against other groups for territory. The case of the chimpanzee murderers Pom and Passion, who killed and ate the infants of other chimpanzees in their group, has been well documented by Jane Goodall's research team. Orangutans have been seen to rape other orangutans. Male lions, when they join a pride, often kill young cubs who were fathered by other lions. Young hyenas, foxes and owls have been seen to kill and eat their siblings.

All is not as humans wish it to be among our evolutionary cousins. One has to sympathise with Jane Goodall's reaction to chimpanzee treatment of one old animal, his legs wholly paralysed by polio, who was lonely, shunned, and sometimes attacked by those who were still healthy. In the hope of inducing companions who were grooming each other to groom him as well, he dragged himself up into a tree:

> With a loud grunt of pleasure he reached a hand towards them in greeting - but even before he made contact they both swung quickly away and, without a backward glance, started grooming on the far side of the tree. For a full 2 min, old Gregor sat motionless, staring after them. And then he laboriously lowered himself to the ground. As I watched him sitting there alone, my vision blurred, and when I looked up at the groomers in the tree I came nearer to hating a chimpanzee than I have ever done before or since (Goodall 1971, 202).

It is hard to romanticise anything this ugly.

It has been a long time since anyone called the lion the King of Beasts, but in recent years dolphins have been portrayed as smarter, kinder, nobler, more pacific, and better at living in groups than people. This ignores the well-known fact that dolphins can be quite aggressive. Recently it has been discovered that some dolphins occasionally rape. At the same time, animal cruelty does not approximate the human standard. It is unlikely that dolphin rape rivals the human figures. One random sample survey in 1977 found that almost half of all the women in one U.S. city had been victims of rape or attempted rape at least once in their lives (Russell 1977, 1982; Russell and Howell 1983). While abuse of young may occur among animals in the wild, nothing compares with more than one in every three girls being sexually abused as children, as another American study shows (Russell 1983, 1986). Physical abuse of others for pleasure is not recorded in the animal world. Torture, as it is performed by humans, is unknown in any other species. No other known animal attacks another merely to see it suffer.

If humans can misunderstand animals by assuming they are more like us than they are, can animals also wrongly project their feelings on to us? Do some animals commit what might be called zoomorphism, ascribing their attributes to humans? A cat who brings a human offerings of dead rodents, lizards and birds day after day, no matter how often these objects are greeted with loathing, commits zoomorphism. This is the equivalent of offering sweets to a cat, as children sometimes do. Similarly, in *The Hidden Life of Dogs*, Elizabeth Marshall Thomas writes, "When a dog with a bone menaces a human observer, the dog actually assumes that the person wants the slimy, dirt-laden object, and is applying dog values, or cynomorphizing" (Thomas 1993, xvi). Were a dog to give a history of the human race, some valuable attributes might be denied us, just as our history of any animal civilisation would doubtless miss many of its signal achievements.

References

Adamson, J., introduction by Julian Huxley. 1961. *Living free*. London: Collins & Harvill.

Anand, K.J.S., and P.J. McGrath (eds.). 1993. *Pain in neonates*. Amsterdam: Elsevier.

Artz, V., and I. Birmelin. 1993. *Haben Tieren ein Bewusstsein? Wenn Affen lügen, wenn Katzen denken und Elefanten traurig sind*. Munich: Bertelsmann.

Bagnall, G., M. Plant, and M. Plant. 1987. Medicine and the media. *British Medical Journal* 295 (6599): 660–661.

Barber, T.X. 1993. *The human nature of birds: A scientific discovery with startling implications*. New York: St. Martin's Press.

Cartmill, M. 1993. *A view to a death in the morning: Hunting and nature through history*. Cambridge, MA: Harvard University Press.
Davies, H., and D. Balfour (eds.). 1992. *The inevitable bond*. Cambridge: Cambridge University Press.
De Waal, F. 1982. *Chimpanzee politics: Power and sex among apes*. London: Jonathan Cape.
De Waal, F. 1989. *Peacemaking among primates*. Cambridge, MA: Harvard University Press.
Eisemann, C.H., W.K. Jorgensen, D.J. Merritt, M.J. Rice, B.W. Cribb, P.D. Webb, and M. P. Zalucki. 1984. Do insects feel pain? A biological review. *Experientia* 40: 164–167.
Emde, R.N., and K.L. Koenig. 1969. Neonatal smiling and rapid eye movement states. *Journal of the American Academy of Child Psychiatry* 8(1): 57–67.
Fisher, J.A. 1991. Disambiguating anthropomorphism: An interdisciplinary review. *Perspectives in Ethology* 9: 49–67.
Gauntlett, I.S., T.H.H.G. Koh, and W.A. Silverman. 1987. Analgesia and anaesthesia in newborn babies and infants. *Lancet* 329(8541): 1090.
Goldfoot et al. 1980. Behavioral and physiological evidence of sexual climax in the female stump-tailed macaque. *Science* 208(4451): 14771479.
Goodall, J. 1971. *In the shadow of man*. London: Collins.
Hall, N. 1992. The painful truth. *Parenting Magazine* (June/July 1992).
Humphrey, N.K. 1976. The social function of intellect. In *Growing points in ethology*, ed. P.P.G. Bateson, and R.A. Hinde, 303–317. Cambridge: Cambridge University Press.
Humphrey, N.K. 1980. Nature's psychologists. In *Consciousness and the physical world*, ed. B.D. Josephson, and V.S. Ramachandran, 57–80. Oxford: Pergamon Press.
Izard, C. 1977. *Human emotions*. New York: Plenum.
Kellert, S.R., and J.K. Berry. 1980. *Phase III: Knowledge, affection and basic attitudes toward animals in American society*. Washington, DC: Government Printing Office.
Kennedy, J.S. 1992. *The new anthropomorphism*. Cambridge: Cambridge University Press.
Lockwood, J.A. 1987. The moral standing of insects and the ethics of extinction. *Florida Entomologist* 70(1): 70–89.
McFarland, D. (ed.). 1987. *The Oxford companion to animal behaviour*. Oxford: Oxford University Press.
Midgley, M. 1973. The concept of beastliness: Philosophy, ethics and animal behavior. *Philosophy* 48(184): 111–135.
Midgley, M. 1978. *Beast and man: The roots of human nature*. New York: Cornell University Press.
Midgley, M. 1983. *Animals and why they matter*. Athens, GA: University of Georgia Press.
Midgley, M. 1992. The mixed community. In *The animal rights/environmental ethics debate*, ed. E.C. Hargrove, 211–225. Albany, NY: State University of New York Press.
Montgomery, S. 1991. *Walking with the great apes*. Boston, MA: Houghton Mifflin.
Moss, C. 1988. *Elephant memories: Thirteen years in the life of an elephant family*. New York: William Morrow.
Orleans, F.B. 1993. *In the name of science: Issues in responsible animal experimentation*. Oxford: Oxford University Press.
Ortega y Gasset, J. 1972. *Meditations on hunting*, trans. H.B. Wescott. New York: Charles Scribner.
Pryor, K. 1975. *Lads before the wind: Adventures in porpoise training*. New York: Harper & Row.
Pryor, K., R. Haag, and J. O'Reilly. 1969. The creative porpoise: Training for novel behavior. *Journal of the Experimental Analysis of Behavior* 12: 653–661.
Rüppel, V.G.A. 1986. "lie" as a directed message of the arctic fox (*Alopex lagopus* L.). In *Deception: Perspectives on human and nonhuman deceit* ed. R.W. Mitchell, and N.S. Thompson, 177–181. Albany, NY: SUNY Press.
Russell, D.E.H. 1977. *The politics of rape: The victime's perspective*. New York: Stein & Day.
Russell, D.E.H. 1982. *Rape in marriage*. New York: Macmillan.
Russell, D.E.H. 1983. The incidence and prevalence of intrafamilial and extrafamilial sexual abuse of female children. *Child Abuse and Neglect: The International Journal* 7: 133–146.

Russell, D.E.H. 1986. *The secret trauma: Incestuous abuse of women and girls*. New York: Basic Books.

Russell, D.E.H., and N. Howell. 1983. The prevalence of rape in the United States revisited. *Signs: Journal of Women in Culture and Society* 8: 668–695.

Schechter, N., C.B. Berde, and M. Yaster (eds.). 1993. *Pain in infants, children, and adolescents*. Baltimore, MD: Williams and Williams.

Staddon, J.E.R. 1989. Animal psychology: The tyranny of anthropocentrism. In *Whither ethology? perspectives in ethology*, ed. P.P.G. Bateson, and P.H. Klopfer, 123–136. New York: Plenum Press.

Symons, D. 1979. *The evolution of human sexuality*. Oxford: Oxford University Press.

Thomas, N.W. 1908. Animals. In *Encyclopaedia of religion and ethics*, vol. 1, ed. J. Hastings, 483–535. Edinburgh: T. & T. Clark.

Thomas, E.M. 1990. 'The old way,' *The New Yorker*, 15 October 1990.

Thomas, E.M. 1993. *The hidden life of dogs*. Boston, MA: Houghton Mifflin.

Tomkies, M. 1987. *On wing and wild water*. London: Jonathan Cape.

Tomkies, M. 1992. *Last wild years*. London: Jonathan Cape.

Turner, E.S. 1992. *All heaven in a rage*. Sussex, UK: Centaur Press.

Tyack, P. 1989. Whistle repertoires of two bottlenosed dolphins, *Tursiops truncates*: mimicry of signature whistles? *Behavioral Ecology and Sociobiology* 18: 251–257.

Wigglesworth, V.B. 1980. Do insects feel pain. *Antenna* 4: 8–9.

Wood Krutch, J. 1950. *The best of two worlds*. New York: William Sloane Associates.

Comment: Between Human and Animal

Henk van den Belt

Abstract As in the other parts of the book, in this part culturally entrenched boundaries and demarcations are also critically re-examined in light of the arrival of the Anthropocene as a new geological era. Here the focus is on rethinking the received distinction between humans and non-human animals. In a long series of discourses making up our cultural heritage, we humans have persistently tried to define the essence of our own humanity and to distinguish ourselves from other animals by laying claim to supposedly unique capacities and achievements like reason, language, morality, religion, technology, law and politics. By now, however, human exceptionalism in its cruder forms has definitely gone out of fashion. None of the contributors to this part of the book subscribes to such versions, but each has different views on how far the critique of human supremacy should be pursued and where the new boundaries should be redrawn, if at all. The issues raised in these contributions are of central importance for the reorientation of animal ethics and environmental ethics in the Age of the Anthropocene.

1 The False Modesty of Biomimicry

A farewell to human exceptionalism is emphatically proclaimed by the influential school of biomimicry in architectural and technological design discussed in *Sanne van der Hout's* chapter. Rather than as conquerors of the Earth, the adherents of biomimicry like to present themselves as humble apprentices and students of living nature, whose ambition is simply to consciously emulate 'life's genius'. There is virtually unlimited opportunity for human designers to receive valuable instruction and advice from the plethora of solutions to daunting technical challenges that the Earth's organisms have already found over the long haul of evolutionary history. The design strategy of following living nature's examples, its proponents suggest, promises to generate inherently sustainable technologies. Van der Hout thoroughly

H. van den Belt (✉)
Philosophy Group, Wageningen University, Wageningen, The Netherlands
e-mail: Henk.vandenbelt@wur.nl

B. Bovenkerk and J. Keulartz (eds.), *Animal Ethics in the Age of Humans*,
The International Library of Environmental, Agricultural and Food Ethics 23,
DOI 10.1007/978-3-319-44206-8_8

analyzes the strengths and weaknesses of the apparently modest claims of biomimicry. She notes that the proclaimed mentorship role of organisms as discussed in the biomimicry literature can be divided into their roles as (a) teachers of values and morals and (b) technical advisers. She criticizes the first mentorship role as being based on a naturalistic fallacy. Van der Hout declares that "apart from human beings, there are no moral agents in nature". Thus in this respect at least, she reaffirms human exceptionalism. She also points out that in purportedly drawing moral lessons from nature, champions of biomimicry are guilty of a very selective, romanticized reading of nature as inherently harmonic and cooperative rather than conflictual and competitive. Neither is there any guarantee that technologies inspired by nature will automatically be nature friendly, as she shows with the example of military body armor that was modeled after the offensive system of the mantis shrimp. It would seem indeed that the language of biomimicry can be used quite easily as a rhetorical device for greenwashing any modern life science technology however invasive it may be. Think about the recent discovery and development of the CRISPR/Cas system. This was originally discovered as a natural system for anti-viral defense in bacteria and archaea, but is already being used as a sophisticated genome editing system in a wide variety of biotechnology applications. The modern bioengineers might try to silence potential opposition to their genome-editing projects by arguing that they are only following Mother Nature's ways.

The proclaimed role of organisms as 'technical advisers' is also in need of further scrutiny. Van der Hout rightly notes that the practice of biomimicry implies more than mere copying or slavish imitation but amounts to a process of "*creative* imitation in which nature's lessons are translated into a technological idea". When we look more closely into what is actually involved in this process of translation, we will also become more wary of the allegedly humble stance adopted by biomimicry advocates. To begin with, it would seem that organisms must already be looked at through technological (or mechanistic) spectacles before they can be set as examples to be emulated by human technology. It was, for example, the technological gaze of the Japanese engineer-cum-birdwatcher Eiji Nakatsu that turned the kingfisher's beak (useful for diving into the water without making a splash) into a relevant model for designing the streamlined forefront of the famous Shinkansen train. In fact, practitioners of biomimicry naturally look at organisms as sophisticated machines. This way of looking might be unproblematic, were it not for the fact that the machine metaphor (and the corresponding mechanistic world picture) is anathema to many environmentalists and animal rights activists. The 17th-century philosopher René Descartes, who so strongly promoted the view of animals as machines, has become, according to science historian Guerrini (2002, 63), the preeminent example of "the soulless, arrogant, uncaring scientist", but also "the scapegoat for our collective guilt" (due to the fact that modern science and technology have partly fulfilled his promise that humans would become "masters and possessors of nature"). Biomimicry's favorite posture of being the humble apprentice of Mother Nature's ways thus appears to be rather disingenuous, more rhetoric than reality. Philosophers of the Enlightenment like Francis Bacon and

Immanuel Kant, who also stressed the need to learn from nature, were less squeamish in this regard, as they admitted that the ultimate aim was to achieve mastery over nature. Bacon minted, in addition to his well-known slogan 'Knowledge is power', the famous aphorism that we can conquer nature only by obeying her. In the Preface to the second edition of his *Critique of Pure Reason* (1781), Kant argued that human reason must not adopt a passive stance towards nature, but take the initiative and "compel Nature to answer its questions". Human reason had indeed to learn from nature, Kant agreed, but he immediately added that we should not be misled by the phrase 'learning from': "Reason must approach nature with the view, indeed, of receiving information from it, not, however, in the character of a pupil, who listens to all that his master chooses to tell him, but in that of a judge, who compels the witnesses to reply to those questions which he himself thinks fit to propose" (Kant 1855[1781], B xiii). Clearly, Kant did not indulge in the false modesty of merely posing as a humble apprentice.

However, Van der Hout would probably not agree that in some sense every technology aims to achieve mastery over nature. In the final part of her chapter she expresses concern that the lessons drawn from nature might be used "in an instrumental fashion". In her view, this means that practitioners of biomimicry would not apply the lessons of their fellow-organisms to live sustainably on planet Earth, but would use them instead for the sake of us humans "to empower ourselves". I think this distinction does not withstand scrutiny. In fact, any human technology involves instrumental use and empowerment. Ultimately, if there is to be any hope of living sustainably on Earth, we will need quite a lot of the "instrumental power" that science and technology can offer us in order to drastically reduce our immense ecological footprint on this planet.

2 The Worlds of Birds

In their contribution to this volume, *Simon Burton* and *Emily Brady* use findings from ornithology and Kant's and Schopenhauer's transcendental idealism to make a plea for epistemic humility as an appropriate ethical virtue in human-animal relations. They reformulate Thomas Nagel's classic question ('What is it like to be a bat?') for a different group of animals: "What is it like to be a bird?' A passionate birdwatcher myself, I very much appreciate the intriguing details they tell about the almost inscrutable 'life-worlds' inhabited by a number of interesting bird species. What lends a special interest to this chapter is that it advocates epistemic humility on the basis of Kant's (and Schopenhauer's) philosophy. This is noteworthy, as Kant is usually considered a prominent representative of human exceptionalism. Earlier, in her book *The Sublime in Modern Philosophy* (2013), Brady already presented a radical reinterpretation of Kant's well-known analysis of the sublime as set out in his *Critique of Judgment*. This is conventionally considered a vindication of the moral superiority of the human subject, but in Brady's view the experience of the sublime is seen as leading up to a sense of humility through an awareness of things in the universe that are ‘bigger than us’. It is Kant’s fundamental distinction between phenomenon and thing-in-itself, Burton and Brady now argue, which naturally leads to a proper sense of epistemic humility when we seriously try to understand what it is like to be a bird. Just to be more specific, we may be talking about a common swift, a willow warbler, a blackcap, a robin, a blue tit, a dotterel, a northern gannet, a kestrel, a bar-tailed godwit or a Caspian tern (to mention the bird species that briefly make their appearance in the chapter). We can attempt to reconstruct how the world looks like for any of these species by collecting sundry facts about their sensory perception, their migration patterns, their ways of getting food, selecting mates and raising offspring, etcetera. Thus we may come to realize that common swifts live on the wing throughout their lives and literally have the air as their habitat; that male blue tits use the ultra-violet part of the electromagnetic spectrum that is beyond the reach of our perceptual apparatus to advertise their true colors to potential mates; or that willow warblers and other migratory birds navigate incredibly long distances across continents by orienting themselves to variations in the Earth’s magnetic field for which we humans lack the appropriate sense.

Through all these efforts we might come to know, at least to some extent, how the world would look like if we were a particular bird species, say, a bar-tailed godwit. This would be a necessary but not a sufficient step to know what it would be like to be such a species. Or as Thomas Nagel would be quick to point out, it still conveys only what it would be like *for us* to live as a bar-tailed godwit, but not for the godwit itself. Burton and Brady hold that it is precisely the growing realization of the inevitable inadequacy of our strenuous attempts to understand what it would be like to be a bird that should lead us to feel a proper sense of epistemic humility.

It is remarkable that Burton and Brady nowhere refer to the pioneering investigation into the species-specific ‘worlds’ or ‘*Umwelten*’ (literally ‘surrounding worlds’) of various animals undertaken by the Baltic-German biologist Jakob von Uexküll in the early part of the 20th century, although his *Umweltforschung* shows a striking similarity to their approach. What is more, Uexküll was also inspired by Kant’s philosophy (and in turn would subsequently inspire several German philosophers such as Heidegger, Scheler, and Cassirer). He followed Kant in sticking to the distinction between phenomenon and thing-in-itself and held that for a biologist (in contrast to a physicist) there are as many worlds as there are subjects. As he was perfectly willing to grant animals the status of being subjects, the implied task for biology was to painstakingly reconstruct how all these different worlds look like for each and every species of animal. This is also what Burton and Brady attempt to do for some bird species.

What proves crucially important in all such exercises is the ability to mobilize scientific insights and technological devices. We were only able to acquire knowledge of the fact that blue tits use the ultra-violet part of the spectrum for selecting mates, because we had earlier extended our own perceptual apparatus by technological means to gain access to it. We could only track the migration route of the bar-tailed godwit by using satellite tracking technology. And the discovery that migratory birds orient themselves to variations of the Earth’s magnetic field was of course only possible on the basis of our own previously developed, technologically mediated apparatus of observation. Much the same holds for the bats discussed by Thomas Nagel. Although some biomimicry enthusiasts like to claim that bats were the first to 'invent' sonar, it is not true that their achievement has inspired the human invention of sonar and radar. It is rather the reverse: our mastery of these new technologies has made it somewhat easier for us to understand the intricate system of echolocation that bats use to navigate the dark. Such considerations might easily lead to a sense of human superiority rather than humility, as Burton and Brady would have it. After all, each species lives in, or is locked into, its own world (its own soap bubble, as Uexküll would say), but it is apparently only given to human beings to transcend the natural limits of their *Umwelt* and thereby to acquire the ability to peer into the many different worlds of other species. Isn't that a reason to celebrate human exceptionalism after all? Burton and Brady, however, would rather emphasize the always provisional and never definitive character of our knowledge and the fundamentally insurmountable obstacles to our effort to gain a deeper understanding of how the world looks like from the point of view of other animals. It is this principled epistemic fallibilism, together with experiences of sublimity and wonder triggered by those aspects of animal lives that challenge our powers of imagination, which in their view should ground a proper sense of humility on our side.

It is clear, nonetheless, that Burton and Brady are performing a delicate balancing act. They advocate humility, but within limits. We are invited to respect the irreducible otherness of different forms of life, but also to attempt 'to grasp nature on its own terms'. In environmental ethics, recognition of the otherness of nature is an important motive for awe and respect. The environmentalists' nightmare would be to live in a world where, to use the words of the German physicist Werner Heisenberg, "we always meet only ourselves" (Heisenberg 1958, 104). Burton and Brady quote a well-known passage from Iris Murdoch, stating that the sudden sight of a hovering kestrel silenced her preoccupation with petty personal concerns. They interpret this passage as illustrating an act of epistemic humility in which we become receptive to the world, but also warn against driving this humility too far. A complete 'unselving' on our part, while recognizing nature's independence from us, would also endanger any serious attempt to engage with other life forms on their own terms and obstruct the 'critical anthropomorphism' that they find commendable. The same passage from Murdoch is also quoted by Sue Donaldson and Will Kymlicka in their book *Zoopolis* (2011, 37). They make the critical point that awe and respect for otherness, though perhaps indispensible for the appreciation of natural beauty, do not exhaust our moral response toward animals. They actually see grave danger in considering animals as our Other: "If we overemphasize animals' separateness from us—their independence, distance, inscrutability, or indifference—we are at just as much risk of moral error as if we overemphasize our similarity by projecting onto them needs, desires, or interests that are distinctly our own" (Idem, 38). The two extremes of this opposition can also be seen as the horns of the dilemma in which Burton and Brady are caught. Their refusal to give up on either of these horns explains why their quest becomes an endless dialectical journey in which knowledge is vigorously pursued to the end but in which the eventual limits of our knowledge also become "openings for developing a less arrogant epistemic perspective".

3 Political Animals?

In her contribution to this volume, *Eva Meijer* mounts a vigorous attack on one of the last vestiges of human exceptionalism, to wit, the idea that politics is the exclusive reserve of human beings and that it is therefore absurd to attribute any political rights and duties to animals. Her assault follows in the wake of the so-called political turn in animal philosophy heralded by Donaldson and Kymlicka's book *Zoopolis* in 2011. While Donaldson and Kymlicka already took a fairly radical and unorthodox position, Meijer tries to develop an even more radical approach. Her chapter starts with reporting some recent findings from ethological research into 'collective decision-making' among various animal species. It appears that African buffaloes, for example, have a sophisticated procedure for deciding to move to new grazing fields, with the females of the herd indicating their preferences by standing up, staring in the preferred direction and lying down again, thus providing inputs for the group decision. This type of behavior is interpreted as 'voting'. Similar findings have been reported about other animal species like red deer, yellow baboons and pigeons. Even the search of a split-off honeybee colony for a new nesting place after the birth of a new queen seems to fit this general pattern. Such a colony, Meijer assures us, needs to make a 'unanimous decision' about the new nesting place within a couple of days. From all these examples Meijer concludes that animals, contrary to the dogma of human exceptionalism, are capable of collective decision-making and are thus political actors in their own right. Once we recognize their political agency, she further argues, we need to rethink notions like 'politics', 'democracy', 'citizenship' and 'sovereignty' in a non-anthropocentric way in order to create the foundations for 'interspecies democracies'.

Meijer starts from the idea that, for democratic reasons, all animals should have a voice in decisions affecting their lives. She holds that animals are perfectly able to express their opinions and concerns, but that they have been effectively silenced by anthropocentric politics. Her aim is to develop a new political framework in which all (human and non-human) animals will have a proper say in the decisions that affect them all. In elaborating this framework, Meijer moves far beyond Donaldson and Kymlicka's position. In the latter's model, she explains, domesticated animals can participate in the political process only if they are represented by human collaborators who have learned how to interpret their expressions of preferences (see also Donaldson and Kymlicka 2011, 153). Meijer therefore attempts to find new models of representation allowing for a more independent role of animals, whether domesticated, wild or liminal. She even suggests that animals should participate in the search for the new political framework itself, as they have the right to 'co-shape' this very framework. Animals will be equal to this task, she claims, because they have already shown on many occasions to be capable of 'changing the rules of the game'. Thus, the effort to 'think and work towards social change' is not an exclusively human pursuit; it is allegedly undertaken together 'with other animals'.

Meijer's suggestions and proposals are bound to strain the credulity of skeptical readers. Various cases that are supposed to prove the political agency of animals are not immune to the charge of over-interpretation and naïve anthropomorphism. Meijer has apparently no qualms to say that animals engage in 'political protest actions' or are 'questioning power hierarchies' when they resist cruel treatment; or that they 'vote with their feet' when they escape from captivity; or that they 'negotiate and bargain' and enter into 'agreements' with humans based on 'mutual recognition'. Her accusation that anthropocentric politics effectively silences animal voices begs the very question that is at issue, namely, whether animals are indeed capable of voicing their opinions and concerns. There is also a curious duplicity in her entire argumentation. We are asked with some vehemence to recognize the political agency of animals, but are not allowed to apply the only standards we have at our disposal to answer such a momentous question, to wit, our existing human standards of politics, democracy, citizenship and sovereignty. Indeed, we are subsequently urged to drastically rethink the very content of these concepts in a non-anthropocentric way. She thus changes the goal posts while the game is on.

4 Emotions, Practices and History

Jeffrey Masson and *Susan McCarthy* open the attack on yet another version of human exceptionalism in their contribution to this volume. They criticize the idea that it is only humans who can have feelings and emotions (at least of the more noble sort). The strict taboo on anthropomorphism in behaviorist animal research often amounts to an unwarranted denial that animals also have emotions. However, they admit that there are also bad or delusional forms of anthropomorphism, e.g. when people imagine "ravening wolves, saintly dolphins or crows who follow parliamentary procedure". The last example suggests that Masson and McCarthy are not willing to attribute political agency to animals.

Michiel Korthals does not directly confront the issue of human exceptionalism head-on, but polemicizes against a principles-based ethics. He advocates an alternative pragmatist approach with a primary focus on the variety of practices in which humans and non-human animals encounter each other. In principle, Korthals argues, such interactions can be enriching both for humans and for animals insofar as they have world-disclosing effects. In his view, animals act as purveyors of new symbolic meanings. A skeptic would nonetheless retort that in most if not all cases these meanings are attributed by humans to animals and not created by the animals themselves.

In her contribution, *Anita Guerrini* targets human exceptionalism in the writing of history and criticizes the idea that it is only humans who make history. The advent of the Anthropocene, she argues, requires a new concept of history that radically de-centers humans and admits non-human animals as historical agents. It would also expand the time-frame from a more narrow human history based on written documents to a wider evolutionary history that encompasses the histories of animals alongside those of humans. It would thus exhibit the 'co-evolution' of animals and humans. This new concept of history is also to underpin a new ethics for the Anthropocene. As yet, Guerrini admits, this ideal is still to be realized, a promise for the near future. Meanwhile, some synthetic biologists (who according to Guerrini represent "very much a science for the Anthropocene, assuming human control over nature and its processes") already make radical proposals to reconstruct various extinct animals. However, de-extinction of animals "erases their history in favor of a human-made one". So there is a definite risk that we will come to live in a world where, as Heisenberg said, "we always meet only ourselves". In that event, history would become even more human-centered.

5 From Humility to Hubris

There thus seems to be something paradoxical with the attempt to overcome human exceptionalism. No sooner is the end of the supremacy of our species proclaimed, than it reenters through the back door. Only a thin line separates humility from hubris. Our ability to probe deeply into the wonderful life-worlds of other species may not only instill a proper sense of modesty but also give rise to self-congratulation. We may admire the intelligent solutions of life's genius only to ruthlessly exploit them in our own technical designs. We may de-center humans from our conception of history only to re-center them with a vengeance. This same ambivalence is implicit in evolutionary biology. After Darwin had dealt a severe blow to our self-esteem by suggesting that we humans evolved from apes, evolutionary biologist Julian Huxley (the grandson of "Darwin's bulldog", Thomas Huxley) healed this narcissistic injury by asserting that we are still at the pinnacle of evolution and that the universe is becoming conscious of itself in us. Man's responsibility and destiny, Huxley prophetically declared in the 1950s, is to be defined as "an agent for the rest of the world in the job of realizing its inherent possibilities as fully as possible" (Huxley 1957, 13). In a sense, Huxley already anticipated the idea of the Anthropocene, as he noted that this new responsibility has actually taken us by surprise: "It is as if man had suddenly been appointed managing director of the biggest business of all, the business of evolution—appointed without being asked if he wanted it, and without proper warning and preparation" (idem.). In the further elaboration of this vision, however, Huxley appeared exclusively concerned with realizing the inherent potential of humankind alone and altogether indifferent to the fate of the rest of the living world. The human species would have to pass a new threshold of existence, which he dubbed "transhumanism" (idem, 17). It can hardly be doubted that the Brave New World that his present-day transhumanist followers and their allies, the synthetic biologists, want to create is bound to be a lonely place for lovers of nature.

References

Brady, E. 2013. *The sublime in modern philosophy: aesthetics, ethics, and nature*. New York: Cambridge University Press.

Donaldson, S., and W. Kymlicka. 2011. *Zoopolis: a political theory of animal rights*. Oxford: Oxford University Press.

Guerrini, A. 2002. The rhetorics of animal rights. In *Applied ethics in animal research: philosophy, regulation, and laboratory application*, ed. J.P. Gluck, et al., 55–76. West Lafayette, IN: Purdue University Press.

Heisenberg, W. 1958. The representation of nature in contemporary physics. *Daedalus* 87(3): 95–108.

Huxley, J. 1957. *New bottles for new wine*. London: Chatto & Windus.

Kant, I. 1855[1781]. *Critique of pure reason*. London: Henry G. Bohn.

Part II
Between Wild and Domestic

Climate Change, Ethics, and the Wildness of Wild Animals

Clare Palmer

Abstract This chapter is a first attempt to consider ways in which anthropogenic climate change may reduce the wildness of wild animals, and explores ethical and value concerns such a reduction might raise. The chapter begins by outlining key impacts a warming climate may have on animal lives. Then it explores different meanings of "wildness" in the context of wild animals, concluding that climate change might reduce both "constitutive" and "self-willed" animal wildness—forms of wildness that many people value. Ethical responses to these potential value losses are shaped by whether a deontological or consequentialist approach to wildness value is adopted. The chapter suggests that consequentialists about wildness value have a range of positive policy options, because they can pursue the creation of new, future animal wildness. Deontologists about wildness value have fewer options, because their focus is on not compromising existing animal wildness, even when doing so would create more wildness in the future.

1 Introduction

One central element of the claim that the Earth has now entered the Anthropocene is that human influence on Earth has become all-pervasive. This influence is most obviously and globally manifested in anthropogenic climate change. As is widely accepted, climate change will have significant impacts on human individuals and societies. But it will also lead to major changes in Earth's ecology—involving individual organisms, populations, species and ecosystems. In particular, anthropogenic climate change will have substantial impacts on wild animals; indeed, many impacts are already being detected. The prospect that such impacts will expand and intensify is a source of deep concern to many conservation organizations. A UK report jointly compiled by the Royal Society for the Protection of Birds, Natural England and the Worldwide Fund for Nature-UK, published in 2012,

C. Palmer (✉)
Texas A&M University, Texas, USA
e-mail: c.palmer@tamu.edu

B. Bovenkerk and J. Keulartz (eds.), *Animal Ethics in the Age of Humans*,
The International Library of Environmental, Agricultural and Food Ethics 23,
DOI 10.1007/978-3-319-44206-8_9

discussed the "devastating impacts of unchecked climate change on nature". The report includes a claim by the biologist Chris Thomas (Watts et al. 2012, 24) that "If the global temperature rises by 4C, we will end up being reviled as the generation that failed to act, that let nature die, even when we knew broadly what would happen".

There seems to be both something right and something wrong about this claim. Whatever the outcome of 4C of climate change might be, it's highly unlikely that it would result in a *dead* nature (though it's not clear exactly what's meant by "nature" here). All the evidence suggests that while 4C of warming would certainly result in some deaths and extinctions, it would also result in new lives. What's more likely than a dead nature is a very *different* nature from the one that would have existed without 4C of climate change. Indeed, this different nature will be one of the hallmarks of the Anthropocene.

To say that a climate-changed nature would be different, not dead, however, is not to deny other parts of Thomas' claim. Perhaps causing or allowing the effects of anthropogenic climate change—even if climate change isn't deadly to nature—should still be regarded as a moral offence, even one worthy of a kind of moral revulsion, as he suggests. This, though, raises questions about who is supposed to have morally offended against whom or what, in what way, and how serious this moral offence should be thought to be.

While there's much debate about the ethical issues raised by the impacts of climate change on human beings, there's relatively little that explores ethical questions raised by the impacts of climate change on the non-human world directly (though see Palmer 2011; Nolt 2011). Because climate change is anthropogenic, rather than a natural phenomenon (discussed below), its impacts on the non-human world seem to be open to ethical scrutiny; but because these impacts are so wide-ranging, such ethical questions are, of necessity, highly varied. In addition, what questions it's thought important to ask will depend on what is thought to *matter* about the non-human world: for instance, sentient or living individuals, species, ecosystems, biodiversity, states of flourishing—and so on.

In this chapter, I will discuss just one ethical concern raised by the impacts of climate change on the non-human world: its impact on wild animals, and, specifically, on the *wildness* of wild animals. I'll begin with a brief explanation of some key terms and background assumptions, and provide a short overview of some of the likely effects of climate change on wild animals. I'll then introduce wildness as a value concern in the context of wild animals and climate change, outlining some of the different things wildness might mean, and considering whether the impacts of climate change should concern us because wild animals will become less wild, and wildness in animals is a value we should try to protect. In the first part of this paper, I'll be concerned with the *direct impacts* of climate change itself on wild animals—not with the potential impacts of climate *policy* on wild animals. I will, however, consider the impacts of some policy responses to climate change on animal wildness towards the end of the paper, where I'll consider how such policy responses will vary, based not only on what's valued about animal wildness, but also on which ethical theory is adopted.

2 Key Terms and Assumptions

To set the context for this chapter, I will begin with some clarifications and by explaining some assumptions.

(a) When talking about "wild animals" here, I'm primarily referring to sentient animals. However, since the focus is on wildness value, rather than on issues such as suffering or death, much of what is said here could equally apply to animals that lack sentience.
(b) The idea of wildness is much contested. Part of the purpose of this paper is to explore different ideas of what wildness might mean in the context of "wild animals"; no particular meaning is assumed at the outset.
(c) The starting point for an argument that there's something morally problematic about climate change is first that humans are *causally* responsible for it, and second that (some or all) humans are morally responsible for its effects. I will just assume causal responsibility here. Establishing moral responsibility is much more complex, in particular because many of the anthropogenic greenhouse gases (GHGs) now in the atmosphere were emitted by individuals who did not know of their possible effects. For simplicity's sake, I'll accept Nolt's (2011) argument that to have moral responsibility for a harm, we must be able (a) to cause or to prevent the harm, (b) to recognize it as morally significant, (c) to anticipate it with some reliability, and (d) to act in less harmful or more beneficial ways. Although human GHG emissions did not always meet these four conditions (so long-dead individuals are not individually morally responsible) some existing people do now meet these four conditions. However, many of those currently living in poverty produce few emissions, and could not survive without the emissions they do produce, giving them few real alternatives in terms of (d)—that is, acting in less harmful ways. This suggests that not everyone now alive meets these conditions; so there's clearly uneven moral responsibility for climate change between individuals and between nations. However, given the subject of this chapter, important as they are, the details of this uneven distribution don't need to be resolved here.

3 Likely Effects of Climate Change on Wild Animals

The extent to which climate will change, and the effects of climate change on animal habitats, is highly uncertain. In particular, we don't know:

(a) What levels of future global GHG emissions there will be, and on what timescale;
(b) The effects of particular emission levels on climate;

(c) The effects of changes of climate on different ecosystems, and whether we are close to particular ecosystem or biospheric thresholds or tipping points (Barnosky et al. 2012);
(d) How changes in climate will interact with other anthropogenic environmental impacts such as deforestation and habitat fragmentation;
(e) The effects of ecosystem changes on different species, and on individual species members.

Despite these difficulties, I'll briefly outline some significant impacts that climate change is likely to have on wild animals (in some cases, these effects are already occurring). While these studies have limitations (for instance, many focus only on North America, Europe and Russia: see Parmesan 2006), this is a fairly conservative account, one that most ecologists and climate scientists accept.

3.1 Changing Temperature and Precipitation

In many places, the climate is likely to become hotter; in particular, minimum temperatures at night and in winter will rise. Different species (and populations within species) are variably susceptible to changing temperatures, since they have physiological differences (Bernardo et al. 2007). Some animal species will expand their ranges. Others will shift their range by moving towards the poles or upwards in elevation; for instance, red fox populations in Canada have been advancing north, while Arctic fox populations have been retreating (Walther et al. 2002). As Parmesan and Yohe (2003) note: 'On average, the spatial distributions of a substantial set of studied species over different taxa has shifted 6.1 km. per decade towards the poles or 6.1 m in elevation per decade.' Success in range shifts, though, depends on the absence of physical barriers to migration (for instance, roads and cities, rivers, oceans and mountain ranges); on suitable soil types; and on the presence of appropriate food species together with the absence of deadly predator species or more successful competitors. For some species, an inability to move as temperature rises has already resulted in range contraction (Bernardo and Spotila 2006); this has been particularly true of polar species and montane amphibians.

Alongside general temperature changes, more extreme weather events are predicted: more heat waves are 'virtually certain' and the frequency of heavy precipitation is 'likely' to increase (IPCC 2014). There may be changes in monsoon patterns, though there is little consensus about what form these changes will take (idem.). These shifts have a number of ecological implications, including changing fire regimes and the distribution of parasites, viruses and other diseases. For example, 80 % of yellow-eyed penguin chicks died of diphtheria in 2004 as virus carriers proliferated in wetter springs and summers (Root and Goldsmith 2010).

3.2 Changes in Phenology and Migration

Climate change is already impacting on the timing of seasonal activities (phenology). In many locations, spring activities already occur earlier, reproductive seasons end earlier, and there are some records of later onset of autumnal events (Hawkes et al. 2007). Seasonal changes may also affect different species located in the same place differently. One effect of this is that the reproductive cycles of predators and prey become asynchronous (Parmesan 2006). In the UK, for instance, newts are reproducing earlier, but frogs are not. So frog embryos and larvae are more exposed to newt predation than they were before (Walther et al. 2002).

Relatedly, climate change will significantly impact on wild animal migration. Migrating species are particularly vulnerable to habitat changes, such as wetland loss, from sea level rise. Climate induced changes in summer or winter habitats, or in stopping places along migration routes and flyways, can have significant impacts on migrants. Warmer winters may reduce migration distances [for instance, studies already suggest that increasing numbers of rufous hummingbirds are establishing resident US populations as far as 400 km from the Gulf Coast, rather than migrating to Mexico (Parmesan 2006; Howell 2003)]. Longer springs and summers may also allow migrants to arrive earlier and spend longer in their breeding area, giving earlier occupation of breeding sites and the possibility of more broods.

3.3 Changes in the Oceans

Climate change will have significant effects on oceans; some effects are already being recorded. The upper ocean is warming (Gleckler et al. 2012), potentially affecting populations of invertebrates such as krill, which are critical food sources for seabirds, penguins, seals and whales (Atkinson et al. 2004). Changes in upper ocean temperature may change the ranges of various species, in particular oceanic mammals. Studies indicate, for instance, that off north-west Scotland, the range of common dolphins (a warmer water species) has expanded, and the range of white-beaked dolphins (a colder water species) has contracted (MacLeod et al. 2005). Decline in Arctic sea ice will significantly impact on the abundance of some Arctic vertebrate populations adapted to living on sea ice for significant portions of the year, including walruses, seals and polar bears. Years where there is low sea ice correlate with low body condition and low ovulation rates in female ringed seals, for instance (Kovacs et al. 2011); while polar bears in the Beaufort Sea have reduced body size and reduced adult survivorship. Conversely, there's evidence that populations of some whales—such as bowhead whales—are expanding or shifting range further north into areas that are now ice-free as warming occurs (idem.). Rising sea levels due to thermal expansion and ice melt will impact on sea marshes, potentially transforming them into shallow open waters. Sea level rises may also turn freshwater wetlands saline, as in Kakadu National Park in Australia, where freshwater-dependent species such as magpie geese, barramundi and freshwater turtles are already being displaced in some locations by saltwater intrusion (Commonwealth of Australia 2010).

This brief overview indicates some of the principal ways in which climate change will impact the lives of wild animals. Hopefully, then, it provides a useful backdrop for thinking about the values at stake in such changes, and what matters ethically about them.

4 Values and Ethical Theories

Wild animals can be valuable in many different ways. Some values concern *states that matter to the animals themselves*. Other values relate to *capacities, properties or relations that (some) animals have*, even though the animals themselves are not normally subjectively aware of them. Concerns about climate change causing wild animal suffering, or killing wild animals prematurely, mostly fall into the first category. Suffering and death are (dis) values that, broadly speaking, matter to animals themselves (even if some animals may not consciously value their own continuing lives). But other kinds of concerns about climate impacts on wild animals concern properties or relations of which animals are subjectively unaware. One of these is wildness, on which I will focus here. Animals themselves can have no conception of their own wildness, however wildness is understood; and loss of their own or others' wildness is not something that animals themselves directly regret. But many *people* value the wildness of wild animals, and would regard any reduction in animal wildness caused by climate change as a loss of value in the world. So two key questions here are whether climate change is likely to reduce, or to compromise, the wildness of wild animals; and if so, whether we should be ethically concerned about it.

Answering these questions, though, is far from straightforward. First, a number of different interpretations of "wild" in the context of "wild animals" exist and may be relevant here. I will return to this in the next section. Second, even people working with the same idea of wildness, and who value it equally, might argue for different actions or policies if they adopt different approaches to ethical theory. Ethical theory, essentially, guides what we do with our values—and here approaches can diverge. Given constraints of space, I will just focus on two approaches to ethical theory here.

Standardly, one principal difference between ethical theories is captured in the distinction between utilitarian and deontological rights theory. So, for instance (in admittedly simplistic terms) a classical utilitarian aims to maximize happiness net of pain. In contrast, a deontological rights theorist aims to prevent the violation of rights, and this trumps maximizing happiness. Even where a value is shared, these different theories entail very different approaches as to what we should *do* with our values. So, for instance, both a utilitarian and a rights theorist may prioritize the value of happiness. But a classical utilitarian will aim to maximize happiness, while a rights theorist will respect and protect individuals' rights to pursue happiness—very different approaches that will often produce different outcomes.

This kind of distinction is relevant for this paper, but we can't use exactly these terms in the context of wildness. First, wildness is not straightforwardly the kind of value that utilitarians have traditionally aimed to maximize. And second, more significantly, it's difficult to claim that there are rights in the context of wildness—that for instance, animals have the "right to be wild". (It may be possible to argue for some version of this view, but I'm not going to do it here.) Better labels for the two views I'm considering here are just "consequentialist" and "deontological" about wildness value, though admittedly these terms are rather vague. A consequentialist approach, as I'm using it here, is one that understands wildness value as something we should try to promote or to maximize. On this view, the goal with respect to wildness value is to get more of it; and (for instance) it's permissible or even required to sacrifice some wildness in one place if by doing so we can get much more wildness elsewhere or in the future. A deontological approach to wildness value, as I'm using it here, aims to respect and protect what wildness there is, and to prevent its reduction or loss. On this view, it is not permissible to sacrifice wildness in one place in order to increase wildness elsewhere or in the future. Wildness is a value we have a duty to protect, not a duty to maximize.

While consequentialists and deontologists about wildness value will agree on what we should do in the context of climate change in some cases, there will be significant disagreements in others. I'll consider these disagreements in Sect. 8 later in the chapter.

5 The Meaning of Wildness

In order to think about the effects of climate change on wild animals, we need to have some sense of what's meant by "wild animals". But we talk about "wild animals" in rather different ways. For instance: wildness here might refer to *dispositional* wildness, that is, to animals that are not tame, where by "tameness" we mean individual animals' lack of fear or aggression towards humans. Wildness might refer to *constitutive* wildness, where by wildness we mean animals that belong to undomesticated species. But since domestication can be differently interpreted, this sense of wildness is complicated. Wildness might refer to something more like *resource independence*—where individual animals don't require humans to meet their needs, and can flourish independently. Or wildness could refer to something more like *self-willed* wildness, where individual animals are essentially free from human-originating or human-imposed constraints or controls to, as it were, "do their own thing". Recently, Vogel suggested an interpretation of wildness as the 'operation of forces in an object or organism that operate unpredictably and beyond the grasp of any human actor' (Vogel 2015, 112). Wildness in Vogel's sense, then, can be found in the built and cultivated environment, not just in the wilderness; a house cat as well as a snow leopard can be wild in terms of unpredictability.

Given so many disparate interpretations of wild animal wildness (and this is by no means a definitive list) an animal can clearly be wild in one sense and not in another. In addition, not only do we use wildness in so many different ways, but in all these senses, wildness comes in degrees. For instance, animals can be more or less tame, more or less domesticated, more or less self-willed. Wildness, however interpreted, describes a scalar relationship between humans and animals: the wilder an animal is, the further it is, in one of these senses, from integration into or control by human actors and human society. So concern about the impact of climate change on the wildness of wild animals is not generally one about complete loss of animal wildness, but rather one about a loss of some degree of wildness.

For the sake of clarity in this chapter, I'm going to put aside some of these interpretations of wildness, and narrow down the meanings I do discuss. The intention is not to dismiss other meanings, but rather to make this paper more manageable in scope. So, I won't further discuss *dispositional wildness*, *wildness as resource independence*, and Vogel's sense of *wildness as unpredictability*. It seems unlikely that climate change—while it may impact on animals' dispositions—is directly going to make wild animals tamer (though some of our *policies* could do this). So, it will have no impact on dispositional wildness. It also seems unlikely that climate change will generally make animals less resource-independent (though there could be a few instances of this, and again, some of our policies in response to climate change could increase animals' dependence on humans). And in Vogel's understanding of wildness as the elusive unpredictability of things, climate change is itself a form of wildness (Vogel 2015, 104). After all, no-one intended climate change to happen; and while in a broad sense we may be able to predict some features of a climate-changed world, in terms of anything more fine-grained, climate change is clearly elusively unpredictable, including with respect to its impacts on particular animals and animal populations. In Vogel's sense, climate change is not likely to be wildness-reducing; it might even be argued to be wildness-increasing! In any case, its loss is not likely to be a concern in the context of this paper. This leaves two senses of wildness to consider more closely here: *constitutive wildness* in terms of animal domestication, and *self-willed wildness*.

5.1 Constitutive Wildness

Constitutive wildness lies at the one end of a spectrum with domestication at the other. However, since domestication may be interpreted in many ways, there are multiple versions of this spectrum (See Swart in this Volume; see also Cassidy 2007). Most (but not all) definitions of domestication agree that something about the genetic adaptation of animal populations or species in relation to people is important. But beyond this, different features and purposes are taken to characterize domestication.

One possible interpretation of domestication refers to species or populations 'intentionally controlled by humans with respect to breeding, in particular deliberate, selective breeding' (Palmer 2010, 66) This picks out *human breeding practices, especially selective breeding* as key to defining domestication. However, this definition raises difficulties. It suggests that just the process of breeding control, especially selective breeding control, constitutes domestication. On this view, captive breeding programs for wild reintroduction, for instance, actually domesticate animals. So does back-breeding animals to create versions of now extinct wild breeds, such as breeding Heck cattle for rewilding projects (Swart in this volume). And this definition may raise historical difficulties, too, by suggesting that domestication didn't begin until people deliberately bred animals. This conflicts with accounts of domestication that maintain some species, at least, partly domesticated themselves before people started breeding them.

Alternative definitions of constitutive wildness take domestication to refer to animals' adaptation to live alongside people, not to whether they have been selectively bred or not. But even here definitions have importantly different emphases. Are house mice and squirrels—the kinds of animals Donaldson and Kymlicka (2011) call 'liminal'—domesticated animals? They have, after all, become adapted to live alongside humans, even though they are widely regarded as pests. On some views, they should be considered domesticated, just because of this adaptation to people. But other views maintain that to be domesticated, the ways animals are adapted to live alongside humans must be for human use or purpose. So, the adaptation of dog breeds to human houses is domestication, since dogs have a human uses and purposes; but the adaptation of mice to human houses doesn't count as domestication—since mice, after all, are the opposite of useful.

Constitutive wildness, then, can refer to (at least!) three different things: (a) not being selectively bred, (b) not being adapted to live alongside humans, or (c) not being adapted for human use or purpose. Zoo animals, for instance, might be domesticated in terms of (a) but not (b) or (c). Happily, for the purposes of this paper, there's no need to argue that one of these is the "correct" view of constitutive wildness. We can think about how climate change might affect constitutive wildness in all these senses.

5.2 *Self-willed Wildness*

The idea of "self-willed" wildness is rather vague, though it has become increasingly widely used of land, ecosystems, animals, and even people (Foreman n.d.; Monbiot 2013). It is possible, though, to isolate an interesting meaning for "self-willed" wild animals. Monbiot (2013) describes self-willedness as where organisms and systems are '*governed not by human management* but *by their own processes*' (Monbiot 2013, italics mine). This doesn't seem to require "agency" in any conscious sense, but rather refers to things changing and developing independently of human control. However, in the case of animals, self-willed wildness can be related to an idea of animal agency—specifically animal agency in the context of freedom from human constraints. More specifically, for an animal to be self-willed is for it to be *free* from human or "civilizational" constraints in expressing or fulfilling essential capacities, behaviors (often species-specific ones) and, in species where this applies, purposes and desires.

So, for instance, wolves hunt and live in family groups; being free to do this is part of what it is to live as a wolf. Wolves in zoos that are unable to hunt and don't live in family groups aren't able to express these essential capacities and behaviors, and to that extent aren't wild in a self-willed way; their behavioral options are limited. For animals to express their self-willed wildness, people need to 'stand back… rather than be in control' (Browning and Gorst 2013, 273).

This sense of "wild animal" refers to human-originating constraints on animals' behaviors and capacities, whether or not humans *intend* to be constraining in what they do. So, for instance, suppose a housing estate is constructed in an area that was formerly uncultivated, the habitat of self-willed wild animals, doing their own thing unconstrained by people. After the development, some of those wild animals may be able to continue to live in that area, but they are now constrained, and their options are limited. For instance, they can no longer move around freely, and access places that they were once free to enter. Such constraints were likely not intended by those people who built the housing estate—they were probably not even considered. But nonetheless, the construction has made animals less wild in a self-willed way because humans have limited their capacities, behaviors or fulfillment of their desires.

6 Climate Change and Animal Wildness

Now that I've outlined senses of wildness as constitutive and as self-willed, let's consider what they might mean in the context of climate change.

6.1 Climate Change and Constitutive Wildness

In the previous section, I identified three senses of constitutive wildness: (a) not being selectively bred, (b) not being adapted to live alongside humans, and (c) not being adapted for human use or purposes. Is climate change likely to reduce wild animal wildness in any of these ways? First reflections suggest not. Certainly, climate change may influence which animals survive to breed, which animals mate with which, and therefore which genetic individuals come into existence. However, humans don't *intend* these impacts, so they don't constitute human selective breeding. The genetic changes that climate change may bring about also does not better fit animals to live alongside humans, nor to be used by them. So, at first sight, it looks as though, at least in terms of meanings (a)–(c) of constitutive wildness, climate change won't have an impact at all.

However, this doesn't deal with all the possibilities here. Suppose particular populations of a shrew species are threatened by climate change. These shrew populations are rather specialized, and can only function well in a narrow temperature range and moisture regime, which is being altered to become warmer and wetter by climate change. Nearby, there's a small population of feral cats. These cats do well in the warmer, wetter climate, and their growing population is partly fueled by feasting on members of the threatened shrew population (though the shrews would still be threatened, if less quickly, by climate change alone). So, the wild shrew population is declining, and the feral cat population is growing, both partly because of climate change. And since cats are more domesticated than wild shrews (in all three senses of domestication) climate change might seem as though it is, in this case, helping to precipitate a decline in wildness.

This case picks up on several ambiguities in my discussion so far. First, it shows that "loss of constitutive wildness" could mean different things. The most obvious meaning would be that a formerly wild population becomes increasingly domesticated [in one or more of senses (a)–(c)]. But it could also mean that a constitutively wild population dwindles, and a different, more domesticated population of another species replaces it. So, even though climate change is not acting directly as a domesticating force, it could have the more indirect effect of reducing or eliminating constitutively wilder populations, and opening up space for the expansion of more domesticated populations.

Second, this case also picks up on a kind of ambiguity about what makes an animal count as domesticated. One possibility is that we call *all* feral cats domesticated because they belong to a species that's domesticated (domestic cats are domesticated in all senses: they were deliberately and selectively bred, they are adapted to live alongside people, and adapted for human purposes). But the domesticated nature of feral cats, in some cases at least, might be questioned. After all, many individual feral cats, or populations of feral cats, have not been deliberately or selectively bred; some of them don't live alongside people (for instance, the feral cats in Australia's outback) and it might be argued that they are not adapted to human purposes (at least, if the main purpose for a domestic cat is to be a family pet, or even a household mouser). And it seems implausible to argue that, in principle, 'all descendants of a domesticated species are domesticated.' We don't maintain that 'all descendants of a wild species are wild'; since wild species are, after all, the origin of domesticated species and populations. So, it would be odd to insist that the reverse could not be the case. It seems more reasonable to suggest that feral cats retain *some elements* of domestication, in terms of being descendants of selectively bred cats adapted for human purposes, and often living close to people (to different degrees). So, let's assume the feral cats in this case remain *somewhat* domesticated. Then if climate change causes the expansion of feral cat populations and the contraction of constitutively wild shrew populations, then climate change *is* contributing to a reduction of constitutive wildness.

Climate change might also have the effect that particular places lose constitutively wild animal populations, so that they just contain fewer wild animals altogether. Suppose, as is likely, that warming at altitude means a reduction in American pika populations on mountains in Colorado, and the harsh mountain environment is not subsequently successfully colonized by any other animal species. Then there just would be fewer wild animals in existence, on Coloradan mountains at least. This seems to be another way in which climate change might mean there are fewer constitutively wild animals in existence. However, we also have to consider other climate effects here: while some places may, as it were, "empty out," of wild animals it's likely that other places will "fill up". Where vegetation replaces ice, for instance, habitats that support more wild animals are likely to be created. So: while it's at least *possible* that there may be fewer constitutively wild animals in total, over time, because of climate change [I'll take "time", in contrast to Nolt (2011), to mean the next couple of thousand years, rather than long evolutionary time] it's also at least possible that numbers of wild animals will remain roughly the same, or even increase—although they will be differently distributed both across species and across places. On some views, as I'll discuss later, the existence of new constitutively wild populations in one place wouldn't compensate for their loss in another. But in terms of total numbers of constitutively wild animals, it's very difficult to say what effect climate change might have.

So, overall, this is a confusing picture; it's difficult to judge the effect of climate change on animals' constitutive wildness. Climate change itself is not directly a domesticating force. But it may—indeed, this is likely—have indirect effects that lead to the contraction of constitutively wild populations and the expansion of more domesticated ones. And it's also possible that climate change could reduce the number of constitutively wild animals in some places; equally, however, it may lead to expansion of their populations in other places.

6.2 Climate Change and Self-willed Wildness

In Sect. 5.2, I described self-willed wildness as concerning animals' freedom from human constraints in expressing their natural behavior or other capacities. Since climate change is anthropogenic, if climate change does constrain animals that were previously wild in a self-willed sense, then it does reduce their wildness. But *is* climate change likely to be constraining in this way?

For many animals, it will be. Climate change will constrain some animals' behavior by, for example, making finding food and water more difficult, or requiring them to spend more time sheltering from the heat. The increased physiological stresses that these changes create, along with susceptibility to new diseases, will restrict some animals' capacities; climate change will therefore reduce their self-willed wildness.

In the case of *constitutive* wildness, while climate change might reduce the numbers of wild animals in some places, there was at least the possibility of expanded numbers in other places. However, self-willed wildness will, in contrast, very rarely be increased by climate change. To explain this, let's consider an example: the wild red foxes that have already benefited from a warming climate. In the past, the foxes were constrained by the natural climate. They had to spend significant amounts of energy keeping warm, prey was difficult to locate, roaming behavior was limited by heavy snowfall, and significant numbers died each year. But nonetheless: because these constraints were not of human origin, the foxes were fully wild in a self-willed sense.

Now these constraints have receded, due to climate change. The foxes can do more, hunt more, and move around more freely. But this does not make them wilder in a self-willed sense, because the relevant definition of wild here is "being governed not by human management but by their own processes". And while they were previously constrained by various *natural* forces, these foxes were never governed by *human* management. So while they are now better off because of climate change, it has not made them wilder by freeing them from human constraints. Indeed, it might be argued that even though these foxes now have more behavioral freedoms they are actually *less* wild, since the climate-changed conditions might be understood as a form of anthropogenic management. However, since loss of wildness is interpreted here in terms of anthropogenic *constraints* rather than just anthropogenic *effects*, climate change doesn't reduce wildness in such cases either; it is neutral.

There will, though, be cases where climate *policy* in response to climate change increases animals' self-willed wildness. Suppose in response to climate change a human community stages a "managed retreat" along a coastline, abandoning roads and buildings and moving back from the shore. Animals previously constrained by human development along this coastline now find these constraints lifted, as traffic disappears, fencing is removed and human presence diminishes. In this case, climate change does indirectly, through human policies, increase self-willed wildness, because it has the effect of reducing anthropogenic constraints on animals. But such effects will more usually be manifest more through climate policy than through climate change itself.

In this section, then, I've suggested that the impacts of climate change on wildness will depend on how wildness is interpreted. In terms of *constitutive* wildness, climate change is not directly a domesticating force, though it may mean an expansion of domesticated animal populations in some places. In terms of *self-willed wildness*, where climate change constrains animals' behaviors, it will reduce wildness. So, some loss of wild animal wildness on account of climate change is likely to occur.

7 The Value of Lost Wildness

Should we, then, be concerned about this loss of wild animal wildness that climate change is likely to bring? Is this a loss of something valuable that we should regret?

Thinking first about constitutive wildness: certainly, many people *do* value animals' constitutive wildness, because they value human-independent "natural" processes (Preston 2012); and "natural" genetic inheritance is one such process. Deliberate human animal breeding on this view, in particular selective breeding for human purposes, is an intrusion of human intentional structure into the process of genetic transmission. The genes that constitute feral cats, for example, bear the hallmark of human intention, even if this hallmark becomes fainter over generations. In comparison, constitutively wild shrews have no such genetic history of human intention. If human-independent processes are what's being valued, then feral cats carry less value, in this regard at least, than wild shrews.

That human-independent genetic transmission *is* widely valued isn't, of course, an argument *for* its value. However, it seems reasonable to follow Sandler (2012) in accepting that if many people have a deep, strong and stable subjective value for something, this provides at least a prima facie reason for protecting it (since it subjectively *matters* to people).

Moving beyond this recognition of existing value to actually formulating and defending arguments that constitutive wildness is valuable is more difficult. Most arguments that defend the value of any kind of wildness, including constitutive wildness, cluster around the idea that we should 'acknowledge[s] that limits to human mastery and domination of the world are imperative' (Hettinger and Throop 1999, 13). This kind of explanation would apply here, inasmuch as the anthropogenic origin of genetic transmission and/or the adaptation of animals such as cats for human purposes can be construed as examples of human mastery; while, in contrast, the continuing wildness of the shrews can be valued as an absence of such mastery. Of course, this idea that wildness *should* be valued as a limit to human mastery and domination is certainly open to objection. But developing a full defense of this view moves beyond the scope of this paper; for my purposes here, I'll just accept that the constitutive wildness lost in cases like the shrews and feral cats is valuable.

Now let's consider cases where climate change constrains animals' self-willed wildness by constraining their natural behavior. The most obvious concern about this loss of wildness arises from concerns that such constraints impinge on animals' welfare. But animal welfare can be interpreted in different senses; and constraints on self-willed wildness are more directly relevant to some senses than others. Welfare may be understood in terms of animals' positive and negative subjective *experiences*, and/or in terms of their satisfied or frustrated *preferences*; and/or in a perfectionist sense, in terms of their ability to carry out *natural behaviors*, often in a species-specific way (Appleby and Sandøe 2002). Obviously, self-willed wildness is close in meaning to this perfectionist sense of animal welfare, though with the specific sense that the animal behaviors concerned aren't constrained by *humans*. (If, say, some animals were trapped by a rock-fall that prevented them from carrying out natural behaviors, there would be a decline in their welfare in a perfectionist sense, but not in their self-willed wildness, since these behavioral constraints wouldn't have a human origin.) If animal welfare matters ethically in a perfectionist sense, then a loss of self-willed wildness in this sense must matter too.

It's certainly contested whether we *should* value animal welfare in this sense. Much work on animal welfare argues that what matters is how animals subjectively experience the world, not whether they are free to perform natural behaviors. If animals' natural behaviors are constrained, but this constraint doesn't negatively impact on how the animals concerned actually feel, then these constraints don't matter (for example, Duncan 1996). On this view, we shouldn't worry about the effects on self-willed animal wildness caused by climate change, because nothing valuable has been lost. This doesn't, of course, mean that behavioral constraints aren't *indirectly* important, because such constraints may cause suffering and other negative experiences for the animals concerned. But then the value at stake here would be the animals' negative subjective experiences, not the loss of the self-willed wildness in itself.

But while there are difficulties in maintaining a *solely* perfectionist account of animal welfare, again for the sake of argument, let's assume that there is loss of value when animals' self-willed wildness is diminished. So, this allows us to conclude: climate change is likely to reduce animals' wildness value, at least in terms of constitutive and self-willed wildness. What, then, should be done about it?

8 Responses to the Loss of Wildness Value in Animals

One immediate response here, for those who value constitutive and self-willed wildness, would be to argue for reductions in GHG emissions so that valuable wildness is not lost in the first place. But given that we're already committed to some degree of climate change, even were such an argument likely to have an effect and emissions to fall, value losses would still occur. So we still need to think about other ways of reducing, preventing, or making up for such value losses. And one key factor here, as outlined in Sect. 4, concerns what approach to ethical theory is adopted. There I outlined two possible kinds of theoretical responses: a broadly consequentialist, and a broadly deontological approach to wildness value. The consequentialist understands wildness value as something we should try to promote or maximize, and accepts that it's permissible or even required to sacrifice wildness in one place or time if by doing so we can get much more wildness elsewhere or in the future. A deontological approach to wildness value on the other hand, respects and protects what wildness there is, and attempts to prevent its reduction or loss; on this view, it's not permissible to sacrifice wildness in one place or time in order to increase wildness elsewhere or in the future.

8.1 *Response to the Loss of Constitutive Value from Climate Change*

How might these two positions work out in terms of constitutive animal wildness, given the context of climate change? Let's start with the consequentialist approach.

The first thing to note—although this is not exactly a *response*—is that, from a consequentialist perspective, wildness lost in one place could be substituted by increased wildness in another, a process we noted earlier. So, if climate change led to the vegetating of former Arctic ice-fields, and these provided a habitat that supported more constitutively wild animals, constitutive wildness in the former ice-fields could substitute for constitutive wildness lost elsewhere, say in the Coloradan mountain tops. While we have little information as to how this "wildness sum" could work out, it's at least important to lay this possibility on the table. And this also suggests that, in terms of policy, *rewilding* projects could help. If land not currently occupied by constitutively wild animals were set aside as new wild animal habitat, this may lead to more constitutively wild animals coming into existence, and thus an increase in wildness in the world. While this would not prevent a consequentialist trying to minimize loss of wildness, such gains elsewhere would be relevant to the overall total of wildness value in existence.

What about cases where one less wild population replaces a wilder population—for instance, where feral cats replace wild shrews? In *predation* cases, where climate change gives a feral population an enhanced opportunity to prey on members of wilder species, the obvious response would be to try to prevent the feral population from advancing into wilder territory. Whether and how this is carried out, though, would depend on how constitutive wildness weighed against other values (such as minimizing feral animal suffering). However, attempting to prevent such situations would not be effective in a simple *replacement* case, where one wilder species is struggling and becoming scarce in the context of climate change, and the feral species is merely taking advantage of unoccupied habitat. In this case, allowing the feral species to expand might actually *increase* wildness (if the alternative is no animals of any kind) rather than diminishing it. In this case there seems no reason to do anything, unless it's likely that other constitutively wilder species would expand into the area if the feral species were kept out of it.

Other possible policy responses for constitutive wildness-valuing consequentialists turn around how domestication is defined. If intentional breeding counts as domestication—my definition (a)—then (for instance) captive breeding of threatened shrews would not help to increase constitutive wildness, since projects involving breeding wild animals would undermine the very value that they were trying to create. On this understanding of domestication, the best way of maximizing constitutive wildness would be just by protecting populations that manifest it. However, if domestication is interpreted as just being adapted to live alongside humans, or as adaptation for human use—my definitions (b) and (c)—captive breeding for wild reintroduction wouldn't undermine constitutive wildness. From this perspective, captive breeding could be an effective way of maximizing constitutive wildness, especially if the animals bred were released in new, even rewilded, habitats in which they would do better (as long as they didn't threaten *other* wild animals already living there).

Someone who adopted a *deontological* approach to constitutive wildness value would, however, be unpersuaded by the idea that wildness generated in one place or in the future could substitute for wildness lost elsewhere and in the present. For instance, a Coloradan hiker who values American pika for their constitutive wildness is unlikely to be persuaded that this particular loss of wildness is more than compensated for by an increase in the number of constitutively wild red foxes thousands of miles away. What matters from a deontological perspective is to protect the wildness value carried by particular wild animals that currently exist, rather than to maximize the amount of wildness value in the world over time. In practice, though, there would be some coincidence in policy responses with value consequentialists. For instance, both approaches would try to prevent constitutively wild animals, struggling on account of climate change, from becoming prey to feral and domesticated animals whose flourishing climate change has promoted. But, in general, deontological policies would promote the protection of existing value, rather than the generation of new value.

As an example, suppose that there are resources available to protect an endangered wild species, but there are two possible ways of spending these resources: either to captive-breed from members of the struggling population, or to initiate protection of the existing population, for instance by providing shelters from heat, provision of water in drought, and so on, instead of captive breeding. Studies predict that, over time, a selective captive breeding program would produce more constitutive wildness [interpreted in sense (b) or (c)] but at the expense of the lives (and the wildness value) of some members of the existing population, either because these individuals die in the course of the captive breeding project, or because they would have survived had their lives been protected by shelters, water etc. Consequentialists would opt for the breeding project, as it could be expected to maximize wildness over time. But from a deontological perspective, what would matter would be protecting the valuable animals that actually exist, not maximizing the amount of constitutive wildness in the world. So, the provision of shade, water and so on would be the preferred policy.

8.2 *Response to the Loss of Self-willed Wildness from Climate Change*

Let's move to the second kind of value loss: constraints on animals' ability to perform certain natural behaviors and to express their "self-willed wildness" on account of climate change. A consequentialist will seek to minimize such constraints, and to maximize animals' ability to perform natural behaviors in the face of climate change. This clearly mandates interventionist policies, since such policies could be wildness-restoring over time. These policies could include provisioning (where, overall, this increases rather than reduces, the performance of animals' natural behaviors); assisted migration (moving animals to live in environments where they are less constrained) facilitated adaptation (genetically modifying animals to be less constrained by climate impacts while continuing to live in the same location) and so on. Even though some of these practices, such as facilitated adaptation, may in the short term result in the loss of individual animals' self-willed wildness (for instance, by confining animals for selective breeding purposes) over time (after these initial confined generations had successfully reproduced and been released) such practices could be expected to maximize self-willed wildness.

From a deontological perspective, the concern here is that constraints on animals' natural behavior, on their self-willed wildness, *harms* them (given a perfectionist account of welfare). So, the goal would be to stop the harm, and (perhaps) to offer compensation to those beings that have been harmed. Given that the source of the harm is climate change, however, stopping the harm is difficult to do directly. So, responses entail finding other ways of lifting constraints on the natural behaviors of wild animals. However, many of the approaches available to consequentialists seem unavailable to deontologists, because they involve imposing further restrictions on the self-willed wildness of some animals in order to increase it for other animals later. Assisted migration, for instance, may involve significant constraints on the first generation moved. They would be captured, confined, potentially used in breeding programs, transported, and released into an alien environment. This wouldn't be permitted from a deontological perspective that valued self-willed wildness, since such constraints harm animals' welfare; it is not permissible to cause poor welfare to individual animals now, in order to create more animals in the future whose welfare is better.

It's worth pointing out that the argument here isn't that deontologists about wildness value don't care about the wildness of future animal generations. But on this view, it's just not permissible to sacrifice one existing animal's self-willed wildness in order to bring about more wildness value in the future. It's the wildness of the existing individual animal that matters, not the state of affairs in the world where there's more wildness. This is like a human case where a deontologist about human rights would not find it acceptable to violate someone's right to live now, in order to maximize the protection of other people's rights to live in the future. This approach means, however, that for a deontologist about wildness value there are fewer options to respond to climate induced constraints on animals' self-willed wildness than are available to a consequentialist.

9 Conclusion

This chapter has attempted to think through a little discussed issue: what values are at stake, and how to respond ethically, to possible climate-induced loss of wild animal wildness in the Anthropocene. I considered in what ways wild animals might become less wild in the context of climate change, and concluded that a reduction in constitutive wildness (in some senses) and a reduction in "self-willed wildness," (interpreted as human-originating constraints on natural behaviors) were both possible outcomes of climate change. Although I did not offer a detailed defense of these forms of wildness value (in fact I'm somewhat skeptical that there are good reasons to value them), since people *do* value these forms of wildness, exploring them further, and thinking about how to reduce their loss, seems important. Consequentialists about wildness value turn out to have more, and more wide-ranging, policy options to respond to climate-induced loss of wildness than deontologists about wildness value. Consequentialists can focus on pursuing the *creation* of wildness as well as the protection of existing wildness, including the creation of new wild animals (even at the expense of the losing wildness in some existing animals). For deontologists, the focus must be on a value-protecting response; but success in achieving this is particularly difficult, given the pervasive impacts of climate change. There is, of course, much more to be said about wildness value, animals and climate change; this is merely a first attempt to work through some basic ideas.

References

Appleby, M., and P. Sandøe. 2002. Philosophical debate on the nature of well-being: implications for animal welfare. *Animal Welfare* 11(3): 283–294.

Atkinson, A., et al. 2004. Long-term decline in krill stock and increase in salps within the Southern Ocean. *Nature* 432(7013): 100–103.

Barnosky, A.D., et al. 2012. Approaching a state shift in Earth's biosphere. *Nature* 486(7401): 52–58.

Bernardo, J., and J. Spotila. 2006. Physiological constraints on organismal response to global warming: mechanistic insights from clinically varying populations and implications for assessing endangerment. *Biology Letters* 2: 135–139.

Bernardo, J., R. Ossola, J. Spotila, and K.A. Crandall. 2007. Validation of interspecific physiological variation as a tool for assessing global warming-induced endangerment. *Biology Letters* 3: 695–698.

Browning, G., and J. Gorst 2013. Wild cattle and the "wilder valley" experiences—the introduction of extensive grazing with Galloway cattle in the Ennerdale Valley, England. In *Trees, forested landscapes and grazing animals,* ed. I. Rotherham, 269–280. London: Routledge.

Cassidy, R. 2007. Domestication reconsidered. In *Where the wild things are now*, ed. M. Mullin, and R. Cassidy, 1–26. London: Bloomsbury.

Director of National Parks. 2010. *Kakadu national park climate change strategy 2010–2015.* Canberra, Australia: Commonwealth of Australia. http://www.environment.gov.au/parks/publications/climate/kakadu.html.

Donaldson, S., and W. Kymlicka. 2011. *Zoopolis: a political theory of animal rights.* Oxford: Oxford University Press.

Duncan, I.J.H. 1996. Animal welfare defined in terms of feelings. *Acta Agriculturae Scandinavica. Section A, Animal Science* 27: 29–35.

Foreman, D. n.d. Wilderness: self-willed land. http://rewilding.org/rewildit/around-the-campfire-with-uncle-dave-wilderness-self-willed-land/.

Gleckler, P.J., et al. 2012. Human-induced global ocean warming on multidecadal timescales. *Nature Climate Change* 2(7): 524–529.

Hawkes, L.A., A.C. Broderick, M.H. Godfrey, and B.J. Godley. 2007. Investigating the potential impacts of climate change on a marine turtle population. *Global Change Biology* 13(5): 923–932.

Hettinger, N., and B. Throop. 1999. Refocusing ecocentrism. *Environmental Ethics* 21(1): 3–21.

Howell, S.N.G. 2003. *Hummingbirds of North America*. Princeton, NJ: Princeton University Press.

IPCC. 2014. *Summary for policymakers*. http://www.ipcc.ch/.

Kovacs, K., et al. 2011. Impacts of changing sea ice conditions on Arctic marine mammals. *Marine Biodiversity* 41(1): 181–194.

MacLeod, C., M. Bannon, G. Pierce, C. Schweder, J. Learmonth, J. Herman, and R. Reid. 2005. Climate change and the cetacean community of north-west Scotland. *Biological Conservation* 124(4): 477–483.

Monbiot, G. 2013. *Feral: rewilding the land, the sea and human life*. London: Allen Lane.

Nolt, J. 2011. Nonanthropocentric climate ethics. *WIRES Climate Change* 2(5): 701–711.

Palmer, C. 2010. *Animal ethics in context*. New York: Columbia University Press.

Palmer, C. 2011. Does nature matter? The place of the non-human in the ethics of climate change. In *The ethics of global climate change*, ed. D. Arnold, 272–291. Cambridge: Cambridge University Press.

Parmesan, C. 2006. Ecological and evolutionary responses to recent climate change. *Annual Review of Evolution, Ecology and Systematics* 37: 637–669.

Parmesan, C., and G. Yohe. 2003. A globally coherent fingerprint of climate change impacts across natural systems. *Nature* 421(6918): 37–42.

Preston, C.J. 2012. Beyond the end of nature: SRM and two tales of artificity for the anthropocene. *Ethics, Policy & Environment* 15(2): 188–201.

Root, T.L., and E. Goldsmith. 2010. Wild species and extinction. In *Climate change, science and policy*, ed. S. Schneider, 44–55. New York: Island Press.

Sandler, R. 2012. Intrinsic value, ecology, and conservation. *Nature Education Knowledge* 3(10): 4.

Vogel, S. 2015. *Thinking like a mall*. Cambridge: MIT Press.

Walther, G.-R., et al. 2002. Ecological responses to recent climate change. *Nature* 416(6879): 389–395.

Watts, O., M. Morecroft, J. Phillips, and J. Taylor (eds.). 2012. *Climate change: biodiversity and people on the front line*. Natural England and WWF-UK: RSPB.

Animal Captivity: Justifications for Animal Captivity in the Context of Domestication

Bernice Bovenkerk

Abstract The central question of this chapter is whether keeping animals in captivity is morally justified. Captivity could be considered inherently wrong when animals are perceived to have an interest in liberty. I argue that this is the case to a certain extent, provided that we use a less stringent notion of autonomy than we do for humans. Next, I address two possible general moral justifications for keeping animals in captivity: (1) it is in the interest of humans to keep animals in captivity and (2) it is in the interest of the animals themselves. Whether these justifications are successful is to a large extent an empirical matter. In general, however, we could say that either animals of a specific species do not have sufficient adaptive capacity to be able to deal with conditions of captivity—in which case harm to the animals' welfare occurs—or they do have this capacity, but this means that their genetic make-up changes over the generations, and they ultimately become domesticated. This aspect of domestication raises a whole new set of questions regarding the justifiability of animal domestication. I argue that the questions raised by animal domestication cannot be completely dealt with within traditional animal ethical approaches, that take individual animals as the sole unit of moral concern. Objections that people voice regarding animal captivity and in particular regarding the often resulting domestication and interfering with an animal's genetic make-up, are more properly directed at the species-level.

1 Introduction

We keep animals in captivity for many different purposes: for recreation or entertainment (zoos, aquariums, rodeos, circuses), for their products (meat, milk, wool, fur), for medical experimentation, for scientific research, for companionship, for their labour (police horses, guiding dogs, transportation) and for their own benefit (sheltering and treating lost and sick animals). The central question of this chapter is

B. Bovenkerk (✉)
University of Wageningen, Wageningen, The Netherlands
e-mail: bernice.bovenkerk@wur.nl

B. Bovenkerk and J. Keulartz (eds.), *Animal Ethics in the Age of Humans*,
The International Library of Environmental, Agricultural and Food Ethics 23,
DOI 10.1007/978-3-319-44206-8_10

whether keeping animals in captivity is morally justified. First, I will briefly address the question whether animal captivity could be considered categorically wrong. Next, I will address two possible general moral justifications for keeping animals in captivity: (1) it is in the interest of humans to keep animals in captivity and (2) it is in the interest of the animals themselves. Whether these justifications are successful is to a large extent an empirical matter. In general, however, we could say that either animals of a specific species do not have sufficient adaptive capacity to be able to deal with conditions of captivity—in which case harm to the animals' welfare occurs—or they do have this capacity, but this means that their genetic make-up changes over the generations, and they ultimately become domesticated. This leads to a situation where we decide 'the way animals are'. This aspect of domestication raises a whole new set of questions regarding the justifiability of interferences in animals' genetic make-up. In this context, I will first address the argument that animals (tacitly) consented to a so-called 'domestication contract'. Next, I will address the objections that exist in society against changing the genetic make-up of animals. I will argue that the questions raised by animal domestication cannot be completely dealt with within traditional animal ethical approaches, that take individual animals as the sole unit of moral concern. Objections that people voice regarding animal captivity and in particular regarding the often resulting domestication and interfering with an animal's genetic make-up, are more properly directed at the species-level. Moreover, some justifications for animal captivity (most notably in the context of zoos) also are directed at the species-level. However, directing our concerns at the species level also raises problematic issues; it is difficult to conceptualise 'harm to species', as the concept of species is a human construct that does not correspond to a particular being 'out there' that can be the object of (experienced) harm. As a way out, I will suggest that the evaluation of animal captivity does not only rest on an assessment of harm done to particular animals, but also on moral views regarding the relationship between the human and other species.

2 Inherent Objections to Animal Captivity

It has been argued that it is inherently wrong to keep animals in captivity, as this commodifies them, or turns them into property (Francione 2010). Others hold that captivity is generally wrong, because most practices for which animals are kept in captivity fail to show animals the respect they deserve on the basis of their inherent value. Regan (1995), for example, argues that limiting the freedom of animals in zoos fails to treat them with respect and would only be justified if this were in the animal's best interest, such as when this would save the life of this particular animal (but not when it would save this animal's species). It remains unclear, however, why limiting an animal's freedom in itself fails to treat it with respect. Does captivity necessarily mean that an animal is used solely as an instrument for our goals or can the animal's own goals also be taken into account? How about animals that are kept in wildlife refuges or as pets? These animals are also not entirely free, but it is at least not immediately clear that they are solely used as a means to our ends.

It also remains unclear why animal captivity by definition would commodify them. While animals in most countries are legally regarded as property, this does not mean that keeping animals commodifies them in a moral sense. True, keeping a companion animal does deprive it of certain freedoms, but many people tend to regard pets as friends or even family members, rather than as property.[1] Neither do most people hold that we can just do whatever we want with animals, as we can usually do to our property. We can raise the question of whether there is something inherently denigrating about being deprived of freedom. In the case of human beings we do seem to think so; taking away a criminal's liberty is regarded as quite a harsh punishment. However, depriving prisoners of their freedom does not automatically turn them into our property, nor does it necessarily fail to show respect for their inherent value.

Could we still not argue that animals have an interest in freedom as such? There is discussion about the question of whether animals have an intrinsic—as opposed to only an instrumental—interest in freedom, which would be violated when we keep animals in captivity. Freedom is an instrumental interest for animals in so far as it facilitates "other goods, such as the avoidance of suffering" (Cochrane 2009, 661). However, in an instrumental conception of interests, captivity would not by definition always violate an animal's interest. In a (perhaps only imaginary) captive situation where all of an animal's interests are met, the animal does not suffer, and has a good life, according to Cochrane (2009) the animal has no interest in freedom.[2] On the other hand, he argues, most human beings do have an intrinsic interest in freedom, or liberty, because they have a stake in being free from control and being the author of their own lives. They have this interest due to the fact that they are autonomous agents who have the 'capacity to frame, revise and pursue their own conception of the good' (idem, 665). This type of autonomy is a 'second order capacity' which requires one to be able to reflect on one's own desires. Except for perhaps cateceans and great apes, Cochrane argues that no animals have this type of autonomy.

The interest that humans have in liberty cannot simply be reduced to an instrumental interest, where liberty would promote happiness or preference satisfaction. In order to illustrate this point, Cochrane refers to the movie *The Truman Show*, in which the main character leads a happy and fulfilling life, but unbeknownst to him he is starring in a TV series about his life. Even though he does not realise that the life he leads is inauthentic, and therefore he is not suffering, we would still say that his interests have been violated. After all, he is not

[1]This at least was the outcome of interviews I held with stakeholders in the Netherlands about pedigree dog breeding in 2015 and 2016.

[2]Cochrane (2009, 662) follows Feinberg (1984) in taking interest to mean "to have some kind of stake" in something. One has a stake in something when one's well-being is affected by it. Well-being is taken to be a prudential value, in other words, "well-being relates to how well things are going *for the individual whose life it is*" (italics in original).

self-governing, he is not the author of his own life. Similarly, someone who is a slave, but is happy and treated very well by her enlightened owner, is still regarded as violated in her intrinsic interest in liberty. One of the questions that Cochrane's account gives rise to, is whether animals do not also have an interest in being the author of their own lives.

In medical ethics it is now common to regard autonomy as a gradual notion: one person is more autonomous than another and someone can be autonomous to make certain decisions, but not others. For example, someone in the first stages of dementia can still decide what she wants to have for dinner or how she wants to entertain herself, but this person can no longer make important financial decisions. Could something similar count for animals? Do animals perhaps have an interest in making certain decisions for themselves, even though they may not be able to foresee the implications of larger decisions, for example about their far future? This would suggest that perhaps the distinction between instrumental and intrinsic interest is not as strict as Cochrane suggests. Gruen (2009), on her part, doubts Cochrane's view that animals would not be able to frame their own conception of the good life. She cites evidence that shows that many animals at least have concepts and she argues that many animals can devise their own plans in life:

> "All sorts of animals make choices about what to do, when to do it and who to do it with. Many animals make plans, by making and saving tools for future use or by caching food to collect at a later time… Social animals often engage in manipulation or deception to try to get what they want and to prevent others from getting it. So it certainly seems like these sorts of behaviors could be considered autonomous in the sense that animals are controlling what they do" (Gruen 2009, 148).

Gruen, in other words, adopts a different account of autonomy than Cochrane. She argues that many animals pursue certain ends, shape their own lives, adapt to changing circumstances, make choices, and try to improve their environment, not only on their own but also by means of collective action (idem, 150). Is all of this sufficient to be able to say that animals have an interest in being free in the sense that they are the author of their own lives? And would this be sufficient to argue that they should never be held in captivity? Or does this toned down version of autonomy merely mean that it is important to let animals make their own—limited—choices, even if they are held in captivity? Should we perhaps let animals decide for themselves whether they have an interest in liberty? Many examples exist of animals trying to escape captive situations (Hribal 2011). This could be interpreted as a simple fear- and flightresponse, but it could also mean that limitations on freedom do not fit into the lifeplan of these animals. In summary, I am inclined to believe that many animals have an intrinsic interest in liberty, although perhaps not to the same extent as most human beings do. Perhaps this interest admits of gradations and perhaps the distinction between intrinsic and instrumental interest is a bit overdrawn. Animals may not have very well articulated conceptions of the good life, which they can reflect on and revise in a very structured way, but they do have concepts and make autonomous decisions in a less demanding interpretation of autonomy.

They seem to value freedom, although we must admit that it is hard for human beings to understand exactly what animals attach value to.

3 General Justifications for Captivity

Even if animals can be said to have an intrinsic interest in liberty, this does not have to lead to an absolute obligation on the part of humans to liberate all animals. This interest would still have to be weighed against other interests. Even though I hold that interests have to be weighed, I do not necessarily adopt a utilitarian style of reasoning. In a deontological argumentation—if we would argue that intrinsic interests form the basis of rights—rights can in some circumstances be trumped by other rights (see also Cochrane 2012). Therefore, we also have to take a look at possible justifications that could be given for animal captivity and weigh these against animals' instrumental and intrinsic interests in liberty. Several justifications for keeping animals in captivity can or have been given, sometimes more and sometimes less explicitly. I will now briefly consider two general arguments: (1) it is in the interest of humans to hold (other) animals in captivity, (2) it is in the interest of animals themselves. I will argue that both arguments often depend on empirical claims and will differ on a case by case basis. I will also argue that the benefits of animal captivity will have to be weighed against the loss of freedom. Freedom entails having space to roam around, having control over decisions and being able to carry out natural behaviour. Whether loss of this freedom causes harm will depend to a large extent on whether the animals will be able to adapt. If animals cannot adapt, harm will be caused, but if they can, there is a good chance that they become domesticated. As we will see later in this chapter, this raises a whole host of new ethical questions.

3.1 Human Interests

Obviously, human beings benefit considerably from keeping animals in captivity, as humans use animal products, use them for research, entertainment, labour, and companionship. As these uses come with a cost to the animals in question—and it is now commonly accepted that many animals have an experiential welfare which can be frustrated or harmed—they need to be justified. The—often implicit—argument to justify these uses is that the interest of humans in enjoying these benefits outweighs the interest of animals in avoiding harm. This argument can be based on three underlying assumptions:

(1) the harm done to animals by keeping them in captivity is not regarded as very serious harm in practice. Note that this is an empirical claim that needs to be supported by observation and research. Moreover, certain practices for which animals are kept in captivity are more harmful than others.

(2) the benefits to humans are so great that this justifies captivity. Again, this is at least in part an empirical claim, the truth of which will differ from practice to practice;
(3) whether or not the harm done to the animals is serious, the interests of human beings simply matter more than the interests of animals.

The first two assumptions are in principle compatible with the view professed by many animal ethicists that the interests of humans and (other) animals should be weighed equally (e.g. Singer 1975; Regan 1983). After all, even if their interests should be given equal weight, it is very well possible that the content of the interests in question differs. It is generally accepted that equal interests should be weighed equally and unequal interests unequally. The third assumption, however, is based on the view that there is an in principle moral divide between humans and other animals. This so-called speciesism is notoriously hard to defend, because animal ethicists can always point to the 'marginal cases argument': there is no morally relevant characteristic that sets humans apart from other animals, since we can always find specific animals that do have the characteristic in question (bonobos that have a rudimentary form of morality, chimps that can learn the basics of language, crows that use tools, etc.) and more importantly, there are specific humans that don't (for example babies, comatose patients, severely mentally retarded people or Alzheimer patients: the marginal cases) (Dombrowski 1997).[3] Yet, many people do have the intuition that when humans and animals have equal interests the interests of humans should outweigh those of animals (VandeVeer 1979). What could be the basis for this intuition and is it justified? I will discuss the in my view two most plausible candidates here.

Firstly, we may think that human and animal interests are simply not comparable. In other words, we cannot speak of equal or unequal interests in the first place; the interests of humans and animals simply cannot be measured along the same scale. The underlying assumption is that because of their different physical and mental make-up humans and animals experience the world differently. Of course, the important question is to what extent these experiential differences would be morally relevant differences. These supposed differences between humans and other animals must be differences in kind, or qualitative differences. If they were gradual, or quantitative differences, the marginal cases argument would apply again. For if they were gradual, this would imply that the interests of, say, a healthy adult human being should also always outweigh those of a baby or an Alzheimer patient. In other words, their interests would not be comparable either and this is a bullet that not many ethicists are willing to bite.[4] What grounds do we have to

---

[3]This marginal cases argument has been criticized, for example by Carruthers (1992) for leading to a slippery slope where in the end we cannot attribute moral status to any human beings either and by Cohen (1986) and Scruton (2000) who argue that regardless of their mental capacities humans are the 'kinds' of beings who deserve moral status. These arguments have successfully been countered by Tanner (2006, 2009).

[4]There are notable exceptions, such as Frey (1988).

conclude that the morally relevant difference between humans and other animals is one not of degree but rather a difference in kind?

Since Darwin we have come to understand that we share a lot more characteristics with animals than we previously thought and that the differences between us and animals tend to be gradual rather than absolute. Still, it may be objected, humans have a much greater ability to shape their environment to suit their own needs and through their superior intelligence and creativity they have developed highways, buildings and mobile phones. How do we explain this if we do not accept an absolute difference between human and animal brains? One way in which current thinking about evolution theory (by cognitive philosophers) tends to explain such differences is through the hypothesis of extended cognition: cognitive processes are not only located in the brain, but they are also constituted in the interaction between the brain and the physical and cultural environment. One cultural factor that is highly constitutive of human cognition is their use of symbolic language. The functional architecture of our brains is not fixed by birth, but develops in accordance with the routines we engage in and skills we acquire. Consequently, external artefacts can play a crucial role in the engineering of our maturing minds.[5] The question of what moral conclusion we should draw from this explanation is contested. However, in my view, this underlines the idea that there is nothing categorically or absolutely distinctive about human brains, but rather that a difference in degree between human and nonhuman brains has accumulated due to environmental/cultural factors.[6] Ultimately, however, the problem with this basis for the view that human and animal interests cannot even be compared is that we simply do not know how animals experience the world and whether, for example, animal pain is qualitatively different from human pain. Yet, we would be committing the fallacy of ignorance if we were to conclude that because we don't know whether animals experience the world the same as us, the opposite therefore has to be the case.

The second option to account for the intuition that humans' interests matter more than animals' is to revert to an evolutionary explanation. Because it was evolutionarily advantageous for human beings to cooperate with other human beings our moral sympathies have evolved and these sympathies have evolved to be stronger for next of kin. This explains, for example, why we feel more moral obligations towards our family members or towards our fellow countrymen than towards more distant others. Care ethicists in this context sometimes invoke affective arguments to put forward that we have more obligations to the ones we are closest to (Noddings 1984). However, an evolutionary debunking argument can be used here to counter this view: we have developed certain intuitions or moral emotions because they were advantageous from an evolutionary point of view (Greene 2007).

---

[5]Thanks to Jeroen Hopster for pointing this out to me.

[6]This is supported by current thinking by biologists who posit the idea of 'inclusive inheritance'. Still, one can wonder, of course, whether the distinction between qualitative and quantitative differences is itself not one of degrees. Thanks to Jeroen Hopster for bringing this point to my attention.

And while, surely, we still rely on other people for our survival, this does not apply to all people in general. Hence, the intuition that people matter more than animals is undermined when we consider the origin of this intuition and the fact that these circumstances do not apply anymore.[7]

We can conclude, then, that both of these initially plausible bases for the view that the interests of humans should by definition outweigh those of other animals are shaky at best. Whether the general argument that keeping animals in captivity because it is in the interest of humans succeeds, therefore, is dependent on the assessment of the gravity of the harm done to the animals vis-à-vis the seriousness of the human interests involved. The outcome will differ from practice to practice. It should be noted here that while this assessment definitely has an empirical component, its outcome also very much depends on the particular concept of animal welfare that is used. Animal welfare is not just a biological term; rather it is an evaluative term that combines biological knowledge and moral norms (Bovenkerk and Meijboom 2013). In general, three views on animal welfare can be distinguished: (1) the view that animal welfare is guaranteed when animals can cope with the conditions they find themselves in (the functional view); (2) the view that it is guaranteed if the animals experience subjective well-being (the affective view); and (3) the view that the extent to which an animal can carry out its natural behaviour is constitutive of its welfare (the nature-based view) (Fraser 2003). These three views are by no means mutually exclusive. In zoos and in animal husbandry, stereotypical behaviour, such as pacing, self-biting, and excessive self-grooming, are found, and these point to welfare problems according to all three views on animal welfare. Stress and social isolation are the most common causes of these behaviours.[8] However, the three views on animal welfare can also be in tension with each other. For example, an animal in intensive farming conditions can be kept healthy and may be able to deal well with its situation, but it may also be bored and therefore not experience subjective well-being. When it is not free to carry out its natural behaviour, according to the nature-based view, it will have diminished welfare as well. Whether human interests outweigh animal interests in a particular practice of captivity is therefore not a straightforward assessment, but will be open to discussion about how we weigh animal interests and what we mean by animal welfare in the first place.

[7]Of course, another argument against this view is simply that we are making an is/ought fallacy if we are basing normative conclusions about the moral status of humans on the fact that we feel more sympathy towards other humans. See also Palmer (2010, 51–54) for a critical discussion of the care ethical move to base obligations on affection.

[8]A study has shown, for example, that gorillas in zoos that are confronted with many visitors are more likely to show stereotypic behaviour—consisting of autogrooming, abnormal behaviour (such as teeth clenching, body rocking, and more banging on the separation barrier) taking less rest, and aggression towards other gorillas—than those who are confronted with low visitor density. This indicates that visitor density leads to stress in the animals. See Wells (2005), who also cites other studies with similar conclusions.

3.2 Animal Interests

Animal captivity may be justified alternatively, by pointing to the benefits for the animals themselves. As nature is 'red in tooth and claw' animal lives in the wild are no picnic. When they are held in captivity, they receive shelter from the elements, and protection from predators, disease, and hunger. Whether these benefits outweigh the harm that is done to the animals in captivity, again, is an empirical question that will differ from practice to practice and that is also dependent on whether one regards liberty as only instrumentally or also intrinsically valuable to animals. In the wild many animals die at a young age; this in fact is part of evolution. On the other hand, one fact at least makes it difficult to establish whether a life in the wild is better for animals than a life in captivity: Apart from companion animals, many captive animals live shorter lives than their natural counterparts, due to the purposes for which they are held, which require killing them at a relatively young age.[9] This is most notably the case in meat production and animal experimentation.[10]

While animals in captivity generally are protected from predators and often from disease and hunger, there are also many cases where captive animals do in fact contract diseases as a result of the way in which they are kept (for example in intensive farming systems) or bred (for example as a by-effect of breeding for pedigree dogs or cats, or as a model for human disease in animal experimentation). Moreover, according to the affective and nature views on animal welfare, captivity will often not be in an animal's interest, as it may be bored and/or may not be able to carry out its natural behaviour. No unequivocal conclusion can be drawn, then, about the benefits of being held in captivity for animals and the argument as it stands is too simplistic. What I have noted here is that it is at least not self-evident that animals benefit from being held in captivity. All this argument does seem to suggest is that *as long as* captivity benefits an animal more than it harms it captivity may be justified.

However, in order to determine whether animal captivity is justified, we need not only look at the net animal welfare benefits in practice, but we also need to weigh these against the loss of freedom that captivity brings about. If we consider freedom as an instrumental good for animals, what are its benefits for an animal? Perhaps the main benefits are control over one's own decisions, space to roam around, and being able to live according to one's own nature or carrying out one's natural

---

[9]Despite claims to the contrary. For example, as the documentary Blackfish shows, Sea World advertises that orcas live up to the age of 40 in captivity, while they live much shorter lives in the wild. In fact, male orcas live up to 50 years in the wild and female ones can even live up to 90 years.

[10]Disagreement is possible about the question whether this premature killing of captive animals can count as a harm and the question rises how to offset lifespan against other benefits that the animals may enjoy by being held in captivity. Some (for example Haynes 2008) argue that killing should be regarded a welfare problem while others argue that shortening an animal's life in itself is not problematic (for example Singer 2011).

behaviour. If we consider at least some animals as purposive agents (Kaldewaij 2013), this means that their purposes can be frustrated and this could be considered a harm (Carruthers 2005). Having limited movement or not being able to carry out natural behaviour both lead to diminished welfare, at least in the nature-based view of welfare. For example, loss of freedom and lack of control over an animal's life can lead to what animal welfare scientists refer to as this animal's 'learned helplessness'.[11]

4 Adaptation

Whether loss of freedom is also experienced as diminished well-being by the animals (according to the affective view) is dependent on the extent to which animals can adapt to their captive circumstances. Animals from different species have a different adaptive capacity to a human-made environment, be it a zoo, a human household or a stable (and even intra-species differences exist). When an animal adapts to the human-made environment, its genetic make-up also tends to change. The animal in effect becomes domesticated, which often involves changes to the animal, such as reduced brain size, reduced aggression, reduced seasonal behaviour, smaller teeth and horns, and neoteny (retention of juvenile characteristics in adults) (Fiby 2012). While many of these characteristics are intentionally bred for, domestication also happens unintentionally. A side effect of keeping wild animals in zoos, for example, is that they become semi-domesticated. This side-effect is often unintended, as people want to exhibit wild animals in zoos, and is especially problematic when a goal of a breeding program is reintroduction into the wild (idem.).

A question that can be raised regarding the value of freedom for animals, however, is whether it is still problematic to keep animals in captivity if their genetic make-up is changed to the effect that they experience no negative welfare problems as a result. In other words, is loss of freedom still problematic if an animal's natural behaviour is altered to the extent that it does not suffer from limited movement or loss of control anymore?[12] Looking back at the distinction between an intrinsic and instrumental interest in liberty, we could answer this question affirmatively if we are willing to adopt a less stringent form of autonomy. On the other hand, if animals were to benefit from being held in captivity, they might in fact

[11]The theory of learned helplessness was first put forward by psychologist Seligman (1972), who showed that dogs that have been given electric shocks become so passive that they fail to avoid further adverse stimuli, even if there is an easy possibility of escape. They are in effect conditioned to believe that harm is inescapable.

[12]I am assuming here that the animals' welfare is not violated in either of the three views on welfare. If it's genetic make-up is altered to the extent that it can cope well with loss of freedom, that it experiences no negative feelings from this lack of freedom, and its natural behavioral repertoire is changed so that it can carry out its 'new' natural behavior in captivity, there is no clear welfare problem.

autonomously choose such a life over a life in the wild.[13] But could animals that have been bred to live in captivity really make such a choice?

A problem is caused by the fact that different types of freedom are at play here. We can think of freedom for animals in the sense of being able to make their own decisions or have control over their lives (such as freedom to choose their own food and sexual partners) or in the sense of freedom of movement. These are instrumentally beneficial to animals if we assume that animals themselves are usually the best arbiters of what is good for them or what gives them pleasure or wellbeing. Of course, this is not always the case, as animals (and humans) sometimes make decisions that are not in fact good for them or do not lead to pleasure or wellbeing in the long run.[14] Nevertheless, wild animals that have had freedom of movement and freedom of choice have in general managed to sustain themselves in their environment and seem to lead fulfilling lives, and moreover, animals often reject attempts to take away their freedom, so it seems at least plausible to assume that freedom is beneficial to them. We can also conceptualise freedom as somehow belonging to the 'wild' state of an animal; as a characteristic of an animal which makes it wild. In the case of some of these types of freedom, it seems that the question of whether we judge captivity to be in the interest of an animal depends on whether we think we should be allowed to alter the genetic make-up of animals in the first place. After all, when we think of freedom as wildness or as having control over one's own life, altering animals' characteristics seems more problematic than when we think of freedom as simply freedom of movement and of simple decisions (see also Palmer's chapter about different conceptions of wildness in this volume). The question of whether changing the genetic make-up of animals is problematic will be the topic of the remainder of this chapter. While I do not pretend to know the answer to this complex question, I will raise some issues that need to be discussed in the context of domestication.

5 Domestication

In the previous section I pointed out that if animals cannot adapt to captive circumstances, they will be harmed by captivity. Otherwise they become domesticated and then we have to ask whether or not it is morally problematic to interfere in the genetic make-up of animals. Before we discuss issues raised by domestication, however, it should be noted that domesticated and wild animals are not clear-cut categories, but rather that we could posit animals on a continuum between wild and domesticated. There is a blurring boundary, then, between wild and domesticated

---

[13]Needless to say, there is disagreement about this. For example, Singer (1975) argues that freedom is more important to animals than the protections offered by domestication, but Budiansky (1992) takes the opposite position.

[14]Think for example of experiments with rats that could self-administer drugs and stopped eating entirely.

animals. As becomes clear in other sections in this volume, there is also a blurring boundary between wild and domesticated *freedom*: due to the fact that wild populations of animals are often too small a continuous exchange between in situ and ex situ conservation takes place. In a sense we could say that wild nature itself is becoming more and more like a zoo.

Many different definitions of both 'domesticated' and 'wild' animals have been proposed. One of the most influential definitions of animal domestication is that by Clutton-Brock (1989, 7): a domesticated animal is an animal 'bred in captivity, for purposes of economic profit to a human community that maintains complete mastery over its breeding, organisation of territory, and food supply'. This definition focuses on human intentions of domesticating animals. However, as mentioned before, animals can become domesticated unintentionally as well, when the captivity, or even the mere presence of human beings, leads to changes in the animal population. Moreover, this definition does not seem to acknowledge the gradual nature of domestication (and of the reverse process of dedomestication). An alternative definition that does take these points into account is given by Swart and Keulartz (2011) who make a distinction between wild and domesticated animals on the basis of two characteristics: the degree to which an animal has adapted to its human environment and the degree to which it is dependent on it. The more an animal has adapted and the more dependent it is on humans, the more domesticated it is. Other than the definition by Clutton-Brock this remains neutral on the human intentions by which animals were domesticated. Moreover, this definition emphasizes the fact that wildness and domesticity are matters of degree. However, while this definition remains neutral on human intentions, we should bear in mind that many of the moral issues raised by domestication focus exactly on intentional human agency.

6 The Domestication Contract

When we are considering the question whether we can change the characteristics of animals to the extent that they become domesticated, a first argument that we need to discuss is that animals have (tacitly) consented to being domesticated. A case that is often brought up is that of wolves, that started following human settlements, scavenging for left-over food, and that benefitted from contact with humans. Gradually, the characteristics that were most conducive to cohabitation with humans—most notably less fear of humans—prevailed and wolves evolved into the ancestors of today's dogs.[15] The accuracy of this theory is still debated, but whether or not this is an accurate depiction of historical reality, we can raise the question

[15]This depiction of the domestication of dogs is disputed by Coppinger and Coppinger (2001) who argue that 'the canid family tree split, and wolves and dogs went along their separate branches. The wolf displays specialized adaptation to wilderness, and the dog displays specialized adaptations to domestic life'.

whether the supposed co-evolution of humans and dogs means that dogs agreed to a type of domestication contract. One of the most influential accounts of the 'domestication contract argument' is given by Stephen Budiansky, who in *The Covenant of the Wild* has posited that animals have voluntarily domesticated themselves by choosing to associate with humans in order to gain benefits and that this has given them an enormous evolutionary advantage (Budiansky 1992). Indeed, the global population of domesticated animals is many times larger than the population of wild animals. However, when animals in a specific practice are made worse off in captivity than they would have been in the wild, according to Budiansky, human beings have broken the contract and they can no longer point to a domestication contract to justify their practices. From another angle, environmental philosopher Callicott (1992) has posited that the social relationships that have evolved between human beings and animals in a 'mixed community' could be regarded as an unspoken social contract. This contract is broken when the relationship is undermined and this is the case when animals are depersonalized and mechanized by humans, as they are for example in intensive husbandry conditions.

Palmer (1997) convincingly argues against the comparison of the relationship between humans and animals with a social contract, as animals cannot be regarded as the type of individuals that could enter into a free and equal contract with human beings.[16] They cannot give informed consent and even if they could, we could only say that the ancestors of currently living domesticated animals entered a contract voluntarily, not the current animals themselves. This, of course, is a problem of historical reality facing social contract theories in the human domain as well, and therefore the idea of 'tacit consent' has been proposed, suggesting that if someone benefits from the political structures of society he or she has tacitly agreed to them. Budiansky's account that animals chose to associate with us and thereby dropped their defence mechanisms, which ultimately led to domestication, could be read as a version of the tacit consent view. However, the historical evidence supporting this view is highly contested. Therefore, some have argued for a hypothetical contract, suggesting that if animals were to understand that domestication would offer the many benefits that it does, they would consent to it (Narveson 1983).[17] The discussion then comes back again to the question whether animals benefit from being held in captivity. In answering this question it is important to distinguish between benefits on the individual and on the species level. While it is true, as Budiansky proposes, that in terms of numbers on the species level domestication has been a successful strategy, this is not necessarily the case on the level of the individual animal. Moreover, even on the species level we can doubt whether more always

[16]It should be noted that while this is a difference between (most) humans and animals, according to most animal ethical theories this is a morally irrelevant difference; it is no adequate basis for treating animal and human interests differently (see Sect 3). After all, marginal cases cannot give informed consent as free equals either. Only for contractarian theories this would be a morally relevant difference between humans and animals, but they cannot give an adequate answer to the marginal cases argument.

[17]Cited by Palmer (1997).

equates with better. Is putting higher numbers of individuals of a species on the planet necessarily a good strategy for that species? Species flourishing does not only have to be measured in quantitative terms.

Furthermore, an underlying problem of speaking about animal domestication as a hypothetical contract is that animal domestication is in most cases irreversible—the animals are bred in such a way that most of them could not even return to a life in the wild, even if they could choose to. There is no exit clause, so to speak. Palmer drives this point home very well in the following quote:

> The nature of the 'animal contract' is such that once in, it is impossible to get out. This makes a mockery of the idea of either tacit or hypothetical consent, since if there are no alternatives consent is meaningless. It would be rather like maintaining that by being born I have given tacit or hypothetical consent to being a human being (1997, 421).

But even if we would support the argument that keeping animals in captivity could be justified by a tacit or hypothetical domestication contract, in many practices in which animals are held in captivity there are grounds for breach of contract, as either they are worse off than they would be in the wild[18] or the relationship between humans and captive animals is undermined by the way in which they are treated. Again, we will need to look at the realities of everyday practices of animal use, in order to judge whether captivity is justified.

7 The Limits of Individualist Approaches in Animal Ethics

I have noted that animal captivity often leads to domestication and, therefore, questions surrounding animal domestication arise. Animal captivity cannot be justified on the basis of a domestication contract. This is not only because animals cannot freely consent to being domesticated, but also because domesticated animals today are what they are as a result of the domestication process and it makes no sense to give hypothetical consent to being born in a certain way. It is the choices we humans have made, and the actions we have undertaken, that have intentionally or unintentionally determined the characteristics of domesticated animals. When keeping animals captive we do not only take away their liberty, but we also change them, sometimes even to the extent that they can no longer survive on their own in the wild.[19] By selective breeding and genetic modification we are at least partly creating 'what animals are like' (Palmer 2010). Domesticated animals are bred to

[18]One could wonder why 'the wild' should be the reference point for how animals should be treated. The assumption that social contract theories make is that tacitly agreeing to a hypothetical contract marks the transition from a state of nature to a state of culture and the 'wild' here refers to the situation before animals were domesticated.

[19]It should be noted that the inability to survive in the wild is often the result of the lack of an apt environment or of a careful rewilding process, and is not always irreversible. Thanks to Franck

emphasize specific capacities, such as playfulness and gentleness and avoid others, such as aggression and hunting instincts.

Moreover, by our breeding decisions we decide whether domesticated animals will exist in the first place. These latter points show what is at stake in domestication. When we are considering the question of whether animal captivity is justified, it is important to note that the sheer existence of many captive animals and their particular characteristics are determined by human beings. This means that we cannot even realistically compare the benefits and costs of captivity for domesticated animals, because without it they would not even exist, since they would be a different individual. Palmer (2011) considers the paradox that this so-called non-identity problem causes for domesticated animals: If an animal that has been bred with characteristics that harm its health and welfare, such as the English bulldog, still has a life worth living, we cannot say it was harmed, because if we had bred the animal without these characteristics it would no longer be this particular dog.

This raises a conundrum for animal ethics. Many people in society object to interfering in animals' genetic make-up. They have moral qualms about genetically engineering animals or to breeding animals with severe health problems as a side-effect, such as English Bulldogs or Ridgebacks. However, traditional approaches in animal ethics are not able to accommodate such objections very well. After all, these changes occur before an animal is born; the individual animal itself is not harmed, even if it experiences welfare problems, because if we were to create an alternative animal without welfare problems it would not be this animal anymore. The harm was done before the animal was born, so to speak. Now, it could be argued that the objections in society are not warranted, and people simply react intuitively to welfare problems without reflection on the soundness of their objection. However, similar objections could be made to the reverse situation, in which no welfare problems are present. If we label the breeding of animals with welfare problems an instance of 'disenhancement', we could label genetic interferences with animals so that they experience better welfare 'enhancement'. For example, it has been argued that we could create farm animals that do not experience pain, so that they can deal better with the intensive husbandry environment (Shriver 2009). A first response of many people is that this is upside down reasoning; shouldn't we adapt the farm environment to the animals rather than the other way around? But we need to ask why this is the case. Why would it be problematic to create such animals? Not because they experience harm to their welfare and neither because we fail to respect their inherent worth. After all, the inherent worth of this particular animal has not been disrespected by bringing it into existence with particular characteristics. The animal is what it is due to our decisions and it has inherent worth based on the characteristics it has. If we had given it different characteristics it would be a different animal.

(Footnote 19 continued)

Meijboom for pointing this out to me. Yet, in the case of pedigree dogs, they would most likely not be able to survive in the wild or be able to go back to a wild dog-like state.

In my view, the problem lies in the fact that the changes that we find objectionable take place not on the level of the individual animal, but on the level of its species or population. However, species and populations lack morally relevant capacities. Most theories about moral status of animals have the following general form: an entity belongs to the moral community if it has morally relevant capacity x. This is because if it has capacity x it has interests. All equal interests should be treated equally. When we decide how to act morally, either we should maximize the good of all those with interests (the utilitarian variety) or we should respect the rights or inherent value of those who have interests (the deontological variety). The capacities fulfilling x are usually sentience or certain mental capacities, such as preference autonomy, having desires, beliefs, memory, etc. (see also Bovenkerk and Meijboom 2012). This is because when these capacities are present, it actually matters to the entity what we do to it. In other words, a being has to have an experiential welfare before it is perceived to have morally relevant interests. Species do not have experiential welfare, however.

8 Beyond the Individual

The points made in the previous section raise the question on what level the justifiability of animal captivity should be examined. Discussions about animals in captivity tend to switch between the individual and the species level, without clearly demarcating them (see for instance Norton et al. 1995). When we are considering captivity, we contrast this not only with welfare, but also with values such as liberty, and perhaps even wildness, and authenticity. These appear to make more sense on the species level than on that of the individual level; after all, individuals do not experience their wildness or authenticity, and as discussed in Sect. 3, while freedom appears to be in animals' interest, it is unclear what *exactly* are the benefits of liberty or freedom for an individual animal. Furthermore, when we consider objections that are often raised in society against, for example, genetic engineering these make more sense on the level of species.[20] After all, the changes to these animals take place on the level of the genome and have already taken place before the animal is born. It is therefore not the individual animal that we are changing but the species (or more properly population) to which it belongs. If we create an animal with health or welfare problems, or even with enhanced welfare we cannot talk of "harms and benefits as if they were available comparable states of the same individual" (Palmer 2011, 47).[21] Moreover, justifications for animal captivity tend

[20]Macnaghten (2001) reports on research into public attitudes towards animals and biotechnology and finds that for example the argument that genetic modification is unnatural is raised. This unnaturality seems to refer more to changing a whole species than only an individual animal.

[21]We could make judgments about whether the animal's life is worth living in absolute terms, but as long as the animal's life is deemed worth living we cannot consistently object to harms done to the animal before it was born. Utilitarians could argue that from an impersonal viewpoint we could

to take place oën the level of the species as well. For example, zoos are often defended because of the role they play in wildlife conservation (Hutchins et al. 2003; see also chapter "Captivity for Conservation? Zoos at a Crossroads" by Keulartz in this volume). While it may not be in the interest of an individual tiger belonging to an endangered species to be kept in a zoo, it is in the interest of the species of tiger in case the zoo has a breeding programme and manages successfully to reintroduce tigers into the wild.

Yet, the problem is that species don't have experiential welfare and therefore they don't have interests in the normal sense of the term. In fact, the concept of species is a human construct that does not neatly correspond to an entity in the real world; there are at least 27 different ways in which species can be demarcated, leading to 27 different sets of individuals.[22] Moreover, species are dynamic, their characteristics and composition change over time. So how could we argue against interferences on the species level? Palmer (2009) considers two possible routes by which species could be said to have interests, but rejects both of these. Briefly, one could argue that species have interests if you regard them as an organism in their own right. However, this begs the question whether these are morally relevant interests. The step to moral status from these type of interests will need a separate argument and will in the end come down to the status of individual members of the species, as only they can be aware of the harms inflicted on them.[23] Secondly, we could consider that species have a type of group interest, if you could speak of groups harms. However, group harms in Palmer's view ultimately come down to the aggregated harms to the interests of individuals now and in the future of a particular species.[24]

So what basis for attributing moral considerability do we have if this is not based on experiential welfare? The idea behind taking experiential welfare as a basis is that only for beings that can have experiences it matters what we do to them.

(Footnote 21 continued)

say that for example a world with a group of dogs with welfare problems due to inbreeding is worse than a world without. However, this would only justify certain objections to interferences with animals. If enhancements would reduce the total amount of suffering in the world, a utilitarian would not object.

[22]Robert and Baylis (2003) discuss the difficulties of demarcating species boundaries.

[23]Johnson (1992), for example, argues that species have interests separate from its members, and he illustrates this by stating that while it is in the interest of a species (or more properly: population) of deer that weaker animals are killed by predators, this is obviously not in the interest of the individual animals that are killed. However, this does not show in itself that the species has a separate moral standing that is not derived from the moral standing of its members now and in the future.

[24]In chapter "The Flights of the Monarch Butterfly: Between In Situ and Ex Situ Conservation" of this volume Marcel Verweij and I actually argue that there are collective dimensions to groups that exceed that of the aggregation of all the individuals of the group. However, these dimensions do not lead us to attribute direct moral status to groups.

Environmental ethicists have also proposed non-experiential bases of welfare, but it remains contested whether these are morally relevant (see for example Rolston III 1991). In chapter "Blurring the Boundaries Between Individualistic Animal Ethics and Holistic Environmental Ethics" of this volume Marcel Verweij and I address and reject the claim that moral status could be attributed to groups in more detail. What we are left with, then, when we want to address questions about captivity and domestication on the level of animal species, is to rely on an *indirect* account of the moral standing of species. What could such an indirect account be based on? First of all, it could have an aesthetic basis. This basis is necessarily limited, however, because what is considered aesthetically pleasing is subjective, or at best inter-subjectively shared by certain groups. An aesthetic basis for moral standing cannot explain objections to interferences with animal species, because perhaps these interferences are aesthetically pleasing. We should bear in mind that aesthetics have played a big part in creating the kind of breeds of domesticated animals that we have today.

An alternative basis could be the following: if we consider ourselves inextricably connected to a nature of which other animal species are a part, we cannot really separate our own species from that of other animals. If we would take a type of biocentric outlook in which we are only part of the wider web of interactions that constitute natural entities, we would consider ourselves as but one species between the species.[25] This means that if we have reason not to want to change our own species, we should pause and think about why we would change the characteristics of other species. This type of argument refers not so much to characteristics of a species that grant the species standing, but rather to the relations we have with species other than our own. It is about how our interferences with other species reflect back on ourselves. Perhaps it also appeals to a virtue-ethical account of what sort of attitudes we are allowed to have. What are we permitted to will by changing other natural entities in this world? For example, should we even will to create farm animals that can deal with overcrowded, factory like environments? This is not a question about harm to the welfare of individual animals, but rather about the role we give ourselves as a species within the rest of nature. This alternative view about dependent, rather than independent, moral standing is of course still rather tentative. Whether this basis would successfully be able to deal with considerations regarding captivity and domestication, is the subject of further research. It should be noted however, that such a view would supplement rather than substitute traditional animal ethical theories.

[25]What I am suggesting here is a type of biocentric outlook as Taylor (2011) has proposed. However, Taylor only considers individual natural entities to be morally relevant and my proposal would see ourselves not only as part of a group of individuals in nature, but as but one species between the species.

9 Conclusion

The central question of this chapter was what justifications we have for keeping animals in captivity. I started by discussing inherent objections to animal captivity and concluded that the objections that captivity commodifies animals or turns them into property were problematic. I suggested that animals may have an intrinsic interest in liberty, however, if we accept a less demanding conception of animal autonomy. Still, the interest of animals in liberty need to be weighed against the possible benefits that animal captivity brings. Therefore, I addressed two general justifications for captivity: (1) it is in the interest of humans to hold (other) animals in captivity, (2) it is in the interest of animals themselves. It is at least dubious to hold (according to the first argument) that human interests by definition outweigh animal interests. Whether it is in the interest of animals themselves to be held in captivity and whether human interests in particular cases outweigh those of animals is largely an empirical question that can differ from practice to practice, but I have at least raised doubts about them. In general, we could say that either animals do not possess sufficient adaptive capacity to be able to deal with captive conditions and in this case their welfare is bound to be harmed unacceptably. Or animals do possess adaptive capacity, in which case their genetic make-up changes and they ultimately become domesticated. I have addressed and rejected the argument that animals have in a sense consented to being held in captivity and to being domesticated, through a (tacit) domestication contract.

Next, I have pointed out that many people raise objections to interfering with an animal's genetic make-up, which happens in the context of domestication, for example through genetic engineering or breeding. However, traditional approaches in animal ethics cannot easily accommodate such objections. This is because their unit of concern is the individual animal that can have experiential welfare, and therefore has interests. Interferences with an animal's make-up—either as enhancement or as disenhancement—take place before an animal is born and as long as we could say that the animal still has a life worth living animal ethicists cannot object to them. After all, there is no comparable state of the same animal without these interferences. The animal is what it is at least partly as a result of our choices. This is a conundrum resulting from domestication, as the process of domestication takes place on the level of species. Yet, it is difficult to argue that we have harmed a species, as species are not the kind of entities that can experience harm. Besides welfare considerations, I have suggested that animal captivity and domestication also touch on other values, such as liberty, authenticity and wildness. These values tend to play a role more on the level of species than on the level of individual animals. Finally, I have suggested that if we want to accommodate these considerations, we may need to grant species a type of dependent moral standing, rather than independent moral status. What exactly such an argument for dependent moral standing would look like is the topic for further research.

Acknowledgments I would like to thank Franck Meijboom, Frederike Kaldewaij, Aaron Simmons, and Joel Anderson for their helpful comments.

References

Bovenkerk, B., and F.L.B. Meijboom. 2012. The moral status of fish. The importance and limitations of a fundamental discussion for practical ethical questions in fish farming. *Journal of Agricultural and Environmental Ethics* 25(6): 843–860.

Bovenkerk, B., and F.L.B. Meijboom. 2013. Fish welfare in aquaculture: explicating the chain of interactions between science and ethics. *Journal of Agricultural and Environmental Ethics* 26 (1): 41–61.

Budiansky, S. 1992. *The covenant of the wild: why animals chose domestication: with a new preface*. London: Yale University Press.

Callicott, J.B. 1992. Animal rights and environmental ethics: back together again. In *The animal rights/environmental ethics debate: the environmental perspective*, ed. E. Hargrove, 249–261. Albany, NY: SUNY Press.

Carruthers, P. 1992. *The animals issue*. Cambridge: Cambridge University Press.

Carruthers, P. 2005. *Consciousness: essays from a higher-order perspective: essays from a higher-order perspective*. Oxford: Oxford University Press.

Clutton-Brock, J. 1989. *The walking larder: patterns of domestication, pastoralism, and predation*. London: Unwin Hyman Inc.

Cochrane, A. 2009. Do animals have an interest in liberty? *Political Studies* 57: 660–679.

Cochrane, A. 2012. *Animal rights without liberation*. New York: Columbia University Press.

Cohen, C. 1986. The case for the use of animals in biomedical research. *The New England Journal of Medicine* 315(14): 865–870.

Coppinger, R., and L. Coppinger. 2001. *Dogs: a startling new understanding of canine origin, behavior and evolution*. New York: Simon and Schuster.

Dombrowski, D.A. 1997. *Babies and beasts: the argument from marginal cases*. Urbana/Chicago: University of Illinois Press.

Feinberg, J. 1984. *The moral limits of the criminal law, volume 1: harm to others*. New York: Oxford University Press.

Fiby, M. 2012. The future of wild animals in 50 to 100 years. Future of zoos symposium. Buffalo, NY: Canisius College.

Francione, G. 2010. *Introduction to animal rights: your child or the dog?*. Philadelphia, PA: Temple University Press.

Fraser, D. 2003. Assessing animal welfare at the farm and group level: the interplay of science and values. *Animal Welfare* 12(4): 433–443.

Frey, R.G. 1988. Moral standing, the value of lives, and speciesism. *Between the Species* 4(3): 191–201.

Greene, J.D. 2007. *The secret joke of Kant's soul*. Cambridge, MA: The MIT Press.

Gruen, L. 2009. *Ethics and animals. An introduction*. Cambridge: Cambridge University Press.

Haynes, R.P. 2008. *Animal welfare: competing conceptions and their ethical implications*. Dordrecht: Springer.

Hribal, J. 2011. *Fear of the animal planet. The hidden history of animal resistance*. Oakland, CA: AK Press.

Hutchins, M., B. Smith, and R. Allard. 2003. In defense of zoos and aquariums: the ethical basis for keeping wild animals in captivity. *Journal of the American Veterinary Medical Association* 223(7): 958–966.

Johnson, L.E. 1992. Toward the moral considerability of species and ecosystems. *Environmental Ethics* 14(2): 145–157.

Kaldewaij, F.E. 2013. The animal in morality. Justifying duties to animals in Kantian moral philosophy. PhD Dissertation, Utrecht University, Utrecht.

Macnaghten, P. 2001. *Animal futures. Public attitudes and sensisbilities towards animals and biotechnology in contemporary Britain.* Report by the Institute for Environmental, Philosophy and Public Policy for the Agricultural and Environmental Biotechnology Commission.

Narveson, J. 1983. Animal rights revisited. In *Ethics and animals*, ed. H.B. Miller, and W.H. Williams, 45–60. Clifton, NJ: Humana Press.

Noddings, N. 1984. *Caring: a feminine approach to ethics and moral education.* Berkeley and Los Angeles: University of California Press.

Norton, B., M. Hutchins, E. Stevens, and T. Maple (eds.). 1995. *Ethics on the ark: zoos, animal welfare, and wildlife conservation.* Washington, DC: Smithsonian Institution Press.

Palmer, C. 1997. The idea of the domesticated animal contract. *Environmental Values* 6(4): 411–425.

Palmer, C. 2009. Harm to species-species, ethics, and climate change: the case of the polar bear. *Notre Dame Journal of Law Ethics & Public Policy* 23: 587–603.

Palmer, C. 2010. *Animal ethics in context.* New York: Columbia University Press.

Palmer, C. 2011. Animal disenhancement and the non-identity problem: a response to Thompson. *NanoEthics* 5(1): 43–48.

Regan, T. 1983. *The Case for Animal Rights.* Berkeley: University of California Press.

Regan, T. 1995. Are zoos morally defensible. In *Ethics on the ark: zoos, animal welfare, and wildlife conservation*, ed. B. Norton, et al., 38–51. Washington, DC: Smithsonian Institution Press.

Robert, J.S., and F. Baylis. 2003. Crossing species boundaries. *American Journal of Bioethics* 3 (3): 1–13.

Rolston III, H. 1991. Environmental ethics: values in and duties to the natural world. In *Ecology, economics, ethics: the broken circle*, ed. F.H. Borman, and S.R. Kellert, 73–96. New Haven, CT: Yale University Press.

Scruton, R. 2000. *Animal rights and wrongs.* London. Demos.

Seligman, M. 1972. Learned helplessness. *Annual Review of Medicine* 23(1): 407–412.

Shriver, A. 2009. Knocking out pain in livestock: can technology succeed where morality has stalled? *Neuroethics* 2(3): 115–124.

Singer, P. 1975. *Animal liberation.* New York: Avon Books.

Singer, P. 2011. *Practical ethics.* Cambridge: Cambridge University Press.

Swart, J.A.A., and J. Keulartz. 2011. Wild animals in our backyard. A contextual approach to the intrinsic value of animals. *Acta Biotheoretica* 59(2): 185–200.

Tanner, J. 2006. Marginal humans, the argument from kinds and the similarity argument. *Facta Universitatis - Series Philosophy, Sociology and Psychology* 5(1): 47–63.

Tanner, J. 2009. The argument from marginal cases and the slippery slope objection. *Environmental Values* 18: 51–66.

Taylor, P.W. 2011. *Respect for nature: a theory of environmental ethics.* Princeton, NJ: Princeton University Press.

VanDeVeer, D. 1979. Interspecific justice. *Inquiry* 22(1–4): 55–79.

Wells, D.L. 2005. A note on the influence of visitors on the behaviour and welfare of zoo-housed gorillas. *Applied Animal Behaviour Science* 93(1): 13–17.

Care for the Wild in the Anthropocene

Jac. A.A. Swart

Abstract Animal ethical approaches often focus on certain individual animal features and capabilities for attributing moral standing to them. These features are usually considered from a moral point of view as not differing for wild, semi-wild, and domesticated animals. However, several authors have argued for more relational approaches, in which relationships between humans, human society, and animals are taken into account, implying that wildness may be considered, in a sense, as a morally relevant aspect. This approach is especially relevant in the Anthropocene, since this new geological epoch is characterized by a significant impact on the part of human society on global geological and ecological systems, and thus on many wild and semi-wild animals. In this chapter some conceptual approaches to domestication and wildness are discussed, and it is argued that we should consider wild animals as entities that are highly and critically dependent on the environment, which should be considered as a network of biotic and abiotic elements, whether that environment is natural or human. Accordingly, it is argued that we need a contextual care approach, as an environmental virtue ethics, implying an attitude of care for the threatened natural environment of wild animals in the Anthropocene.

1 Introduction

Animal ethics traditionally focuses on certain features or abilities of animals, when it comes to attributing moral standing to them. For example, according to utilitarian reasoning, sentience, and especially ability to suffer, is necessary and sufficient for moral standing. However, moral standing, based on sentience, is only a precondition here: It is the suffering itself that really counts (Singer 1990). Rights-based approaches also focus on sentience, or, better yet, on being a subject-of-a-life, implying an intrinsic value or inherent value for animals, and therefore a moral

Jac.A.A. Swart (✉)
University of Groningen, Groningen, Netherlands
e-mail: j.a.a.swart@rug.nl

B. Bovenkerk and J. Keulartz (eds.), *Animal Ethics in the Age of Humans*,
The International Library of Environmental, Agricultural and Food Ethics 23,
DOI 10.1007/978-3-319-44206-8_11

right; this encompasses not only a right not to be harmed but also to be treated in a respectful way (Regan 1983). Furthermore, the more recently proposed capability approach (Nussbaum 2006, 365) stresses certain animal abilities, needed to flourish in the context of the species to which it belongs.

These approaches do not morally distinguish between wild and domesticated animals, because it is assumed that wild and domesticated animals do not significantly differ with respect to these morally relevant features. However, in practice, domesticated animals such as, for example, companion animals and cattle are, as a matter of fact, much more subjects of human care compared to wild animals. This difference is usually based on the notion that wild animals are probably best off by leaving them on their own, detached from human inference: "Let them be" (Regan 1983, 361). Palmer (2010, 2) labels this approach as the laissez-faire intuition (LFI). Furthermore, the capabilities approach acknowledges that flourishing may result in quite different human roles when it comes to treating wild and domesticated animals: "Part of what it is to flourish, for a creature, is to settle certain important matters on its own, without human intervention, even of a benevolent sort" (Nussbaum 2006, 373). However, both Palmer and Nussbaum argue that there are limitations to the requirement not to interfere in the wild animal's life, since human beings affect the habitats of wild animals strongly, implying a responsibility to bring about conditions for their well-being and/or flourishing.

The latter issue is very relevant in the Anthropocene, that is, the current epoch that is distinguished from the Holocene by the appearance of human society as a geophysical force on a global scale. Climate change is one of the clearest indications of this, but there are many more phenomena such as the appearance of floods, loss of tropical woodland, reduced biodiversity, disturbed ecological food chains, etc. (Crutzen 2002; Steffen et al. 2007, 2011). The rise of the Anthropocene not only threatens the provision of environmental goods and services, it also threatens the living conditions of many wild animals. Their territories are increasingly fragmented, contaminated, and disturbed by transport activities, tourism, fishery, urbanization, climate change, and agriculture. These developments affect their access to grazing areas, water stores, shelter possibilities, migration routes, etc. Worldwide, populations of almost 30 % of vertebrate species have declined (WWF 2010).

This not only challenges the classic LFI approach but it also blurs the demarcation between wild and domesticated animals, as more and more wild species are affected by and have to adapt to these Anthropocenic conditions. Wild animals are in a sense involved in a process of Anthropocenic domestication. In this paper, some conceptual approaches to domestication and wildness are discussed, and it is argued that we should consider wild animals as entities that are highly and critically dependent on the environment, whether that environment is natural or human. Accordingly, discussing recent contributions by Palmer (2010), and Donaldson and Kymlicka (2011), a virtue ethics-based contextual care approach, implying an attitude of care for the threatened natural environment of wild animals in the Anthropocene, is advocated.

2 Domestication

According to Dutch law (Staatsblad 2002), a wild species is a species that naturally lives in the wild. This rather general view seems to be in accordance with the more detailed description by Palmer (2010), who considers wildness in the sense of the constitutive (in the sense of bodily) and locational (or physical) distance between humans and animals. Constitutive wildness refers to animals that are not domesticated in the sense of not being intentionally bred and controlled by humans (idem, 63–65). Locational wildness refers to the wild-urban spectrum where a wild animal is considered to live far away from human settlements.[1]

On the other hand, many authors argue that the essential dimension of domestication is the development of particular relationships between humans and animals, which may consequently affect the animal's genetic constitution, resulting in the appearance of certain morphological and behavioral traits such as, for example, depigmentation, shorter muzzles, docility, floppy ears, curly tails, smaller brains, juvenile behavior, etc. (Wilkins et al. 2014).

In a review, Russel (2002) distinguishes dominant perceptions concerning the qualification of the human-domesticated animal relationship: a controlling and a symbiotic relationship. According to the 'control approach' (my term) the human-animal relationship is considered to be suppressive: Animals are forced into the human system. For example, Clutton-Brock (1981, 21) characterizes a domesticated animal as an animal "… that has been bred in captivity for purposes of economic profit to a human community that maintains complete mastery over its breeding, organization of territory, and food supply". Elsewhere she states that domestication should be considered as a biological and cultural process, through which "tamed animals are incorporated into the social structure of the human group and become objects of ownership" (Clutton-Brock, 1989, 7). Domestication is here seen as a practice, in which the animal is considered as an instrument or an economic good for human benefit.

However, O'Connor (1997) argues that the control approach assumes a wild-domestication dichotomy, which is inadequate for describing the different relationships between humans and wild animals that can be found in reality. He applies a symbiotic approach, making use of biological concepts to describe the wide range of interactions between different species. For example, mutualism implies that both species receive benefits from their interactions, whereas competition may lead to negative consequences for both. In addition, we find modes of interaction, where only one of the species is affected positively (commensalism) or negatively (amensalism). Finally, he distinguishes contramensalism, where the interaction benefits one species and harms the other, as is the case with predation and parasitism.

[1]Palmer (2010, 63) also distinguishes dispositional/behavioral wildness, which refers to the wild–tame spectrum and especially to the animal's behavior: A tamed animal shows non-aggressive behavior toward humans (or its owner). This aspect is, however, set-aside in her approach.

According to O'Connor, these different relationships between species can also be applied to qualify animal domestication. From this perspective, the control approach may be considered as a form of contramensalism, as it attributes a passive role to the animal in the process of domestication: Animals are removed from their natural geographical range, controlled in their breeding, and embedded in human society. However, O'Connor interprets domestication historically rather as a mutualistic and commensal relationship, in which both humans and animals actively operate. In this view animal domestication is "a form of behavioral coevolution by which mutualistic and commensal relationships developed between people and other species, either because it was to the benefit of both species, or because it was beneficial to one and neutral (or at least not strongly detrimental) to the other" (1997, 153–154). According to this view domestication should be seen as a gradual phenomenon where some species have—so to say—chosen to live in or closely to human settlements, for example, rat and mice species, garden birds, and feral animals.,[2,3,4] Wild animals are thus considered, according to this view, as rather opportunistic entities. They just make use of the resources available. Migration, colonization, and the establishment of new species assemblages just happen in nature.

3 Rewilding

The symbiotic perspective on domestication may also imply that the process of domestication can be turned around, and may lead to the 'rewilding' of formerly domesticated animals. The term rewilding was coined by David Foreman, who established the Wildlands project (Foreman et al. 1995) in the United States. The aim of this movement is to restore natural processes and wilderness areas, along with the preservation and reintroduction of wild fauna.

In this context, the case of the Oostvaardersplassen (OVP) in the Netherlands is interesting (Vera 2009; Kolbert 2012; Lorimer and Driessen 2014). This natural area of around 5600 ha was developed in the 1980s after a large-scale reclamation project in the eastern part of the IJsselmeer, a freshwater lake close to Amsterdam.[5]

[2]Feral animals are animals that have gone wild after being domesticated.

[3]Thus, according to the symbiotic view, humans did not intentionally domesticate animals in ancient times: Instead, animals became adapted to a human environment because of reproductive benefits. Domestication is considered here as the animal's answer to the rapidly changing environment in the late Pleistocene and early Holocene (O'Connor, forthcoming).

[4]However, we cannot consider all forms of domestication from such a coevolutionary perspective, since it is clear that modern breeding technologies, such as, e.g., artificial insemination and embryo selection (Swart 2014), cannot be considered as examples of the animal's answer to a changing environment.

[5]The IJsselmeer was a result of the closure of the Zuiderzee by a sea wall in the early 20th century. Its original surface was 590,000 ha. However reclamation and diking projects reduced and divided the lake into two lakes (IJsselmeer and Markermeer), with a total surface area of 180,000 ha now.

In the 1980s and 1990s, small herds of Konik horses (27 animals), Heck cattle (35 animals), and red deer (54 animals), were introduced into the area. The Konik horse is a Polish breed descended from the extinct European Tarpan. The Heck cattle are a result of a German project in the early 1930s to breed back the extinct aurochs. Although both the Konik horses and Heck cattle were bred lines, the animals were considered to be able to survive under natural circumstances.

By introducing these animals, a natural process of ungulate grazing could be realized in this 'new wilderness,' as the area is often called. These small animal populations have increased exponentially since their introduction but have stabilized more or less in the last 10 years. Currently (2014–2015) there are around 2750 red deer, 950 Konik horses and around 200 Heck cattle (Staatsbosbeheer 2015). In the meantime, many other wild animals have found their own way to this area: geese, foxes, buzzards, muskrats, goshawks, gray herons, kingfishers, and white-tailed eagles. The area is now recognized as an important protected nature reserve (in Dutch: Staatsnatuurmonument) and is characterized by extensive management, where the introduced animals are considered to be wild animals.

The populations of the animals in the OVP are primarily regulated by natural food availability affecting the reproduction and mortality rates of the animals. Only under extraordinary circumstances is additional feeding provided. In late winter periods, food availability can sometimes be so low that a double-digit percentage of the population dies.[6] Such mortality rates of large herbivores, which are regulated by bottom-up regulation (primary production, water quality, etc.) (Sinclair et al. 2003; Hopcraft et al. 2010) can also be found elsewhere in nature (ICMO 2006). Although the case of the OVP may be considered as a great success from a rewilding perspective, it has also met with much criticism from the public, the Dutch Animal Protection Society, animal ethicists, and veterinarians, since they consider the introduced animals still as kept animals, living, after all, in a fenced area.

The widespread deaths from starvation that happen from time to time—often broadcast by modern public media—especially give rise to societal protests. However, in a lawsuit by the Dutch Animal Protection Society against Staatsbosbeheer, the agency managing the OVP, the Court concluded that it was reasonable that these animals could no longer be considered as kept animals (Court verdict 2007). According to the verdict, the Dutch state had lost its ownership when it introduced the animals. In addition, the subsequent process of rewilding of the animals was considered a reason supporting the loss of the state's ownership of the animals. Nevertheless, according to the critics, humans still have a responsibility to take care of these animals. The controversy has ultimately lead to a practice of

[6]Most of the animals do not die from starvation but are killed by early-reactive culling, as a response to public concern concerning the welfare aspects (Staatsbosbeheer 2013).

management, in which the animals are monitored and culled, if they are seriously suffering and/or it is expected that they will not survive a winter period of food shortage.[7]

4 Human-Animal Relationships

The rise of the Anthropocene blurs the difference between the wild and the human world, and stresses the role of the human-animal relationship and interactions, also in wild areas. Relational approaches may therefore be increasingly relevant. A number of authors consider relationships between humans and animals as additional grounds for the moral standing of animals and thus human responsibility in addition to the animal's traits and capacities. For example, Palmer (2011, 713) states that if animals "have deliberately been made vulnerable or dependent, whether by domestication, captivity, or serious prior harm such as habitat destruction, 'we' should compensate, protect, care for or assist those animals in ways that relieve the burdens 'we' have created." Thus, according to this view we have no special obligations toward wild animals that have not been affected by us (according to the LFI), but we do have them towards those animals that have been made vulnerable and dependent on us such as, for example, domesticated animals. This also implies that we have special obligations to wild animals as a compensatory justice for past harm caused by humans. Palmer (idem.) describes the case of wild coyotes that are seriously hurt by losing their territory to housing developments. Since these animals are considered as constitutively wild, we should compensate them in one or another (e.g., by habitat restoration). This obligation not only applies to the housing developers but, according to Palmer, also to some extent to people that share in the benefits of house construction, for example, residents.

On the other hand, according to the approach of Palmer (2010, 2011), we do not have special obligations to truly wild animals, that is, animals that are physically and bodily detached from humans: They are not domesticated and live outside human settlements. Thus we do not have the obligation to help truly wild animals that have run into trouble through natural disasters such as flooding, storm, or fire. This position can be recognized, for example, in the Dutch governmental guidelines for seals stranded on the beach, stating "that people should respect the inherent individuality of animals living in the wild. This means that interventions involving the intrinsic value of the animal can only be permitted in extraordinary circumstances. Hospitalization of animals that are ill, wounded, or otherwise in need is an intervention affecting their intrinsic value. After all, the inherent individuality of an ill, wounded, or dying animal living in the wild must also be respected" (LNV 2003, 7, my translation).

[7]According to the management plan, the aim is to cull 90 % of the animals requiring culling, while they are still capable of standing. However, recent reports show that this objective is not achieved every year (Beheeradviescommissie Oostvaarderplassen 2014).

Thus, we have, according to Palmer, no obligations towards truly wild animals but special obligations towards wild animals that suffer from human actions, even if we are not causally involved but nevertheless profit from past harm to them.[8] In a response to a critique by Weber (2015) that we have only voluntary obligations in such cases, Palmer argues (2015) that special obligations also arise for community members sharing the attitudes that have led to animal harm, even if they were not actually involved and have not agreed to accept such an obligation. Considering the rise of the Anthropocene through the emission of greenhouse gases with its benefits for so many people (car driving, house warming), this reasoning would imply special obligations towards the animals that are negatively affected by it. However, we may wonder if the concept of special obligations is applicable in the Anthropocene, where the causal link between cause and effect is so unclear and indirect. This issue is also recognized by Palmer (idem, 702–703) in her response to Weber and accordingly she suggests political activity in order to change political and legal frameworks.

Such a political move is also suggested by Donaldson and Kymlicka, who consider relationships between humans and animals in terms of political theory. They reject suppressing forms of domestication, but consider the existence of domesticated animals in our society as a matter of fact and argue that they are members of a shared community of sentient animals and humans. Referring to recent insights on the citizenship rights of mentally disabled people, they argue for citizenship and political rights for domesticated animals: "The citizenship status of animals—just as in the human case—is determined not by their cognitive capacities, but by the nature of their relationships to a particular bounded political community" (Donalson and Kymlicka 2011, 61). This implies, according to the authors, that domesticated animals should have the right of residency, recognition of their subjective good, and political agency (e.g., by representatives).

By contrast, wild animals are not and should not be considered as members of our shared community and cannot be entitled to such citizenship. Rather, the authors consider these animals as members of sovereign communities that should not be disturbed: "Respect for animal sovereignty (the right of wild animal communities to lead autonomous, self-directed lives) places a strong check on human activity and on interventions in the wild" (idem, 205). According to the authors, this vision implies that wild animals communities "cannot be invaded, colonized, or robbed by others" (idem.) and that we should keep our hands off their territories: "Wild animals have legitimate interest in maintaining their social organization on their territory, they are vulnerable to the injustice of having alien rules imposed on them and their territory, and sovereignty is an appropriate tool for protecting that interest against vulnerability to injustice" (idem, 174). This does not, however, imply a total ban on assistance: "Some forms of positive care are consistent with

[8]The concept of special obligations refers to obligations "to those with whom we share a special relationship, understood as placing the demands and interests of one subset or group of persons above those outside such relationships" (Murphy 2012).

respect for sovereignty. A duty to assistance can be triggered by natural disasters which undermine the viability of sovereign animal communities (and which we are in a position to relieve), or it can be triggered by external threats to wild animal communities from a destructive invader (e.g., a rogue bacterium, or giant meteor, not to mention human invaders)" (idem, 206).

In addition to wild and domesticated animals, Donaldson and Kymlicka (2011) distinguish so-called liminal animals: Animals that live in or close to human settlement but that are not considered as being domesticated. This is a very heterogeneous group of species consisting of, for example, park geese, garden birds, feral animals. These species lack, according to the authors, their own territory. They do not belong to the shared community of humans and animals, and cannot be entitled to citizenship, nor do they have wild animal sovereignty, since they are nevertheless dependent on human settlements. But, like human denizens and migrants, they have a basic right to be present in our society, to travel, and to basic forms of assistance in order to maintain their autonomy.

Another relational approach is suggested by Mark Coeckelbergh. According to this author, current dominant approaches of animal ethics are trapped in Platonic and Cartesian reasoning by individual property thinking. According to the author, this type of reasoning neglects the myriad relationships that constitute human and nonhuman animals. Such thinking implies—just as happens in laboratories—that animals "are taken out of their usual context in order to determine their properties, their essence." Accordingly he states: "The modern scientist, who forces nature to reveal herself, is now accompanied by the moral scientist, who forces the entity to reveal its true moral status" (Coeckelbergh 2012, 17). Instead, Coeckelbergh considers relationships as the basic condition for moral standing. Relationships should, however, according to this author, not be seen as properties but rather as an a priori given, in which we are already engaged, making possible the ascription of moral status to entities. Nevertheless, Coeckelbergh also cannot escape property thinking himself, when he distinguishes humans and animals by referring to cognitive and emotional features such as having the feeling of hope (idem, 131).

5 A Contextual Care Approach

The approaches of Palmer, and of Donaldson and Kymlicka, focus especially on relationships and interaction between humans and animals. The approach by Coeckelbergh (2012) focuses much more on a network of constituting relationships covering both the cultural and natural sphere. In this paper, I will elaborate on the concept of constituting relationships, based on earlier work by Jozef Keulartz and myself, in which a care approach is proposed (Swart 2005; Swart and Keulartz 2011; Keulartz and Swart 2012).

According to the approach outlined in this chapter, the concept of relationships should not be restricted to human-animal relationships alone but also to animal-environment relationships and interactions, whether in a natural or human societal domain. With respect to wild animals, we may think of kinship relationships, herd-individual interactions, food-web relationships (between vegetation and herbivores, and between prey and predators), parasitism, and pathogens, etc. Wild animals should be seen as nodes in a challenging and complex dependency network of abiotic and biotic factors. Imagine the environmental challenges for a knot (*Calidris caudatus*) that breeds in the Northern hemisphere. It flies back in the late summer or early autumn to the Southern hemisphere and returns in spring to its breeding place. The bodily storage of fat, the availability of resting and foraging places on the flyway, the danger of being predated upon or shot during the trip, the different weather and climate conditions it will meet—these are all critical, sometimes deadly factors in its life. Only animals that are able to cope with the heterogeneous natural environment can survive. Now, if we care for wild animals we should also care for the natural environment on which wild animals are so highly dependent. Following Fisher and Tronto, caring can be considered

> as a species activity that includes everything that we do to maintain, continue and repair our 'world' so that we can live in it as well as possible. That world includes our bodies, our selves and our environment, all of which we seek to interweave in a complex life-sustaining web (cited in Tronto 1993, 103).

The proposed approach in this chapter may be characterized as a *contextual care approach*, since it considers taking care of interactions or relationships that animals have with the natural and/or non-natural environment as a pivotal point of view. This contextual care approach extends care beyond personal spheres, and covers not only personal relationships between humans and animals but also dependency relationships of animals with their natural environment. According to this perspective, the analytical unit of moral consideration is thus the animal in relation to the environment on which it is dependent. Furthermore, recognizing the animal's intrinsic value, the relationships on which it is dependent are inseparable and constitutive for the animals, and should also be considered as intrinsically valuable. Thus abandoning a dog in a forest because one cannot take it along on a vacation trip is not only wrong because of its cruelty, but also because it demonstrates a lack of respect for the animal's constitutive relationship with the human community. Similarly, capturing a wild animal is wrong, because the animal loses its relationships with the natural world on which it depends, and the natural world is replaced by an environment that the animal is not adapted to and is not used to.

In this context human responsibility is increasing, as we increasingly affect the natural environment and the wild animals living there. It implies 'non-specific care,' that is, care for the animal's natural environment, not specifically tailored to the individual traits or circumstances of the wild animal but rather to those ecosystem elements that will contribute to the survival, subsistence, and natural flourishing of the animal. Building eco-viaducts, restoring migration routes, reducing human disturbance, recovering vegetation are all examples of non-specific care measures. They often fit in with practices of preservation, conservation, and ecological restoration (Swart 2005; Swart and Keulartz 2011). However, wild animals are not the only creatures dependent on their environment. This is also true for domesticated animals. For them, the environment is the human society on which they are highly dependent, for example, for food, shelter, reproduction, etc. Research animals, zoo animals, cattle, etc., are all highly dependent in this way on human society and are in this respect domesticated. Thus, in contrast to wild animals, humans should give specific care to these animals: care that is adjusted to their species-specific features. Thus, it is not that wild animals should be provided with less care as compared to domesticated ones, but rather that the type of care that should be provided to these animals differs, that is, specific or non-specific care, depending on the context and the sort of relationships they are involved in. In addition to specific care, individual care may also be distinguished. This refers to the required care for animals with very personal relationships with humans, for example, companion animals (Swart and Keulartz 2011; Keulartz and Swart 2012).

Considering domestication as a gradual phenomenon (Klaver et al. 2002) we can also distinguish semi-wild animals: animals having a position between fully wild and being domesticated. These animals are moderately dependent on the human environment. This is a very heterogeneous group consisting of animals that are called liminal animals by Donaldson and Kymlicka (2011), or as animals living in the ‘contact zone’ by Palmer (2010). Humans should give moderate and mixed levels of specific and non-specific care to these animals; see Fig. 1. As each group of animal species in Fig. 1 is rather heterogeneous, the actual form of care may differ for different species and situations.

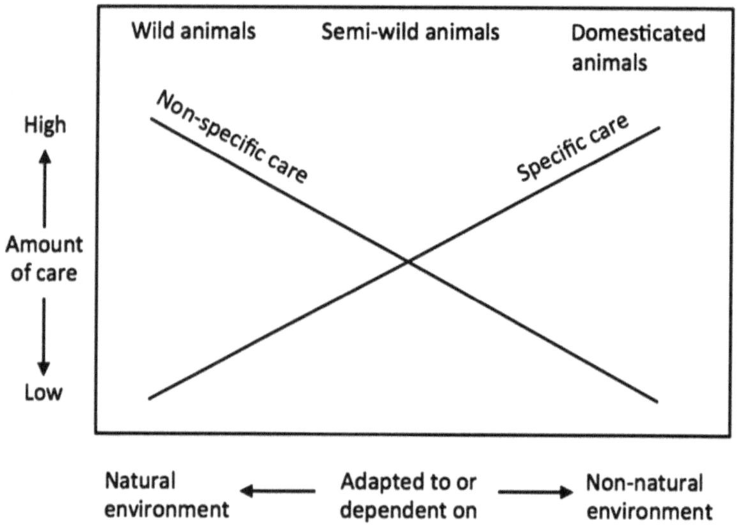

Fig. 1 The model of specific and non-specific care. The amount of care we should provide to animals is similar for different groups of animals. However, the type of care relates to the contexts, which they are adapted to or dependent on. After Swart and Keulartz (2011), ICMO2 (2010), and RDA (2012)

The model in Fig. 1 makes the controversy over the introduced animals in the OVP, described above, more understandable: Proponents of the rewilding project place the introduced animals somewhere in the left region, assuming that they are fairly able to adapt themselves to the natural circumstances where they can flourish as wild animals. Critics, however, consider these animals still as domesticated and vulnerable animals, and place them somewhere on right side of Fig. 1.

The contextual care approach should not be considered as a replacement for the capacity approaches in more traditional animal ethical circles, as was suggested by Coeckelbergh (2012). Rather, it is an additional ground for morally respectful behavior, as it acknowledges the role of relationships for the animal. Thus sentience or the capacity to flourish are still grounds for moral standing, implying a particular care towards the animals given the relationship that it is dependent on. As a consequence, we may distinguish different interpretations of intrinsic value: Naturalistic, species specific, and more individualistic interpretations of an animal's intrinsic value can be distinguished going from left to the right in Fig. 1 (Swart and Keulartz 2011). Respecting the relationships of an animal in its environment implies that moving animals along the horizontal axis in Fig. 1, through rewilding projects, should remain within the range of conditions that animals can still flourish under. It does not forbid rewilding, but such projects should be accompanied by measures that aim to keep harm and discomfort within natural or societally accepted boundaries and to give the animals the opportunity and time to establish new relationships.

The model of specific and non-specific care was adopted as a background philosophy by the Second International Commission on Management of the OVP (ICMO2 2010). The task of this commission was to evaluate the management in the OVP, as a reaction to public concern about the well-being of the introduced animals. The committee concluded that these animals are indeed to be considered as wild but that they were also limited in finding shelter or their ability to migrate to areas with shelter and/or a greater diversity of food types. According to the Committee these animals have an intermediate position in Fig. 1. Consequently, a number of measures, such as the construction of artificial shelter ridges, early reactive culling, and connections to other natural areas, were suggested.

The Dutch Council on Animal Affairs, an independent advisory committee to the Dutch government on animal affairs, consisting of experts from a wide spectrum of societal stakeholders, also refers to the proposed contextual care approach in developing its Assessment Model for duty of care and (un)acceptable suffering (RDA 2012).

6 Why Should We Care?

The contextual care approach differs conceptually from the political rights approach advocated by Donaldson and Kymlicka (2011), who place individual animals in a political and institutional framework as co-citizens, co-residents, denizens, nomads, or foreigners in order to derive animal rights. These authors consider wild animals in this context as foreign and autonomous beings living in their own territories, which we should respect and where we should keep our hands off, as in the LPI approach described by Palmer (2010). However, we may question the concept of autonomy for wild and semi-wild animals, since they are—in my view—intrinsically related to and highly dependent on biotic and abiotic living conditions in their (semi)natural areas for their survival and subsistence.

Furthermore, the concept of territories should be carefully considered in order to avoid the anthropomorphic trap of identifying biological territories as human-political territories. For example, wild animals are not per se anchored in a local context or territory but may make use of several locally distanced natural and non-natural areas. Many mammals, birds, and fish species migrate over long distances and do not limit themselves to wild areas for their rest, feed, and reproduction. Often they also make use of cultivated areas. For example, the black-tailed godwit (*Limosa limosa*) breeds on Frisian pastures in the summer and can be found in wintertime in Spanish and African rice fields. Moreover, when agriculture changes, the godwits adapt their migration (Márquez-Ferrando et al. 2014). These birds simply make use of the available natural or agricultural resources in accordance with the symbiotic interpretation of domestication described above. The concept of politically defined sovereign territories of humans and wild animals, as proposed by Donaldson and Kymlicka (2011), seems therefore rather problematic, especially in the Anthropocene with its blurring demarcation of wild and domesticated animals, since it seems that humans are probably the most liminal animals of all themselves, if we take the perspective of wild animals living in their putative autonomous territories.

The contextual care approach considered in this chapter is based on the recognition that humans do have a duty of care towards entities with moral standing, which are threatened by human causes, and that the type of care depends on the animal's position in Fig. 1. According to this care framework, the duty of care towards domesticated animals is primarily based on direct and causal relationships between humans and animals, generating a responsibility for their well-being. However, as noted earlier in this chapter, such direct and clear causal relationships are usually lacking in the case of wild animals. The link between causes and harm due to the rise of the Anthropocene is therefore not or only barely identifiable on the level of human individuals, due to the global, multi-scaled, and longitudinal character of the Anthropocene (Jamieson 2007; Sandler 2010). Moreover, individual human behavior usually has an insignificant effect on global phenomena. For example, the causal effect of not driving my car anymore is actually insignificant when compared to the total amount of emitted greenhouse gasses, let alone the subsequent effects on living conditions, for example, of the polar bear, sea turtle, North Atlantic right whale, or Australian frog species, to mention just a few of the many species that are threatened by climate change.[9] How should we morally justify a care approach to wild animals if such causal responsibility is so unclear,

[9]http://wwf.panda.org/about_our_earth/aboutcc/problems/impacts/species/.

insignificant, and temporary and geographically far away? Jamieson (2007) argues for a virtue-based environmental ethics: "When faced with global environmental change, our general policy should be to try to reduce our contribution regardless of the behavior of others, and we are more likely to succeed in doing this by developing and inculcating the right virtues than by improving our calculative abilities" (Idem, 167–168).

The care approach advocated here may be considered as such an environmental virtue ethics towards wild animals, since it assumes that humans have the responsibility and obligation of care-taking for the humanly threatened environment of animals in the Anthropocene. Whether or not we are directly causally involved, humans have, as entities gifted with reason, rationality, and agency, the ability and therefore the responsibility to do so, and to develop or to support political structures to implement care-taking initiatives as institutionalized human duties. We are all inhabitants on this earth, and as humans we have to find a morally reasonable way to live together on our shared planet. This perspective is, for example, put forward by the Earth Charter, a document commissioned by the United Nations, and now supported by thousands of civil society organizations and individuals worldwide. Principle 1A of the Chapter states: "Recognize that all beings are interdependent and every form of life has value regardless of its worth to human beings" (Sanders 2014). Thus, although the proposed contextual care approach does not start from a political theory, it recognizes that political institutions are nevertheless significant, since they may force or stimulate their human members to implement non-specific care measures for wild animals as their earthly cohabitants.

The concept of contextual care fits in—given the context of the Anthropocene—with current secular concepts of environmental stewardship (Welchman 2012). For example, Worrel and Appleby (2000, 269) define stewardship as "the responsible use (including conservation) of natural resources in a way that takes full and balanced account of the interests of society, future generations, and other species, as well as of private needs, and accepts significant answerability to society". In discussions on 'novel ecosystems,' which can be considered as a conservationist's or restorationist's answer to the rise of the Anthropocene, ecosystem stewardship is put forward as an appropriate normative stance that reconciles human and natural interests (Seastedt et al. 2013). In the words of Chapin et al. (2010, 241): "This unsustainable trajectory demands a dramatic change in human relationships with the environment and life-support system of the planet". Finally, the term earth stewardship is used in circles of the Ecological Society of America (ESA), which thus stresses the cohabitation of the biosphere, socio-ecological networks, and integrating ethics (Rozzi et al. 2015).

In conclusion, humans are increasingly reshaping our world, affecting the lives of many wild animals, which we should consider as cohabitants living on the same earth. Wild animals often cannot escape the impact of humans and have to cope with human impact any way they can. The contextual care approach justifies the recognition, preservation, and restoration of the myriad relationships that humans and domesticated, semi-wild, and wild animals are involved in and dependent on.

Acknowledgments I would like to thank Henk van den Belt of the Social Sciences Department of Wageningen University for his comments on a previous version of this chapter. I also thank Han Olff of the Conservation Ecology Group, and my colleague Koen Beumer of the Science and Society Group, both at the University of Groningen, for their discussions on this topic.

References

Beheeradviescommissie Oostvaardersplassen. 2014. IMCO2 in de Oostvaardersplassen. Evaluatie implementatie van ICMO2-adviezen en het beheer in het Oostvaardersplassengebied. Website Rijksoverheid. https://www.rijksoverheid.nl/documenten/rapporten/2014/12/02/icmo2-in-de-oostvaardersplassen-evaluatie-implementatie-van-icmo2-adviezen-en-het-beheer-in-het-oostvaardersplassengebied. Assessed 6 Nov 2015.

Chapin, F.S., S.R. Carpenter, G.P. Kofinas, et al. 2010. Ecosystem stewardship: Sustainability strategies for a rapidly changing planet. *Trends in Ecology & Evolution* 25(4): 241–249.

Clutton-Brock, J. 1981. *Domesticated Animals, from early times*. London: Heinemann & British Museum (Natural History).

Clutton-Brock, J. 1989. Introduction to domestication. In *The walking larder. Patterns of domestication, pastoralism, and predation,* ed. J. Clutton-Brock, 7–9. London: Unwin Hyman.

Coeckelbergh, M. 2012. *Growing moral relations. Critique of moral status ascription*. New York: Palgrave Macmillan.

Court verdict. 2007. Website of the Netherlands Judiciary. http://uitspraken.rechtspraak.nl/inziendocument?id=ECLI:NL:GHSGR:2007:AZ9246. Assessed 6 Nov 2015.

Crutzen, P. 2002. Geology of mankind. *Nature* 415(6867): 23.

Donaldson, S., and W. Kymlicka. 2011. *Zoopolis. A political theory of animal rights*. Oxford: Oxford University Press.

Foreman, D., J. Davis, D. Johns, R. Noss, and M. Soulé. 1995. The Wildland mission statement. *Wild Earth* 1: 3–4 (special issue).

Hopcraft, J.G.C., H. Olff, and A.R.E. Sinclair. 2010. Herbivores, resources and risks: Alternating regulation along primary environmental gradients in Savannas. *Trends in Ecology & Evolution* 25(2): 119–128.

ICMO. 2006. Reconciling nature and human interests. Report of the International Committee on the Management of large herbivores in the Oostvaardersplassen (ICMO). The Hague/Wageningen, Netherlands: Wageningen UR - WING rapport 018.

ICMO2. 2010. Natural processes, animal welfare, moral aspects and management of the Oostvaardersplassen. Report of the Second International Commission on management of the Oostvaardersplassen (ICMO2). Wing rapport 039, The Hague: Wageningen.

Jamieson, D. 2007. When utilitarians should be virtue theorists. *Utilitas* 19(2): 160–183.

Keulartz, J., and J.A.A. Swart. 2012. Animal flourishing and capabilities in an era of global change. In *Ethical adaptation to climate change. Human virtues of the future,* ed. A. Thompson, and J. Bendik-Keymer, 123–144. Cambridge, MA: The MIT Press.

Klaver, I., J. Keulartz, H. van der Belt, and B. Gremmen. 2002. Born to be wild: A pluralistic ethics concerning introduced large herbivores in the Netherland. *Environmental Ethic* 24 (Spring): 3–21.

Kolbert, E. 2012. Recall of the wild. *The New Yorker*, December 24–31: 50–60.

LNV. 2003. *Leidraad opvang gewone en grijze zeehond.* The Hague: Ministry of Agriculture, Nature and Food Quality.

Lorimer, J., and C. Driessen. 2014. Wild experiments at the oostvaardersplassen: Rethinking environmentalism in the anthropocene. *Transactions of the Institute of British Geographers* 39 (2): 169–181.

Márquez-Ferrando, R., J. Figuerola, J.C.E.W. Hooijmeijer, and T. Piersma. 2014. Recently created man-made habitats in Doñana provide alternative wintering space for the threatened continental European black-tailed godwit population. *Biological Conservation* 171: 127–135.

Murphy, S. 2012. Special obligations. In *Encyclopedia of global justice*, ed. D.K. Chatterjee. Berlin/Heidelberg: Springer. http://www.springerreference.com/index/chapterdbid/327837.

Nussbaum, M.C. 2006. *Frontiers of justice*. Cambridge, MA: The Belknap Press.

O'Connor, T. 1997. Working at relationships: Another look at animal domestication. *Antiquity* 71: 149–156.

Palmer, C. 2010. *Animal ethics in context*. New York: Columbia University Press.

Palmer, C. 2011. The moral relevance of the distinction between domesticated and wild animals. In *The Oxford handbook of animal ethics,* ed. T.L. Beauchamp, and R.G. Frey, 701–725. Oxford, NY: Oxford University Press.

Palmer, C. 2015. Response to "vulnerability, dependence, and special obligations to domesticated animals" by Elijah Weber. *Journal of Agricultural and Environmental Ethics* 28(4): 695–703.

RDA. 2012. *Duty of care, naturally. On the welfare of semi-captive and wild animals.* The Hague: Raad voor Dieraangelegenheden.

Regan, T. 1983. *The case for animal rights*. Berkely, CA: University of California Press.

Rozzi, R.F., S. Chapin III, J.B. Callicott, S.T.A. Pickett, M.E. Power, J.J. Armesto, and R.H. May Jr. (eds.). 2015. *Earth stewardship*. Dordrecht: Springer.

Russel, N. 2002. The wild side of domestication. *Society & Animals* 10(3): 285–302.

Sanders, S.R. 2014. The dawning of an earth ethic. *Ethics & International Affairs* 28(3): 317–324.

Sandler, R. 2010. Ethical theory and the problem of inconsequentialism: Why environmental ethicists should be virtue-oriented ethicists. *Journal of Agricultural and Environmental Ethics* 23: 167–183.

Seastedt T.R., K.N. Suding, and F.S. Chapin. 2013. Ecosystem stewardship as a framework for conservation in a directionally changing world. In *Novel ecosystems. Intervening in the new world order,* ed. R. Hobbs, E. Higgs, and C. Hall, 326–333. Oxford, UK: Wiley-Blackwell.

Sinclair, A.R.E., S. Mduma, and J.S. Braghares. 2003. Patterns of predation in a divers predator prey system. *Nature* 425: 288–290.

Singer, P. 1990. *Animal liberation*, 2nd ed. New York: Avon Books.

Staatsblad (Bulletin of Acts and Decrees). 2002. Besluit van 19 April 2002, houdende eisen aan het houden, huisvesten, verzorgen en tonen van wilde dieren in dierentuinen (Dierentuinenbesluit), Staatsblad, 2002214.

Staatsbosbeheer. 2013. Managementplan Oostvaardersplassengebied 2011–2015. Uitwerking en implementatie van ICMO2 maatregelen, monitoring en communicatie. Website Staatsbosbeheer. http://www.staatsbosbeheer.nl/over-staatsbosbeheer/dossiers/oostvaardersplassen-beheer/feiten-en-cijfers. Assessed 6 Nov 2015.

Staatsbosbeheer. 2015. Jaarrapportage 2014–2015. Website Staatsbosbeheeer. http://www.staatsbosbeheer.nl/over-staatsbosbeheer/dossiers/oostvaardersplassen-beheer/feiten-en-cijfers. Assessed 6 Nov 2015.

Steffen, W., P.J. Crutzen, and J.R. McNeil. 2007. The anthropocene: Are humans now overwhelming the great forces of nature? *Ambio* 36(8): 614–621.

Steffen, W., A. Persson, L. Deutsch, J. Zalasiewicz, M. Williams, K. Richardson, C. Crumley, P. Crutzen, et al. 2011. The anthropocene: From global change to planetary stewardship. *Ambio* 40(7): 739–761.

Swart, J.A.A. 2005. Care for the wild. Dealing with a pluralistic practice. *Environmental Values* 14 (2): 251–263.

Swart, J.A.A. 2014. Technologies used for Animal Breeding. Ethical Issues. In *Encyclopedia of food and agricultural ethics*, ed. P.B. Thompson, and D.M. Kaplan. Dordrecht: Springer Science+Business Media. http://www.springerreference.com/docs/html/chapterdbid/385461.html.

Swart, J.A.A., and J. Keulartz. 2011. The wild animal in our society. A contextual approach to the intrinsic value of animals. *Acta Biotheoretica* 59: 185–200.

Tronto, J.C. 1993. *Moral boundaries. A political argument for an ethic of care*. New York/London: Routledge.

Vera, F.W.M. 2009. Land-scale nature development—The Oostvaardersplassen. *British Wildlife* 20(5): 28–36.

Weber, E. 2015. Vulnerability, dependence, and special obligations to domesticated animals: A reply to palmer. *Journal of Agricultural and Environmental Ethics* 28(4): 683–694.

Welchman, J. 2012. A defence of environmentasl stewardship. *Environmental Values* 21: 297–316.

Wilkins, A.S., R.W. Wrangham, and W.Y. Fitch. 2014. The "domestication syndrome" in Mammals: A unified explanation based on neural crest cell behavior and genetics. *Genetics* 197 (3): 795–808.

Worrell, R., and M.C. Appleby. 2000. Stewardship of natural resource: Definition, ethical and practical aspects. *Journal of Agriculture and Environmental Ethics* 12: 263–277.

WWF. 2010. *Living planet report 2010*. Gland, Switzerland: WWF International.

The Wolf and the Animal Lover

Martin Drenthen

Abstract For Aristotle a true friendship can only exist between free human beings, because true friendship is based on a shared understanding of the good. Yet today, some animal philosophers argue that friendships can exist between humans and animals, maybe not in Aristotle's sense of the word but in another way, that appreciates how animals are different from us humans, yet also share a certain commonality. Usually, these reflections on human-animal friendship concern human relations with domestic animals, notably pets. But can we befriend wild predators, those animals that by their very nature can be dangerous to us? In this paper, I examine what it might mean to befriend a wild animal, and more specifically whether it would be possible to become friends with wild wolves. I will argue that any friendly relation with wild animals will consists of a paradoxical combination of benevolent involvement and loving detachment.

1 Friendships Between Humans and Animals

Many people today are convinced that friendships between people and animals are possible. That thought is relatively new, especially in philosophy. According to Aristotle, true friendship can only exist between two free citizens, because according to him friendship consists of a conscious form of mutual benevolence.[1] Aristotle distinguishes three kinds of friendship: friendship can be based on mutual benefit, or friendship on the pleasure that friends derive from being in each other's company, but in friendships of the third kind, true friendship, mutual benefit and enjoyment go together with a third factor, namely a mutual admiration for the moral qualities of the other. Real friends wish each other the best because they see the

[1]Friendship with a slave is impossible, according to Aristotle, insofar as he is a slave. '*Qua slave* one *cannot be friends* with him; but *qua* man one can.' (Nicomachean Ethics, 1161 b5).

M. Drenthen (✉)
Radboud University, Nijmegen, The Netherlands
e-mail: m.drenthen@science.ru.nl

B. Bovenkerk and J. Keulartz (eds.), *Animal Ethics in the Age of Humans*,
The International Library of Environmental, Agricultural and Food Ethics 23,
DOI 10.1007/978-3-319-44206-8_12

other for what they are. And since only humans have the ability to recognize moral virtues in one another, for Aristotle a true friendship can only exist between free people.

Yet today, some animal philosophers argue that friendships do exist between humans and animals, maybe not in the true sense of the word but in a weaker version.[2] It is indeed difficult to deny that some aspects of friendship between people and animals do exist. The mutual benefit is evident in the relationship between a dog and his master: the dog gets food and care, the owner company of the animal, and both appreciate the presence of the other for that reason. Most pet owners enjoy the company of their pets. We all know how the company of dogs can enrich people's lives, and also that many people would be willing to sacrifice many things for their pets. Conversely, most pet owners generally believe that their pet is also experiencing pleasure in the company of their owner. We all know the stories about dogs that languish in grief at the grave of their owner—apparently a clear case of selfless love?

Aristotle might notice that such relationships between animals and humans ultimately are based on mutual benefit and pleasure alone, and thus cannot be labeled friendship in the strict sense. Mutual goodwill and benevolence could be reduced to mutual interest and pleasure—the dog helps to fight the owner's loneliness, the owner provides food and shelter to the dog. Something similar can be said about the apparently friendly relations that some animals appear to have among each other.

Videos about friendships between animals of different species that share a household and thereby appear to maintain friendly relationships—dogs and horses playing together, pigs and ducks that keep each other company, dogs that care for cat litter, even cats who play peacefully with a parakeet—, are highly popular on the internet.

Although in such cases there is clearly a relationship that involves more than just mutual exploitation, according to Aristotle these cases can never truly be called friendship in the full sense of the word. True friendship admittedly often starts because of mutual benefit and enjoyment of each other's company—we go to the movies with friends to enjoy a nice evening and enjoy each other's company. But as soon as these activities at a certain point initiate a true friendship, we will become genuinely interested in the wellbeing of our friend. For Aristotle, true friendship is based on a shared understanding of the good, and for that reason friendships with and between animals are out of the question.

The British philosopher Rowlands (2011) acknowledges that in Aristotle's work there can be no question of true friendship between man and animal, because animals lack the ability necessary for true friendship. But according to Rowlands,

---

[2]For example, see the article 'Animal Ethics Based on Friendship' by Barbro Fröding and Martin Peterson in the first issue of the *Journal of Animal Ethics* (Fröding and Peterson 2011). Fröding and Peterson believe that recognition of the mutual benefits that pets and their owners from each other is enough to base a friendship on.

Aristotle pays insufficient attention to the typical nature of the human-animal relationship, and fails to acknowledge the admittedly different but not for that reason inferior friendships that can exist between humans and animals. Rowlands argues that sometimes, relationships between humans and animals can also be based on more than just shared interest and pleasure. These relations can also be fueled by a deeply felt mutual benevolence and can thus be called a kind of friendship. Obviously, such friendly human-animal relationships will differ from those between humans, because humans and animals are obviously different in many aspects, but according to Rowlands there are also similarities. According to him, even the aspect of adoration, which according to Aristotle is so essential to a full-fledged friendship, may play a role in our relationship with animals. "Why do we derive pleasure from the presence of companion animals? The answer is, I think, pretty clear: We admire them in various ways" (Rowlands 2011, 78).

Our sense of admiration for animals is not founded on a common sense of the good—as is the case in friendship between human friends—rather it is based on the significant difference between humans and animals.

> When Aristotle talks of mutual admiration, it is admiration based on a certain kind of sameness or commonality—shared moral virtue. A master and a slave have nothing in common. That is why they cannot be friends. The mutual admiration that forms the basis of Aristotle's friendship in the real and primary sense is admiration of qualities that you could in principle have—that you should aspire to have—even if you do not, at the present time, possess them. But sometimes with animals, the admiration is grounded in qualities that you lack—and that you admire, in part, because you know you could never have them. The mutual admiration constitutive of human–animal friendships may involve an element of difference that has no echo in the Aristotelian account (idem.).

As an example of admiration as the basis for a human-animal friendship Rowlands describes his admiration for a shepherd dog with whom he lived for many years:

> Perhaps I should speak only for myself. I have always admired—although often in different ways—the animals […] who have found their way into my life. Take Hugo: I admire Hugo —a schutzhund from a long line of schutzhund champions—because of his courage. But I also admire him for the extraordinary tolerance, forbearance, kindness, and gentleness he exhibits toward my two young sons. I admire the unerringly amiable way in which he greets other dogs. I admire his extraordinary lust for life and the enormous amounts of excitement he can generate in connection with things—a walk, a thrown ball—that seem to me small things. Aristotle would, at least as I understand him, deny that these are virtues that Hugo exhibits; and that, I think, is a problem for Aristotle, not Hugo. The pleasure I derive from Hugo's company derives from my admiration of these and other virtues; accordingly, it is this admiration that lies at the core of our friendship (idem, 78–79).

What is striking however is that Rowlands mentions exactly those properties of Hugo as admirable that we also regard as virtuous in human beings: forbearance, kindness, patience. Although Rowlands claims that our adoration for animals refers to the way they differ from us, that is not apparent from the concrete example he gives. He appears to admire the dog mainly for those characteristics in which the animal resembles ourselves.[3]

If a difference exists at all between human-animal friendships and human friendships, then this will have to do with the asymmetry of the former. Rowlands implicitly acknowledges, for instance, that it remains very doubtful whether the admiration is mutual, when he indicates that he *hopes* his friendly admiration for the dog Hugo will be reciprocated by the animal:

> Because I admire him, I also try to ensure he admires me in return, and so I make sure that in my dealings with him, I am always fair, consistent, calm, stable, and when necessary, strict. I am more than happy to call Hugo my friend. And the basis of this friendship is admiration that I hope is returned. This, I think, is the best way of thinking about friendships between animals who are human and animals who are not (idem, 79).

Is reciprocity a necessary condition to speak meaningfully of friendship? One can understand that Rowlands hopes that his feelings of friendship are reciprocated. And it is indeed conceivable that Hugo likes it when his owner is reliable, consistent and clear—the welfare of a dog is, after all, largely dependent on a clear leader. But are we therefore also allowed to say that the animal admires his owner for those qualities, and that this admiration forms the basis for a mutual human-animal friendship?

Perhaps we should not be too strict when it comes to a friendship between humans and animals. Even though the relationship is asymmetrical and there are fundamental differences in the meaning that friendship has for humans and for animals, shouldn't we acknowledge that in many human-animal relations a form of reciprocity exists, and that these relationships can give joy and satisfaction to both parties? Dogs have been domesticated and live with together with humans for between 30 and 40 thousand years.[4]

In those thousands of years, the dog has changed from a wild and undomesticated species into a species that is all set to communicatione with humans. It might therefore be argued that in this millennial coexistence of man and dog, a very specific form of interspecies communication has emerged which emerg has created a certain commonality on which also affective relationships can grow. It is because of this shared basis in a common form of communication that dog owners may hope that their feelings of friendship are answered by their dog.

Still, there is also good reason not to think too lightly on this point and move past it. The asymmetric nature of human-animal relationships makes it difficult to speak of real friendship. The difference between humans and animals may be inspiring, but it is also essentially unbridgeable. The friendship between man and

---

[3]Elsewhere, as he describes his relationship to the wolf, Rowlands gives a better example of the way our admiration for an animal may be based on those properties in which it differs from us, and therefore could be the basis for a different kind of friendship. More on that below.

[4]Recent research (Shipman 2015) even suggests that the domestication of the dog played an important role in the evolution of modern humans. According to American anthropologists, working with dogs gave modern humans a decisive advantage over the hitherto more successful Neanderthal.

beast will mean something else for a human being than for an animal. The world of a human being is ultimately and fundamentally inaccessible to an animal—and vice versa. Human friends talk to each other about the meaning of their friendship and share their vision of the good life, animals don't. It is clear that anyone who believes that his dog understands him so well, ultimately falls victim to a nice and comforting illusion; there are many examples of people who lose sight of the fact that animals are different from humans in many important respects. If we want to remain aware of the difference between man and animal we can surely *hope* that our warm feelings for an animal will be reciprocated, and interpret the joy of a dog when the owner is returning home, or the 'good-natured' look and 'comforting' hug it gives when its owner is feeling sad as tokens of friendship, but we can never be really sure of the friendly meaning of such behavior. Undoubtedly, humans and some animals can develop all kinds of emotional ties with each other, but what they cannot do is share the *meaning* of these relationship. It is therefore problematic to speak of friendship between humans and animals, at least in the conventional sense of the word. And as the difference between man and animal becomes larger, the problem gets bigger.

But maybe we should take seriously Rowlands' suggestion that human-animal friendships are of a different kind, not so much based on what we have in common, but rather on the recognition and admiration for what makes us different.

2 Being Friends with a Wolfhound

In his 2008 book *The philosopher and the wolf*, Mark Rowlands reports about his relation with a wolf (or rather with a wolfhound, a hybrid between a wolf and a dog), with whom he lived together for over 10 years. Rowlands points out that wolves are poorly understood in Western culture and philosophy. All things considered, most philosophical reflections about the wolf do not concern the animal, but rather the wolf as a 'shadow of man'; wolves have become a symbol of the dark side of human, something that must be overcome. Famous is Thomas Hobbes' statement "man is wolf to man."[5] Instead, Rowlands describes his relationship to a real animal of flesh and blood.

A wolf differs substantially from a dog. Through its particular evolution a dog is evolved as a being that is all set for coexistence with humans, and is to a large degree able to communicate his needs to human beings and to interpret human signs. Dogs and humans thereby inhabit a common sense world. A dog looks at you and tries to read your intentions. For example, a dog enjoys returning a discarded cane to its owner just to please him; for a wolf it doesn't make sense to chase a stick and return it to the person who cast it. A dog understands what you mean when you

[5]Hobbes (1998, p. 3).

point out where the ball is B; but whereas a dog follows your finger to see where you are pointing, a wolf only looks at the tip of your finger. Where a dog is looking at you and is open to communication, a wolf is observing you as an outsider, and then tries to predict your behavior. A wolf is immersed in his own world, and remains an outsider to our world. That is the reason why a wolf can never be fully trained. People and wolves can learn to deal with each other, but the ease with which dogs and people communicate with each other is missing between humans and wolves. Because of this, our relationship with wolves will be substantially more asymmetrical than our relation with dogs.

Rowlands buys the wolf named Brenin early 90s as a puppy, and he soon discovers that living with a wolf is much more difficult than living with a dog. Initially all kinds of problems emerge—when Rowlands leaves the animal alone in the house, the wolf basically demolishes the furniture—but in the end he succeeds in training the animal, which makes it possible to live together. He has a relationship with the animal that lasts for years; they. Rowlands and the wolf become inseparable: they travel around the world together, and when Rowlands lectures at the University, the wolf is sleeping in a corner of the classroom.

Rowlands compares the training process of Brenin with the way you can train a dog. A dog is essentially focused on human communication, and one can use this fact when training a dog. A dog is prepared to do things to make his owner happy; a wolf on the other hand only does what makes sense to itself. The only way to train a wolf, as Rowlands puts it, is to ensure that the desired behavior of the animal somehow appears to it as necessary and inevitable. At its core a wolf remains an independent animal.

> Fundamentally, Brenin was not my property; and he was certainly not my pet. He was my brother. Sometimes, and in some respects, he was my younger brother. At those times, and in those respects, I was his guardian, protecting him from a world that he did not understand and that did not trust him (Rowlands 2008, 44).

In Rowlands' descriptions, Brenin often appears as an independent being and a stranger, an inhabitant of another world. Rowlands is fascinated by precisely this strangeness and autonomy of Brenin, and admires it:

> On our runs together, I realized something both humbling and profound: I was in the presence of a creature that was, in most important respects, unquestionably, demon-strably, irredeemably and categorically superior to me. This was a watershed moment in my life [...] I can't ever remember feeling this way in the presence of a human being. That's not me at all. But now I realized that I wanted to be less like me and more like Brenin. [...] I think, if you want to understand the soul of the wolf—the essence of the wolf, what the wolf is about—then you should look at the way the wolf moves. And the crabbed and graceless bustling of the ape, I came to realize with sadness and regret, is an expression of the crabbed and graceless soul that lies beneath (idem, 85–86).

Rowlands admiration for the wolf is primarily triggered by Brenin's independence and fascinating strangeness. The wolf is mysterious and strange. Several times, Rowlands calls the animal a graceful ghost apparition. From Rowlands' descriptions Brenin emerges as an animal that inhabits the world in a radically different way than we do—and much more so than dogs.

Firstly, the wolf has a strength, grace and aloofness that people do not have; the contrast leads Rowlands to become aware of his own limitations as a clumsy ape. But conversely Brenin is incapable of behavior that according to Rowlands is typical for ape-like, political animals like us: the ability to tell stories, to please, to mislead and deceive. In the introduction to his book Rowlands writes:

> It took a long time, but at last I understand why I loved Brenin so much, and miss him so painfully now he has gone. He taught me something that my extended formal education could not: that in some ancient part of my soul there still lived a wolf. Sometimes it is necessary to let the wolf in us speak; to silence the incessant chattering of the ape (idem, 8).

Rowlands claim is that admiration for the otherness of the animal can form the basis for a friendship. But what stands out in his description is primarily how the difference between humans and animals leads to a new perspective on ourselves: the core of Rowlands' confrontation with the grace of the wolf is a humbling experience, that leads up to self-reflection.

But is Rowlands not simply projecting his own human weaknesses onto Brenin the wolf? In that case, it would be justified to doubt whether this really is friendship, or whether the wolf is merely a projection of a human being who is unhappy and unsatisfied with his own limitations.

A narcissist is someone who is unable to maintain a genuine friendship because he sees the other merely as a means to affirm himself, to make himself feel better. The other is seen only insofar as it meets the expectations. Ultimately, for a narcissist it's all about himself. In the similar way humans can use animals to satisfy their own desires, ignoring their otherness. Although animals communicate with us, because of the asymmetry in our relationship with them, there is little that precludes when we project our own wishes and desires on them and we ignore their inalienable otherness and the unbridgeable difference between them and us.

Still, a supposedly friendly relationship with an animal does not necessarily have to be narcissistic. Although it appears in Rowlands' description of his relationship to the wolf that Rowlands' focus in that relationship with the wolf is mainly on what it does to himself, the animal appears to him not merely as an object for self-affirmation. On the contrary. Rowlands' relationship with Brenin shows that respect for the individuality of the animal and a growing awareness of self in the relationship with that other being can go together.

The confrontation with the otherness of an animal can lead to a new self-awareness, but that's not the only thing that matters. Rowlands is aware that he cannot speak for the wolf, and that he cannot know for sure what the value of their relationship is for Brenin. Rowlands is, in other words, aware of the irreconcilable otherness of the wolf.

But in the very recognition of the irreconcilable difference it shows that a meaningful commitment between humans and animals can exist, and that we might label such a relation as friendship. But this combination of meaningful connection and recognition of difference is precarious in that our asymmetric relationships with animals and requires a balancing act. And to the degree that animals are further removed from us, friendship between humans and animals will be more difficult. In that respect it is good to remember that Brenin was no wolf but a wolf-dog, a half-domesticated animal that lived with a man in a world controlled by humans. But what if the animal concerned is a truly wild animal?

3 Friendship with Wild Animals?

What can go wrong once we lose sight of the distance between us and wildlife is shown in the brilliant Werner Herzog documentary *Grizzly Man* (USA 2005). Grizzly Man is about the young American Timothy Treadwell, who lived for thirteen summers in Alaska between wild bears and eventually was eaten by a bear. The film gives us a glimpse into the contemporary fascination with wildlife and the idea that friendship with wild bears is possible.

Initially Treadwell is very aware of the danger, but the longer he is able to gain the trust and respect of indigenous grizzlies, the more his quest for peaceful co-existence transforms into an attempt to make friends with the animals. Increasingly, he talks about the animals as if they were his friends, giving them nicknames, trying to touch them, becoming more and more infatuated with the animals. Where the viewer is initially impressed by Treadwell's courage, his ability to decipher bear behavior and his attempt to peacefully live together with them, as the movie progresses, Treadwell's attitude toward the bears slowly but steadily becomes more troubling.

The documentary presents the gruesome death of the bear lover as an inevitable result of a tragic misunderstanding. We see how Treadwell's tormented soul is looking in bears for a kind of friendship that he cannot find with people; and how in his desire to become one with the bears, Treadwell slowly loses sight of the world in which bears live: a world in which moral considerations and friendship play no role.

From his supposed friendship with bears Treadwell increasingly turns against people. Park managers warn against accustoming bears to humans, they prohibit feeding the bears and call people to stay at a distance. Bears without fear for people are dangerous and must be killed; people and bears should therefore keep distance from each other. According Treadwell, however, park warden and conservationists are merely interested in bears as a resource: he believes they want to shoot the bears; and sees himself as the only true friend of the bears, with a moral duty to protect them against evil mankind.

The film features Sven Haakanson, a native resident of Alaska and curator of the local museum. He criticizes Treadwell's attempt to become one with the bears and close friendship with them. He explains that the indigenous inhabitants of Alaska realize that there is an unbridgeable gap between the worlds of humans and that of bears. That gap poses a limit that must be respected, not so much because of reasons of security, but mainly out of respect for "the bear and what the bear represents."

Haakanson criticizes people like Treadwell because, although they pretend to bridge this gap, they are actually just projecting their own needs and desires onto the bears; they use bears to supplement their own sense of emptiness. Haakanson considers this as the ultimate form of disrespect and typical of the way modern city people interact with nature—if they want something they just take it. Above all, modern man wants to give into his desire for pure untouched nature and in doing so kills that which he desires.

The problem, of course, does not just occur in our relation with wild bears, but with all the wild animals. Wild animals inhabit their own world, and when we pretend to be friends, we risk losing sight of the inalienable otherness of those animals. The otherness of wildlife goes along with their wildness and uncontrollability—it belongs to the essence of wild animals that they do not bother about our rules and considerations, but rather inhabit their own world. By pretending that we can befriend them, we are either fooling ourselves and we're overlooking the fact that wild animals cannot be controlled, or we smuggle away that in our relationship with them we are mostly interested in ourselves and that we are actually using them for our own desires and impose our will onto their wildness.

4 The Wolf in the Netherlands

How difficult it apparently is to not lose sight of the otherness and unruliness of wild animals in the admiration we feel for them, and in our attempt to live together with them in a peaceful way, is also evident from the debate which arose when on March 8, 2015 the Netherlands was visited by a wild wolf for the first time in over 150 years. It was a young animal, that probably grew up on a German military training site, which and was looking for a new habitat. The animal crossed the border near the town of Emmen and then walked many kilometers through the northern parts of the Netherlands, to eventually disappear back into Germany after four days. A one year earlier, a dead wolf was found in the North East Polder—but on closer inspection that animal turned out to be deposited there by people, presumably as a prank. But now, finally, there was a real report of an actual wolf living in the Netherlands.

The arrival of the animal caused a lot of commotion. For a long time, most Dutch people thought it impossible that a wild animal such as a wolf would ever return to Dutch soil. Many believed that the modern Dutch landscape is actually unsuitable for a wolf—and that any wolf arriving would inevitably cause all kinds of problems. Many saw their suspicions confirmed when the wolf—contrary to what was predicted by experts (a wild wolf would be timid, afraid of people, invisible)—seemingly at ease walked on the sidewalk of the small village of Kolham, where it was filmed by a surprised motorist. The images went around the world; "Terrifying footage of wolf prowling city streets looking for its next meal," the British newspaper Mirror headlined, with a sense of exaggeration.

Many felt that there had to be something weird going on with the animal. Someone on the internet said what many thought: "All fairy tales. This is not a wild wolf but an animal that has escaped from a private zoo. A real wolf would never walk the streets in between humans."[6] Even though wolves have been living in European cultural landscapes for many centuries, many people deep down still believe that a wolf belongs in the uninhabited wilderness.[7]

The distinction between culture and wilderness is reminiscent of the kind of distinctions that the British anthropologist Mary Douglas has analyzed in her influential classic *Purity and danger* (1966). Douglas' study shows how people create their own reality through a systematic ordering and classification of matter. Something appears unclean soon as it is in the wrong place, "dirt is matter out of place," says Douglas. The distinction between culture and wilderness also functions as such a symbolic distinction between clean and unclean. The unclean wilderness appears as a threat to the pure culture.

In the 1959 Russian novel *Life and Fate* by Vasily Grossman, this old European view is nicely phrased by main character Viktor Pavlovich:

> Man never understands that the cities he has built are not an integral part of Nature. If he wants to defend his culture from wolves and snowstorms, if he wants to save it from being strangled by weeds, he must keep his broom, spade and rifle always at hand. If he goes to sleep, if he thinks about something else for a year or two, then everything's lost. The wolves come out of the forest, the thistles spread and everything is buried under dust and snow. Just think how many great capitals have succumbed to dust, snow and couch-grass (Grossman 2006).

In other words, the thought that threatening, dangerous wolves belong to the wilderness, is not based on biological fact, but on a symbolic distinction between culture and what lies beyond. The arrival of the wolf to the Netherlands is perceived as a threat because the animal itself seems to ignore and undermine this boundary between culture and wild. Wolf haters fear the unruliness of the wolf that threatens to undermine the meaningful cultural world that was conquered on nature and therefore want to expel the animal to another world.[8]

---

[6]Wolves in the Netherlands Facebook page of 09/03/15, 16:20. Wolves in the Netherlands (*Wolven in Nederland*) is a coalition of conservation groups and other stakeholders, that seek to prepare the Dutch population for the arrival of wolves, by showing how the wolf could be an asset to the ecological health of the Dutch nature, and by providing information about wolf behavior and possible measures to prevent damage to and loss of livestock.

[7]Strangely enough, you will also find that idea in other places in Europe with a lot more space, such as Scandinavia and France. In Norway we even see conspiracy theories about wolves that are secretly introduced by animal activists and the government in a conspiracy against the rural population. See Skogen et al. (2008).

[8]For that reason it will not reassure people when experts point to the statistically low probability that wolves pose a threat to humans and livestock.

Many people nevertheless feel that the wolf is an impressive charismatic animal that deserves admiration and respect and with which we should try to live together peacefully. But for people who want to welcome the wolf too it can be difficult to acknowledge the otherness of the animal. The wolf that roamed the Netherlands in March 2015 for 4 days was not so much bothered by people who wanted to expel or kill it, but mostly by those who wanted to see it with their own eyes. These people did not want to shoot the animal, but wanted to make pictures and share them on social media.[9]

The wolf is one of the most charismatic animals in Europe, it is seen as a beautiful, intelligent and social animal that deserves respect. Many people even feel a deep emotional bond with the animal. But no other animal is so fraught with all kinds of symbolic meanings; sometimes the animal itself almost becomes invisible under all symbolism. In many stories about wolves, wolves are depicted as innocent, pure, honest, and authentic. The wolf is a symbol of the power and wisdom of nature that modern society has lost sight of. It is in this context that many people talk about friendship with the wolf. "I prefer a wolf as a friend to a hunter," one can sometimes hear among animal lovers. "The wolf is not dangerous, people are"; "The wolf is at least honest …"; "It's people, not the wolf, who behave like beasts."[10]

Many of these wolf fans identify with the wolf, not primarily with the real animal with its typical characteristics and behavior, but notably with the wolf as a symbol of the pure, morally pure wilderness. As the innocent victim of a hostile human culture that is just out to subdue and destroy nature, wolves should be protected from the evil of humanity. The risk of such an approach that idealizes wolves is that it blinds people to potential conflicts that could arise when real wild wolves move into a well-ordered, domesticated landscape such as the Netherlands. The romantic idea of a harmonious, equal and mutual friendship with wild wolves, it seems, ends up being illusory. It appears that the romantic wolf is ultimately not much more than a projection of our own desire for purity and of our discomfort with our modern self.

If we as a modern society want to learn to live together with wild animals which also show up in our immediate environment, in the place where our children play and we nurture our gardens, then we will have to learn to deal with the stubbornness that is inherent in wildlife. By denying that an animal is different and our relation to it will for that reason be complicated and difficult sometimes, we tend to lose sight of what is required of us if we really want to live together peacefully with these wild animals.

---

[9]The route of the wolf can be followed in detail on many websites. For example here: http://www.metronieuws.nl/binnenland/2015/03/volg-het-spoor-van-de-wolf.

[10]See the Facebook page of Wolves in the Netherlands.

5 A Tense Friendship

To the extent that a friendly relationship with wild animals is at all possible, the otherness and unruliness of animals therein will have to be given a place. In our relationship with animals, we need to find elegant and graceful ways to deal with each other and grant a place to the significant otherness of animals.[11] When it comes to our treatment of wild animals, that may mean that we need to recognize that humans and animals are best served by maintaining a certain distance from one another.

When it became clear that the Groningen wolf almost fell victim to the human fascination or even love for wild animals, some people argued that we should leave the animal alone, to keep distance out of a feeling of love. One wolf fan put it like this:

> Dear Wolf, quickly go back to a large deep forest somewhere across the border. There clearly is no future for you here: too many thrill seekers with their cameras and wild goose chase stories. I'll miss you, but above all I want you to be safe and live in freedom. I ask all people who see you to leave you alone, to not make pictures or videos and to not chase you, but to respect you for what you are: a free soul, a Gypsy in transit.[12]

This wolf lover recognizes the insurmountable asymmetry in our relationship with wolves, and aims to respect the individuality and unruliness of the wild animal itself, even is this means keeping in check one's own desire for closeness and union with that animal. Maybe this paradoxical combination of loving commitment and recognition of insurmountable otherness can be rightfully labeled as 'friendship' towards wildlife.

A friend of wildlife is aware of the problem that follows from the differences between humans and animals, and does not try to take away these differences, but rather to give them a place, even if that means that he has to keep distance to the animal for which he feels affection and admiration.

6 Close

Friendships between people and animals differ fundamentally from friendships between humans because of the asymmetry of the relationship between human and animal. But meaningful relationships of mutual goodwill between humans and animals do exist: as is the case in our relationship with companion animals and pets, where animals and humans enjoy each other's company and benefit from the relationship, and can even communicate across species borders that they appreciate

[11] Donna Haraway speaks of '*significant otherness*' (Haraway 2008, p. 15).

[12] 'Leave our wolf alone' Facebook page, entry from 03/09/2015 (accessed on 1 May 2015).

each other's company. These 'friendships' are based in a shared common atmosphere of interspecies communication. In those cases our warm feelings of goodwill for an animal may even be answered by that animal in a manner that is appropriate to that animal.

It may be difficult to develop such a friendly understanding with wild animals, because that relationship is asymmetrical in a much more fundamental way. And as an animal gets wilder the problem gets bigger. We may cherish warm feelings of benevolence towards wildlife, even admire wild animals for what they are and how they are different from us, but a wild animal is not really interested in us but only in how we might matter to it. It may be indifferent because it got used to us, or it may feel threatened and flee, or it can attempt to benefit from its relationship with us and therefore seek our company. For potentially dangerous animals like the wolf this is bound to lead to problems.

Because of the fundamental asymmetry in our relationship with wild animals, there is a greater risk that we expect more from our relationships with wild animals than these animals can deliver. In that case we end up mostly with ourselves and our own dreams and desires about harmonious coexistence with animals will get the better of us. Wild animals live in their own world, and any attempt to bridge the gap between their world and ours through friendship is not just based on an illusion, but all too easily turns into a disrespectful denial of individuality and wildness of the animal.

This does not mean, of course, that we could not have a relationship of benevolence towards wild animals, or that we should not seek a conflict poor way to live together with other animals. Such coexistence can mean that we are trying to give these animals the space they need, and give the animals themselves the opportunity to learn how to live with us. And that we recognize that sometimes it will be best to keep a distance—to stay at a distance ourselves or to keep the animals at a distance—even if we actually desire for intimate contact with an animal. 'Friendship' with wildlife precisely consists of the paradoxical combination of benevolent involvement and loving detachment.

For those who regard the wolf as friend and innocent victim of modern society, the possibility of human-wildlife conflicts in itself presents a challenge. Love for wolves cannot be easy; what is more, a too rosy picture of wolves fails to do justice to their very nature as predators. It is easy to love beautiful and innocent animals, but sharing spaces with large carnivores will never be easy. Recognizing wolves as real animals living in the ecological and social context of our landscapes demands we acknowledge them as predators. As long as we lived separate lives, we might could try to ignore their presence, but now that we share the same landscape, we can no longer do so. In this age of the Anthropocene, in which the wild and the human are no longer neatly seperated, we will have to be prepared to enter into a negotiation process in which we both recognize the wolf's agency and accept that its interest can conflict with ours. From the perspective of wolf management, the resurgence of the wolf confronts us with our desire for control, not only control over nature, but also control over nature within us. We cannot but play an active role in organizing our relation with the wolf; we need to find an appropriate *habitus* that allows us to live together, and that will require some degree of management and control. But respecting nature's autonomy also implies a willingness to live with wild creatures, not just when they are charismatic and cute, but also when they are a nuisance. Without practicing tolerance—the virtue of enduring those things that are difficult to endure—wildlife management will inevitably incarcerate wildness. We should challenge our profound but problematic fascination for the vitality of wild animals and remind ourselves that the threat to the their unruly wildness does not only come from our urge for control, but also from our unbounded desire to become one with them.

References

Douglas, M. 1966. *Purity and danger*. London/New York: Routledge.

Fröding, B., and M. Peterson. 2011. Animal ethics based on friendship. *Journal of Animal Ethics* 1 (1): 58–69.

Grossman, V. 2006. *Life and fate* (trans: Robert Chandler). New York: New York Review of Books.

Haraway, D. 2008. *When species meet*. Minneapolis, MN: University of Minnesota Press.

Hobbes, T. 1998. *On the citizen* (trans: Richard Tuck and Michael Silverthorne). Cambridge: Cambridge University Press.

Rowlands, M. 2008. *The philosopher and the wolf. Lessons from the wild on love, death and happiness*. London: Granta.

Rowlands, M. 2011. Friendship and animals: a reply to Fröding and Peterson. *Journal of Animal Ethics* 1(1): 70–79.

Shipman, P. 2015. *The invaders: how humans and their dogs drove Neanderthals to extinction*. Cambridge: Harvard University Press.

Skogen, K., I. Mauz, and O. Krange. 2008. Cry wolf! narratives of wolf recovery in France and Norway. *Rural Sociology* 73(1): 105–133.

Blurred Boundaries in Wildlife Management Practices

Susan Boonman-Berson

Abstract Human-wildlife conflicts have been increasing at alarming rates over the last few decades. Wildlife management practices deal with preventing and disentangling these conflicts. However, which approach should be taken is widely disputed in research, policy, in-the-field-wildlife management and local communities. One key aspect in deciding which approach should be employed concerns the drawing of boundaries, between human and wild animal, between territorial spaces, in the use of certain categorizations and in the implementation of policy regulations. Coexistence between humans and wild animals is propagated worldwide even as human-wildlife conflicts increase. To achieve coexistence, wildlife management practices need to adopt corresponding management strategies. In this chapter it is argued that, in aiming for coexistence, wildlife management practices employ apparently obvious boundaries, which may, however, be adjusted to prevent and disentangle human-wildlife conflicts. Subsequently, it is argued that coexistence between humans and wild animals can only be attained by adopting an integrated approach to human-wildlife-interactions, merging knowledge from the natural and social sciences, and wildlife management practices. This argument is illustrated by exploring the dynamics of boundary drawing in the practice of wild boar management at the Veluwe, the Netherlands and black bear management on the Colorado Front Range, USA. These wildlife management practices display respectively both strategies of confinement and strategies of alignment. In conclusion, wildlife management practices aiming for coexistence require both strategies. Strategies of alignment, however, are prerequisite to opening boundaries between humans and wildlife in order to manage their conflicts as relational endeavours.

S. Boonman-Berson (✉)
Forest and Nature Conservation Policy
Group and Environmental Policy Group,
Wageningen University, Wageningen, Netherlands
e-mail: shboonman@gmail.com

B. Bovenkerk and J. Keulartz (eds.), *Animal Ethics in the Age of Humans*,
The International Library of Environmental, Agricultural and Food Ethics 23,
DOI 10.1007/978-3-319-44206-8_13

1 Introduction

"Unexpected Visitors: Wildlife Invasions in Our Towns and Cities", "Urban Invasion: Wily Foxes Embrace Easy City Life", "Oh, Deer: Wildlife Invades Gold Coast", characterize the headlines of media reports from around the world. How to deal with wild animals invading human territories is a topic of much debate and research (Bradshaw 2009; Knight 2000; Thompson 2002). Wildlife managers must decide what, if any, strategy to take to prevent wildlife crossing the boundaries of human property or to disentangle conflicts arising from these perceived transgressions. However, which approach should be taken is much disputed in research, policy, in-the-field-wildlife management and local communities. Potential wildlife management strategies can range from tranquilizing, relocating or killing individual 'problem' animals, confining them to particular areas by placing physical barriers such as fences, controlling or extirpating entire populations of these 'problem' animals from a particular area to acceptance of the possibility that human-wildlife conflicts might occur while taking no particular safety measures. Other strategies aim at humans rather than animals, such as education and enforcement of rules designed to prevent human-wild animal conflicts. Examples of such human-focused strategies include media attention (TV, radio, newspaper, internet) to reach many people at once (Cassidy and Mills 2012), educating individuals concerning their behavior, and fining individuals for behavior such as not storing their trash 'animal-proof' or for illegally feeding wild animals.

These examples illustrate debates about a variety of the boundaries which come into play when dealing with human-wildlife issues. These include boundaries between human and wild animal, between territorial spaces, between ignorant and competent people, and between individual animals and whole populations. For instance, Milton (2000, 242) argues that conservation is a boundary maintaining exercise; "In order to conserve the things that constitute nature, the boundaries that separate the must be maintained, and in order to conserve nature's 'naturalness', the boundary between the human and the non-human must be preserved". The latter involves not only categorizing wild animals, for example according to problematic or not, or humans versus wild animals, but also requires a spatial ordering to establish where particular wild animals (should) belong. Lulka (2004), discusses that to control certain wild animal populations, wildlife management plans need to shift focus from static boundaries in defining territorial spaces towards dynamic boundaries in which movement is key. He argues that a focus on movement provides a way to link humans and wild animals together in order to better manage human-wildlife interactions. Baruch-Mordo et al. (2011) argue that education as a management strategy should highlight pro-active enforcement, for example by dispensing warning notices, in addition to providing information. Using a combination of educational strategies increases the chance that a variety of people will be affected, and is therefore more likely to produce a change in human behavior when dealing with wildlife. In another debate about boundaries, Bear (2011, 299) says that a focus on individual animals might "offer considerable potential to engage and mobilise wider human interest", and especially "offer an alternative perspective to the often quantitative discussions", which are typically about populations of animals. But, as he warns, a focus on individual animals does not mean that human characteristics should be attributed to animals. This touches upon a currently sensitive issue concerning the attribution of agency to animals in scholarly literature dealing with relations between humans and wild animals (see amongst others Donaldson and Kymlicka 2011; Philo and Wilbert 2000; Wolch and Emel 1998).

In this chapter wildlife management practices dealing with human-wildlife issues are explored regarding how and what kind of boundaries are drawn to manage human-wild animal interactions and how and to what extent coexistence between humans and wild animals is aimed for.

To achieve human-wildlife coexistence, I argue in this chapter that wildlife management practices should continue to make use of boundary drawings when choosing management strategies. At the same time, I question the need for rigid boundaries and rather direct the attention to the creation of more dynamic, flexible boundaries which focus on aligning human and wild animal behavior in order to seek a form of coexistence. The concept of *co*existence is described in terms of human and non-human animals who engage with each other in a collective endeavor. Coexistence can be best understood as a 'co-production' by various humans and wild animals (Francis et al. 2011) and a move beyond essentialized, taken-for-granted representations of nonhuman others (Bear 2011; Philo 1995; Philo and Wilbert 2000). It is particularly stated in this chapter that thinking in terms of management strategies drawing rigid boundaries (as described by the notion of confinement), ignores calls to go beyond the boundaries of humans versus wild animals and instead to co-exist with wildlife (Cassidy and Mills 2012; Hinchliffe et al. 2005; Hinchliffe et al. 2006; Lorimer 2006, 2008; Noske 1997) by using flexible, dynamic boundaries (as described by the notion of alignment).

To make these arguments a number of points have to be taken into account. First, speaking in terms of conflict situations is problematic, as Bradshaw (2009, 246) argues when demonstrating the adversarial use of the term 'human-elephant conflict': "A sympathetic elephant 'bad guy' and an erring but well-intentioned human 'good guy'". It is debatable if human-wild animal conflicts involve a dispute between human and wild animal or between different humans, and it is questionable whether humans and wild animals are worlds apart or co-construct the spaces in which they dwell (Ingold 2005; Johnston 2008). Therefore, in this chapter I instead address human-wild animal *interactions* to emphasize their mutual relationship. These interactions between humans and wild animals entail processes of 'action-reaction'. The processes are not necessarily symmetrical, but both human and wild animal behavior is taken into account in co-constructing their interactions. This means that to avoid problematic interactions, whether for human or wild animal, their mutual relationship, their interaction, is central. Additionally, imposing on wild animals and/or humans the requirement 'to behave' in a particular way by using management strategies of disciplining will never result in an end product that specific people, such as researchers, in-the-field-wildlife managers, policymakers or local residents, had in mind because the focus is on domination rather than on achieving a cooperative relationship and solution (see also Bradshaw 2009, 163 and further). Thinking in terms of mutual relationship also contributes to coexistence between humans and wild animals.

In addition to the problematic division in relationship, Bradshaw (2009, 235) also states that "nature's rhythms and patterns do not conform to those created and imposed by Western humans". In other words, there is no such rule as "one management strategy solves all human-wildlife issues". Anthropogenic influences also play an important role here. For example, humans might intend to manage human-wild animal interactions by placing fences, but, despite these fences, the presence of wild animals in certain areas might rather be regulated by the distribution of natural food resources for wild animals which change on a yearly or larger time scale due to—intentional or unintentional—anthropogenic influences such as global warming, extensive forest fires, logging practices or human population growth. These factors have an effect on the management of human-wild animal interactions and call into question the use of rigid boundaries (Bear and Eden 2011; Lorimer 2006).

This chapter first discusses coexistence in wildlife management practices and focuses on the dynamics of drawing boundaries. Subsequently two wildlife management cases illustrate these dynamics in managing human-wild animal interactions: wild boar management at the Veluwe, the Netherlands, and black bear management on the Colorado Front Range, Colorado, USA. The dynamics described vary at different levels and the boundaries depicted are not only blurred, but also entangled throughout with respect to the described wildlife management practices. The two contrasting cases reveal different ways of approaching wildlife management aimed at coexistence between humans and wild animals and illustrate two concepts that contribute to the attempt to achieve coexistence: confinement and alignment. They also illustrate how living in a multispecies community requires approaching these multispecies communities in a 'multispecies' way. This can be done by viewing humans and wild animals, as well as populations and individuals as aligned, intertwined entities, instead of thinking of them as separate entities that have to be confined by and investigated employing particular scientific disciplinary traditions such as social sciences and natural sciences. This new way of thinking may result in the creation of management strategies that include strategies of alignment, or combine strategies of confinement and alignment, in their aim for coexistence between humans and wild animals in scientific as well as local wildlife management practices (Hinchliffe et al. 2006; Latour 2004; Lorimer 2008; Turnhout et al. 2013).

2 Drawing Boundaries with Wild Creatures

As early as 1995 Philo had expressed his concern regarding the socio-spatial processes of exclusion to which wild animals were so often being subjected. He described them as being excluded from city spaces, where humans are supposed to live and work, expelled for a variety of reasons including being considered polluting, disruptive and discomforting. To understand these socio-spatial processes, it has been argued that these so-called invading wild animals should be (re)included in contemporary research on human (-animal) geography (Philo 1995). This generated a variety of ideas, amongst others the concept of coexistence between humans and wild animals (Buller 2013; Hinchliffe and Whatmore 2006; Knight 2000). Coexistence in this chapter refers to the idea of humans and wild animals as active participants in spatial interactions. Their mutual relation is key, which involves processes of connection and division (Latimer 2013). Specifically, this perception of coexistence makes possible a more dynamic appreciation of differences between areas and between humans and wild animals. For instance, populations of wild animals are viewed as active masses that vary—both on a population level and on an individual animal level—in their behavior in different spatial settings (Hinchliffe et al. 2005), instead of static masses that have to be confined to certain areas to fit pre-defined classifications (Lulka 2004; Urbanik 2012). According to Hinchliffe and colleagues' argument (2005), the involved wild animal (s) come to be seen as fellow subjects in a dynamic process of coexistence. Thus aiming for coexistence assumes that both humans and wild animals actively co-shape the space in which both can dwell. It expressly rejects the idea of wild animals being treated as objects or as simply 'other entities' to be acted upon by humans. As a result, from the perspective of coexistence, preventing and disentangling human-wildlife conflicts implies mutual adjustment on the part of both humans and wild animals as they learn to live together. Relationships between humans and wild animals, in contrast to those between humans-domesticated animals (pets), are intermittent and fleeting (Bear and Eden 2011). Therefore, studying as well as managing human-wild animal interactions becomes a dynamic, situated, relational affair.

Drawing boundaries remains useful in wildlife management practices, since boundaries reduce the spatial extensiveness and complexity of the world to manageable units (Jones 2009). However, hard lines, black-and-white thinking, and deeply rooted dualisms can result in unproductive boundary disputes about, for example, wildlife conflict management (Keulartz 2009). Keulartz suggests moving away from thinking in terms of boundaries towards thinking in degrees employing his notion of "gradualization": "a broad continuum, a hybrid middle ground" (idem, 36). Urbanik (2012) describes the notion of hybridity in (individual) human-animal relationships, denoting that relationships differ depending on place (Lulka 2009). For example, an encounter with a free roaming wild boar differs from an encounter with a captive wild boar at a children's farm. The individual relationship between human and wild animal is emphasized. Subsequently, Urbanik expands this concept by examining the processes of power in these human-animal relationships. These processes of power are important, she argues, since, depending on the place of an encounter, different power relations exist. For example, humans may feel that they have control over the wild boar in the children's farm, but feel less in control when coming across a wild boar in an 'open space'. Similarly, processes of power may be able to explain which actions might be taken in specific circumstances. For example, when a wildlife manager encounters a black bear sniffing in a garbage can, this might result in management actions aimed at capturing and relocating the particular black bear because that bear might create problematic situations, while when a tourist—who is excited to spot a black bear during his/her holiday—encounters the same black bear sniffing in a garbage can, (s)he might grab his/her camera to make a picture, further leaving the black bear alone. The above described variation in the drawing of boundaries is what I call in this chapter 'the dynamics of boundary drawing'.

Specifically, I explore in this chapter the dynamics of boundary drawing in wildlife management practices to clarify the kind of control that is mobilized to deal with human-wild animal conflicts. This is done by focusing on the doings, sayings, and spatialities of both humans and wild animals. These doings, sayings and spatialities have implications for what management strategy will be taken. For instance, Marvin (2000) discusses the killing of foxes in the English countryside and concludes that the dispute surrounding these killings is what these foxes *do* rather than what they *are*. In his research it is the illegitimate killings of the fox—because the fox kills game birds and domesticated livestock which are only supposed to be killed by human beings—that causes the fox to be perceived as a pest and legitimises controlling foxes through fox-hunting. While, he states, the fox as such, was seen as an attractive animal in his research and people spoke positively about the physical body of the fox. Here, it is what the fox does that makes it a 'pest fox'. Cassidy and Mills (2012) depict the 'urban fox' as a central character in transgressing socio-spatial boundaries when it enters 'human spaces', in their case entering a home in East London in 2010. They discuss the consequential confusion concerning the 'correct' space of belonging for humans and 'urban foxes', and the confusion about the 'accompanying' behavior of an animal categorized as wild versus one categorized as urban, such as the fox, or as a domesticated animal, such as a dog (as pet). These ambivalent spatial and representational conceptualizations of human-wild animal relations have implications for which wildlife management strategies are applied in practice. In the case of the 'pest fox' in the rural country side (Marvin 2000), it is hunted, in the case of the 'urban fox' (Cassidy and Mills 2012), it is disputably considered to be both a pest, which legitimizes its killing and confinement, and feral, adapted animal that humans are persuaded to tolerate and align with.

Common practice in dealing with human-wildlife conflicts is to control or discipline the wild animals. This can take different forms, depending on how the boundaries are drawn. In drawing these boundaries, processes of power, as described by Urbanik (2012), are important. For instance, Rinfret (2009) argues that the reintroduction of wild animals involves targeting controlled encounters between human and wild animal. The involved "wild" animal is argued to be a "human-constructed, representation of the wild" that is disciplined by using sophisticated technology such as "shocking collars" for wolves—when they come close to a farm they are shocked—, and GPS monitoring devises for condors—to track daily activities in order to protect them in their "natural" surroundings. These animals are confined to certain locations by the use of technological devices.

Palmer (1996), who particularly defines human-animal interactions as power relations, distinguishes between disciplinary techniques interfering in internal human-animal practices, such as training, taming and teaching in a variety of ways (including the offer of reward and affection), and those interfering in external human-animal practices, such as confinement and physical punishment (including trap—and—relocate and "shocking collars"). Although these practices are intertwined, she argues that these relationships can easily turn into human domination over wild animals, when the desired disciplining of the wild animals is not having the desired effect. As a result the animals revert to being treated as 'things' (objects) rather than 'beings' (subjects) in a relationship.

Both Palmer (1996) and Rinfret (2009) discuss the doings (actions) of wild animals that precede the introduction of disciplinary devises in terms of "acts of resistance", which constitute reactions to human acts. Despite their depiction of wild animals as actors, they do not go on to consider a possible alignment between humans and wild animals that would move beyond targeting domestication and disciplining wild animals. The disciplining of wild animals also brings into question the ambivalent notion of how *wild* animals are defined, and challenges the distinction between domesticated and wild animals.

A particular distinction between domesticated and wild animals is made in policy regulations which impact the management of human-wild animal interactions. In the case of wild animals, no one is assumed to be—in social or legal terms—responsible for the animals. The responsibility for the acts of these animals and their related consequences is debatably a (social) collective issue, one which could be attributed to the wildlife manager, to the 'injured' (human) party or to the wild animal in question. In the case of domesticated animals, the owner is responsible for any attacks that occur. The attributed responsibility is an especially delicate issue in the implementation of wildlife management strategies, especially when potential economic losses play a role. Are these losses to be considered the consequences of nature or can they be attributed to a particular human party?

Last but not least, an important boundary having an impact on the application of management strategies in wildlife management practices dealing with human-wild animal interactions is between the social and natural sciences. Traditionally, the behavior of wild animals is investigated and subsequent wildlife management strategies are proposed by natural scientists. They determine how the wild animals forage, reproduce, and survive in a particular ecosystem and how many of a specific wild animal an ecosystem can sustain (Thompson 2002). It is only recently that the human factor has also been recognized as an important focus for research and management concerning human-wildlife conflicts (Knight 2000). Humans may interpret the behavior of wild animals in a variety of ways. Their observations can be paired with emotions ranging from excitement to intense fear with respect to the particular wild animals under study (Cassidy and Mills 2012). The subsequent management strategies vary accordingly. Consequently scholars suggest that there is a need for (more) social science research to deal appropriately with the increase in human-wildlife conflicts worldwide (Baruch-Mordo et al. 2009; Knight 2000). Although it is acknowledged that both fields of knowledge—the one dealing with the behavior of wild animals and the other dealing with the behavior of humans in areas where both dwell—are necessary to solve human-animal conflicts, the question of how these knowledge fields can be aligned in constructing management strategies aiming for coexistence has been underexplored.

In the following sections two contrasting cases illustrate the dynamics of boundary drawing in managing human-wildlife interactions. Wild boar management at the Veluwe is examined first. The empirical data on wild boar management is drawn from prolonged participant-observations of wild boar counting in the Veluwe area in 2010 and 2011, hunting observations, a total of 30 interviews with in-the-field wildlife managers, researchers, local residents, policy makers, four focus groups in the area, and several key documents as well as media attention given to the subject in the period that the research was conducted and the years immediately thereafter. This case clarifies the notion of 'confinement' by analyzing how and what kind of boundaries are drawn to manage wild boars and how that contributes to coexistence between humans and wild boars in the Veluwe area. Particular emphasis is placed on the drawing of boundaries through the choice of a management strategy and how particular management strategies of confinement are implemented.

Next, black bear management on the Colorado Front Range, Colorado, USA is examined. The empirical data concerning black bear management is drawn from 5 days of participant-observations with in-the field-wildlife managers in the summer of 2012 during a two month stay on the Colorado Front Range, USA, 37 in-depth interviews with researchers, policymakers, in-the-field-wildlife managers and local residents, three focus groups with local residents and several key scientific, policy, and educational documents on black bear management in this region. This case clarifies the notion of 'alignment' by analyzing how and what kind of boundaries are drawn to manage the black bears and how that contributes to coexistence between humans and black bears on the Colorado Front Range. Particular emphasis is placed on the relationship between human and black bear in deciding what management strategy to use and how to implement particular management strategies of alignment.

The main characters are referred to by titles such as in-the-field-wildlife manager, researcher, policy maker, local resident and local activist for reasons of anonymity and analytical clarity. When individual wild animals are discussed, they are usually referred to as '(s)he' (when gender is not known) and 'he' or 'she' when gender is known.

3 Wild Boar Management at the Veluwe

The Veluwe is the second most important nature area in the Netherlands. It is approximately 100.000 ha made up of forests, heathlands, drift sand and some cultivated fields, and is dissected by several roads. Human settlements are found within the area. An important feature of the Veluwe is the presence of wildlife such as wild boar (*Sus Scrofa*). Due to the area's spatial arrangement, current wild boar management is multi-spatial. In principle, wild boars are allowed to roam 'freely' within the Veluwe. However, this 'free-range' is restricted by several fences constructed to prevent the boars from invading certain areas such as human settlements, highways and agricultural zones. These areas are designated as zero-tolerance areas by national and provincial policy. Their boundaries are key to wild boar management, especially in terms of preventing problematic human-wild boar interactions, as maintaining public safety is a top provincial priority (Provinciaal Bestuur van Gelderland 2002). In general then, to achieve coexistence between humans and wild boars while simultaneously maintaining public safety at the Veluwe apparently requires placing physical dividers in the landscape in order to control wild boar movement and regulate the number of wild boars permitted to roam in particular areas.

Physical dividers, for example (electric) fences preventing wild boars from entering certain areas, such as gardens, can be found all over the Veluwe. Wildlife road barriers are another way of separating wild boars and humans. Some of these road barriers have been specifically developed to block wild boars while allowing access to other wild animals, such as red deer. Another strategy for physically separating wild boars from humans is the establishment of so-called 'grazing areas' for wild boars. These grazing areas are resource rich, largely open areas which wild boars are attracted to, as an in-the-field-wildlife manager explains when we pass one: "Wild boars come here because there is food and they know it is safe because it is fenced". The safety of these areas for wild boars is related to a management strategy for population regulation and is applied outside these 'safe-zones': culling. Before examining population regulation in detail, it is important to note that these grazing areas not only control the movement of wild boars, but also control the movement of tourists visiting the area. As an in-the-field-wildlife manager puts it:

> Also tourists know their way to this [grazing] area, because they know they can spot wild boars here and the infrastructure is designed in such a way that it restricts these people to walk on the trails, and thereby limit disturbing these animals.

In other words, the development of these grazing areas, as well as other physical dividers in the landscape, can be seen as strategies of confinement controlling the movement of both wild boars and humans in the area.

Among in-the-field-wildlife managers one physical divider which employs physical punishment is particularly disputed: chasing away. This strategy is implemented when wild boars cross the boundary of zero-tolerance areas. An in-the-field-wildlife manager explains it in this way:

> [wildlife managers] used to scare them away with fireworks [in town centers]. At first, that was effective, but soon they became used to that and weren't frightened anymore. So we stopped doing that. We need to keep shooting these boars, because you know, once they are completely habituated, you don't get rid of them anymore from your town center…

He argues that wild boars are intelligent animals, but don't comply with human defined rules about staying away from zero-tolerance areas. As a result, he argues, extirpation of wild boars in these zero-tolerance areas seems necessary. His understanding of the relationship between humans and wild boars has changed from teaching these wild boars how to behave to domination by killing them.

However, in managing human-wild boar interactions the most significant approach entails regulating the size of the population. This approach assumes that a particular population size will prevent these animals from entering zero-tolerance areas in their search for food. To regulate population level a variety of boundaries are drawn. This regulation is "based on research on wild boar carrying capacity in each indicated wild boar habitat" (Provinciaal Bestuur van Gelderland 2002). Wild boar habitat at the Veluwe is specified by their foraging behavior and ecological characteristics. This wild boar carrying capacity, denoted here as the maximum allowance of wild boars, must first be determined. It is initially derived from a scientific model constructed in 1999 (Groot Bruinderink et al. 1999), based on the actual food availability for wild boars on the Veluwe, without any supplementary feeding. However, in practice, the maximum allowance is in reality based on preventing any problematic situations between humans and wild boars. As a policy maker notes "we just look per area [organizational unit] how many animals we want […]" and his colleague adds "the main aim is safety here, fluctuations are not tolerated". This means that social concerns rather than ecological—food related—concerns determine this maximum—desired—allowance of wild boars at the Veluwe. This maximum is rigidly observed as will be shown. Thus, to determine the maximum allowance of wild boars at the Veluwe a variety of boundaries are effected: rather rigid spatial boundaries (zero-tolerance areas, organizational units), and a blurred boundary between reliance on ecological knowledge about the number of wild boars the Veluwe area can sustain versus reliance on the socially preferred number of wild boars all play a role in the determination of the maximum number of wild boars at the Veluwe. Importantly, in the determination of these boundaries wild boars are treated as objects—units or numbers—rather than animals that move around.

To arrive at this maximum number it is first necessary to ascertain the current wild boar population. To do this wild boars are counted yearly during two evenings in late spring (May-June). Wild observation posts are established to count them and the counts are divided in categories: sex (female, male), and age (young adult, piglet). Wild boars are attracted to these posts by bait laid down there. This means that these wild boars are disciplined in order to count them. To discipline the wild boars, wild boar managers accumulate detailed knowledge about their behavior and employ this knowledge to train the wild boars to come to these observation posts so they can be counted. As one in-the-field wildlife manager explains, they distribute bait on the same spot every day at the same time for three weeks before the counting. At some places this is done by hand, at others it happens automatically. Wildlife managers themselves are trained to count the boars using a variety of camouflage techniques, such as using grease clothing, situating themselves "against the wind", and making no noise. Although wildlife managers indicate that they "try to think like a wild boar", so that the boundary between human and wild boar seems to be blurred, still, the relationship is one of domination. In this interaction between humans and wild boars the count, not the relationship itself, is central.

Once actual wild boar numbers are approximated, the number of wild boars to be culled is calculated. First the wild boar count is corrected for the number of piglets still to be born. The correction factor used is based on expected food availability in summer/autumn, the experience of in-the-field-wildlife managers and data from former years. After corrections are made, the pre-set maximum allowance of wild boars per area is subtracted from the total number calculated for that specific year. This yields the number to be culled. Over the last five years this fluctuated around 3800 or 4000 wild boars to be culled at the Veluwe annually, leaving about 1100. To cull the calculated number, a hunting table is set up that describes the number of wild boars to be culled, divided in categories of sex (female, male), and age (young adult, piglet), and divided per organizational unit as well as for the whole Veluwe. The resulting hunting table shows—again—that, here, wild boars are being treated as objects; they are represented by numbers that do not refer to particular—individual—animals. In this hunting table they are just numbers.

The culling of wild boars corresponds to the procedure for counting wild boars; wildlife managers discipline wild boars to come to observation posts by using bait, and discipline each other (based on their knowledge of wild boar behaviors) by using a variety of camouflage techniques. Hunting follows on an individual basis or through so-called 'situated' hunting where several hunters hunt close to each other, positioned in fixed observation posts, attracting the boars using bait and with the intention to hunt as many wild boars as possible in one evening. The latter strategy covers larger areas than can individual hunting and is used by preference in areas close to human-populated areas, the so-called 'risk' areas. These two culling strategies use different spatial boundaries to achieve the legally defined population size of wild boars and manage human-wild boar interactions. Besides culling that is spatially restricted to observation posts, some wildlife managers "go after the boars", as a wildlife manager explains, by observing where a boar has recently been rooting, putting bait there and waiting near that spot until they return. As with counting wild boars, here too humans discipline the boars in their need to cull them. In culling wild boars during hunting season (July-January, possibly until March), a categorical shift in focus occurs, from hunting piglets in summer, to targeting older age groups and poorly fed boars later in the season. The reasoning behind hunting piglets in summer is to reduce population size relatively quickly. This is a clear indication that wild boars are treated principally as numbers, with the aim being to approximate the desired population size as much as possible.

Thus, in order to regulate the population size of wild boars, rather static boundaries are drawn to manage human-wild boar interactions (zero-tolerance areas, organizational units) and wild boars are primarily treated as objects to be acted upon (counting and culling) and divided into particular categories (sex/age/problematic). In this way, individual wild boars roaming the Veluwe 'disappear' in numbers of maximum allowance and to be culled/not-culled. The relationship between human and wild boar in population regulation can be best described as one of power (dominance) and of numbers.

Current wild boar management as described above is severely criticized. For example, this critical remark is made by a local activist who questions the current way of regulating population sizes: "Management shouldn't focus on numbers [...] we stick too much on models, too much on 'there are too many' wild boars. [...] we need to adapt [to wild boars] or fence no-go areas". He questions controlling human-wild boar interactions through a focus on models and numbers. He touches on the issue of (mutual) relationships between humans and wild boars, saying that we—humans—need to adapt to them in order to live together at the Veluwe. Thereby he specifically addresses the idea that wild boars are wild animals and not just numbers to be acted upon. He does indicate, however, that spatial boundaries might still be necessary and thus a need for physical dividers in the landscape.

The need for—strict—spatial boundaries to control interactions between humans and wild boars at the Veluwe is also supported by most permanent residents. Generally speaking, permanent residents are not too concerned about the numbers of boars present in their surroundings and suggest that they are able to take their own measures to manage their relation with the boars. As a local resident states, "I like to see wild boars, but I wouldn't like to see a wild boar disrupting my yard every night. I would place a fence". When residents place fences as a physical divider around their garden, it can be regarded as a type of confinement, imposing restrictions in the movements of both humans and wild boars.

In contrast, wildlife managers emphasize the need to control wild boar numbers and behavior in order to maintain public safety. Next to regulation on population level, this also includes controlling the individual behavior of wild boars (here interpreted as killing so-called problematic wild boars). For instance, an in-the-field wildlife manager said that he once saw a male wild boar walking along the road several times, and that boar had to be shot, because he was likely to cross the road and could cause an accident. The anxiety over potential accidents is expressed in the next statement by a wildlife manager. He also stresses the importance of regulating population size and meeting the desired population target:

> ...when we don't meet the targets [to hunt a certain amount of wild boars]...when somebody is killed by a crossing wild boar, national media jumps in and points at us that we did a bad job, now somebody is killed.

The above interchanges reveal a distinction in focus concerning current wild boar management between wildlife managers and residents. Wildlife managers justify the applied wild boar management strategies of confinement by emphasizing the need to control populations of wild boars in order to maintain public safety. Thereby they treat wild boars as a group, in plural, with a strict focus on numbers and restricted spatial presences accomplished through the creation of zero-tolerance areas. As a result of drawing and maintaining these tight numerical and spatial boundaries, humans and wild boars are in general separated from each other. Seen in this light, *co-existence* means living next to each other, not co-shaping the environment where both dwell. Residents, on the other hand, look towards relaxing the sharp boundary between humans and wild boars and instead focus on—individual—interactions with wild boars and alternative management strategies to align with wild boars, given the spatial limitations in densely populated areas.

To conclude, current wild boar management at the Veluwe can at best be described in terms of management strategies of confinement. Humans control the wild boars by drawing and maintaining sharp spatial boundaries, focusing on and managing their numbers and assigning them to a variety of categories (problematic, age, sex). By using these management strategies of confinement, the wild boar as animal disappears in numbers. In contrast, the next case illustrates strategies of alignment in which the relationship and interaction between human and wild animal is centralized.

4 Black Bear Management on the Colorado Front Range

Over the last decades, the Front Range of the Colorado Rocky Mountains, USA has become a heavily human populated area. The amount of (sub)urban areas has more than tripled state-wide since 1960 (Center of the American West 2001; Colorado State Demography Office 2014). Considering the increase in humans living on the Front Range, researchers argue that it is not surprising that human-black bear conflicts have rapidly increased in this period (Baruch-Mordo 2007). Since researchers indicate that the black bear population at the Front Range is increasing as well, conflicts between humans and black bears are even more likely to occur (Vieira 2011). There is a growing need to manage interactions between human and black bear on the Front Range, on the one hand to reduce human-black bear conflicts and secure public safety, and on the other hand to achieve the state-wide wildlife aims to enhance and sustain populations of native, desirable non-native and migratory wildlife populations for the use, benefit and enjoyment of Colorado residents and visitors (Colorado Wildlife Commission Policy 2009).

The overall aim in Colorado is to achieve coexistence between humans and black bears. In order to further this aim and manage the interactions between humans and black bears, two (complementary) directives have been introduced. They prescribe strategies concerning how to manage these interactions and how to deal with conflict situations. The first directive (2-strike management directive) is a state-wide directive, the second directive (trash directive) is an—optional—directive for municipalities. The implementation of the two directives with respect to the management of human-black bear interactions is described below.

The 2-strike management directive, introduced in May 1994, is a state-wide policy aimed at guaranteeing human safety. This directive is put into effect when a black bear is considered to be a 'problem bear' by an in-the-field wildlife manager, for instance because (s)he broke into a home searching for food and/or threatened people. The bear in question will then be trapped, tagged, and relocated to 'prime' bear habitat (first strike). A labeled problem bear trapped for a second time (second strike) will be killed. The intention of this directive is to keep black bears out of cultivated areas and give them a second chance by providing them with their own—safe and natural—surroundings. In other words, this policy aims at separating human and black bear dwellings to prevent problematic situations and assumes that black bears are intelligent animals that learn from previous experience. In-the-field-wildlife managers confirm that black bears are intelligent animals and that it is possible that they might be disciplined by implementation of this directive (trapping, relocating) and stay away from certain areas. But most in-the-field-wildlife managers regard implementing the 2-strike directive as problematic, as one manager explains:

> You can take a bear you can haul it a hell of a long ways but if it is a territorial bear its gonna be back. You know and so that is another reason why I can't tell these people [who had a conflict with that bear] with a straight face we are going to trap this bear and your problem is going away.

In fact, in practice most in-the-field-wildlife managers experience safeguarding human safety and black bear safety as a balancing act. One in-the-field-wildlife manager explains it in this way: "Combining the fundamental mandate to protect the public and protecting that bear is tough". One difficulty this balancing act creates involves the drawing of boundaries in that an in-the-field-wildlife manager needs to decide whether a particular bear is to be considered a 'problematic bear' and having done so then needs to decide whether or not to implement the 2-strike directive. These boundaries include spatial boundaries designating where black bears are allowed to roam and where not; did that bear cross a particular boundary? They also include categorical boundaries about what black bear behavior determines if (s)he is a problem bear. In fact, in-the-field-wildlife managers will try to determine what boundaries have been crossed by either black bear or human in order to discover the causes of the conflict they need to solve. Thereby they examine the actual interaction between human and black bear, to see what human behavior might have influenced the 'problematic' behavior of a particular black bear (e.g. feeding bears), what black bear behavior might have influenced humans to act in a specific way (e.g. breaking into a home). Wildlife managers deal with the difficult task of drawing appropriate boundaries that meet the needs of both human and black bear on a daily basis in bear season (approximately April-November each year). In practice, this also implies that implementation of the 2-strike directive is not always the best solution in solving human-black bear conflict situations according to many in-the-field-wildlife managers at the Front Range. They examine individual human-black bear interactions and subsequently determine what strategies to implement with respect to either black bear and/or human. In the following example an in-the-field-wildlife manager explains the difficulty of determining what interactions have occurred and the difficulty of drawing spatial boundaries to decide what management strategies to implement.

He describes a recent event when he received a call from a local resident reporting a black bear break in. The lady of the upper story of the house said it was her fault that the bear came in because she left the door open. The bear could smell food, entered her home and drank coca cola and ate a bunch of eggs and made a big mess in the living room and kitchen area. When the lady heard the noise, she came out and howled at the bear and the bear ran out the door. The wildlife manager talked to the lady the day after to determine what should be done. The main question he needed to answer was whether this bear had been aggressive to human beings. Did (s)he stand his/her ground did (s)he huff, bluff charge which would indicate that (s)he is a bear to be concerned about, in other words a problem bear. In this case it sounded like the bear took off, and in agreement with the neighbor living beneath the lady, they decided to place an 'unwelcome' mat at the front door. This is an electric mat that when the bear returns—which is probable according to the wildlife manager, because (s)he has associated the house with food—and steps on it, (s)he will get a shock and hopefully that will deter her/him from breaking in there again because it will create a negative association with (this) house(s). Within the framework of the 2-strike directive, the in-the-field-wildlife manager needed to separate the dwellings of human and black bear by gathering detailed information about the actual human-black bear interaction. The bear entered a home, which was considered as problematic behavior by the wildlife manager; the bear crossed the line of human property. But since neither the lady of the house nor the neighbors below had been hurt, and the bear took off, it was decided to discipline the black bear and not give her/him a first strike. In other words, (s)he wasn't trapped and the 2-strike directive wasn't implemented as such. In fact, as many in-the-field-wildlife managers indicate implementation of the 2-strike directive to be problematic, they tend to avoid applying the 2-strike directive by disciplining both the respective black bear(s) and the human(s). Therefore, as in the example, information about the behavior of the black bear and the humans in question is needed before deciding what management strategy to use in addition to separating their spatial dwellings as much as possible.

Avoiding implementation of the 2-strike directive is also an applied strategy used by many local residents and Bear Aware volunteers, as one explains: "Every time when there is a bear call we have to figure out how to deal with that bear and it could be several different things that we might try before we hit the one that works the most". She explains that first whatever attracted the bear has to be removed, and this mostly means the people need education (as in the case of the lady who needed to lock her door). If they decide to condition the bear, different strategies are examined:

> Using paint balls or rubber buckshots [rubber bullets and beanbags shot by a shotgun, stings the bear—but it doesn't hurt him—, and makes a lot of loud noise] so the bear will run away, or use pepper spray or an airhorn gun [compressed air that will make a lot of noise after pushing the button]. These airhorns work quite well at first but then the bears get used to that; it scares them if they are close but if they are not very close they just look at you. [We also consider] putting security things on top of the dumpsters and the trash that the bears do not like.

While the 2-strike directive is aimed at disciplining black bears, trash directives in some Colorado towns aim at disciplining humans. The aim of the directive is to get humans to store their trash in black bear resistant trash containers. If people don't store their trash properly they risk a fine. But before fining people, they are educated about trash storage and about how to prevent black bears feeding on anthropogenic food sources in general. This is done by in-the-field-wildlife managers as well as by Bear Aware volunteers (neighbors talking to neighbors). Next to talking with people, they also make use of brochures, bumper stickers ("Keep bears wild", "A fed bear is a dead bear"), newspaper articles, internet websites devoted to "Living with wildlife", TV and radio broadcasts, etc. In other words, disciplining humans to "behave Bear Aware" is diverse. For instance, one Bear Aware volunteer describes her activities in her neighborhood:

> I really worked in my community and south of here where there is a real problem with bears getting in the trash. I did that last fall. I had their brochures, but I also had my own. And I went to over 300 houses and left notices at their doors when they were not at home.

If people still don't comply with the trash directives in the respective towns, they risk a warning notice. If this still doesn't result in black bear proof trash storage, people are fined (the amount of trash-fines differ per municipality, but are generally around 100 dollars). The trash directive specifies the trash can as a boundary that is drawn between human and black bear and if storage is bear-proof, coexistence between human and black bears seems to be possible.

Next to trash-directives some towns also revised their building code to keep bears out of homes. These are generally towns in which an increased number of black bears breaking into homes had been observed. In one such building code, it is stated that it is no longer permitted to build a house with wooden door handles. Door knobs have to be solid and round; hollow door knobs can be squeezed by black bears, and levers can be pushed down so that the door can be easily opened by black bears. Physical dividers are also sometimes used to manage human-black bear interactions. For instance, one in-the-field-wildlife manager experiments with putting a perimeter around an entire camp ground located in prime bear habitat, with electric fence to anticipate any—food conditioned—black bear attacks.

The importance of disciplining humans in managing human-black bear interactions is stressed by the following statements of, respectively, an in-the-field-wildlife manager and a policymaker in a town discussing the introduction of a trash directive:

> The really important part of nuisance bear management is getting the public to understand that their actions impact bear behavior and ultimately what they do will result in the well-being of individual bears. We [wildlife managers] oversee populations of bears and so we can oversee and manage bear populations very well, but we need the public involved in doing things that will help us keep bears out of town and out of trouble.

> It's not that the community doesn't value their wildlife, they're just not having been convinced it's their role to secure their trash to protect their wildlife.

Currently, according to the above descriptions, policy directives on the Colorado Front Range are focused on either disciplining the problematic black bears or on disciplining residents. According to several in-the-field-wildlife managers, residents, and researchers as well as policy makers, black bear management is more about people than about black bears. However, in practice a variety of management strategies emerge, especially in avoiding giving a bear a first strike. As a researcher points out: "Black bear management is a mixture of social science and wildlife science that are in play here [at the Front Range]". The management strategies employed often focus on the relation between human and black bear, on the interaction that occurred and how to align them. These management strategies of alignment address individual black bears and individual humans, and aim at both human and bear safety. Aligning black bears and humans is achieved by disciplining them simultaneously. In other words, the aim is to bring about mutual adjustments, thereby centralising the relationship between human and black bear.

5 Discussion

The first case described above demonstrated that managing human-wild boar interactions at the Veluwe employs management strategies of confinement. These strategies involve humans dominating wild boars by drawing and maintaining rigid boundaries (spatial, numerical, categorical) and by regulating wild boar movements through reducing population size or through the use of physical dividers in the landscape. In contrast, managing human-black bear interactions on the Front Range rests on the employment of management strategies of alignment, with, at its core, the relation between human and black bear and their mutual adjustments to local circumstances. Consequently, managing human-wild boar interactions at the Veluwe results in spatial separation between the two and thus moves away from the principle of coexistence according to which human and wild animal would be expected to co-construct the space in which both dwell. In contrast, while in managing human-black bear interactions the interactions between humans and wild animals are centralized with the aim to create mutual understanding and co-shaping of their environment.

However, the case of black bear management on the Front Range reveals that the 2-strike policy is still aimed at acting upon black bears when black bears don't learn from their mistakes and continue to transgress particular human-set boundaries. In addition to managing black bears, the emergence of trash directives in Colorado towns are representative of the need to manage and educate humans as well as the bears in order to achieve coexistence. In local practices, black bear management is even more promising than management directives suggest in their aim for coexistence. In practice, in-the-field-wildlife managers intend to avoid implementation of the 2-strike directive by gathering detailed information about the 'problematic' interactions, such as what boundaries have been crossed and how by either black bear or human, and simultaneously seeking for the best possible solution for both human and black bear.

In contrast, in wild boar management at the Veluwe static—policy driven—boundaries, mainly physical and numerical boundaries, guide practice. Populations rather than individual wild boars are managed according to these regulations. Traces of alignment between human and wild boar can, however, be found in local practices. Although the relationship between human and wild boar can at best be described as one of domination, building on the detailed knowledge about (individual) wild boar behavior collected by in-the-field-wildlife managers in the process of carrying out their duties may eventually result in thinking in terms of strategies of alignment and in an attempt to co-exist with these wild animals.

Retaining the idea of separating humans from wild animals as much as possible to reduce conflicts between them is untenable in this era of declining wildlife territories and human population growth. In addition, wildlife protection programs around the world are doing so well, that many wildlife populations are booming (Von Drehle 2013). This means that the need for (alternative) human-wild animal management strategies is urgent. A continued focus on either humans *or* animals doesn't lead to ideas about how to live *together*. As the cases illustrated, management can easily slide back to strategies of confinement and disciplining human or wild animal when policy, research or in-the-field management remain focussed on either one of them (Palmer 1996; Rinfret 2009). When approaching the issue of managing human-wild animal interactions as a collective undertaking in which both humans and wild animals have a 'voice', not in the sense of treating humans and animals as equals, but rather treating both as active participants in interaction, alignment seems to be possible and capable of generating solutions for dealing with current human-wildlife issues. This is not meant to suggest that strategies of confinement are undesirable or unnecessary. In fact, particular spatial boundaries may be necessary to safeguard human and/or wild animal safety. For instance, it might be necessary to protect wildlife territories where cultivation could otherwise result in the disappearance of resources wildlife might need for their survival.

To propagate management strategies that make use of flexible, dynamic boundaries in their aim for coexistence, both detailed knowledge about animal as well as human behavior, and their interactions is needed. This requires integrated knowledge drawn from both the natural sciences and the social sciences. It also calls for further exploration of what Lorimer (2005) calls a "multi-sensual approach", since humans and wild animals have different preferences in using their senses. Gaining detailed knowledge about the use of the senses in humans and wild animals contributes to mutual understanding and subsequently to the creation of strategies of alignment. Thereby the treatment of these animals as fellow-participants is key to a deeper understanding of the processes of power taking place between human and wild animal (Urbanik 2012).

Despite the differences in applied management strategies in the two cases, both practices demonstrate that the respective animals are seen as intelligent animals. In the case of current wild boar management in particular this raises several ethical questions, such as how to justify culling wild boars and making use of their intelligence to attract them and subsequently cull them? A continued and strict focus on numbers justified by implementing policy, does not obscure these questions. But when coexistence is aimed at, especially by means of managing interactions between humans and wild animals, these kinds of ethical questions are important to answer. To paraphrase Bradshaw (2009, 235) concerning the management of wild boars at the Veluwe: The end product serves not to prevent and disentangle human-wildlife conflicts, but predetermined policy. In the case of black bear management on the Colorado Front Range, these kinds of ethical questions are indeed questioned by in-the-field-wildlife managers. Especially in the implementation of the 2-strike directive. They argue about the justification of killing 'problematic' black bears after 2 strikes when the cause of conflict was human induced (e.g. leaving out food attractants). In-the-field-wildlife managers tend to deal with these questions by attempting to avoid implementation of the 2-strike directive, instead disciplining both humans and black bears simultaneously when a particular boundary has been crossed.

Finally, both of the described wildlife management practices make use of disciplining wild animals. Disciplining wild animals is ambiguous. Or better phrased, when wild animals are disciplined, the line between domesticated and wild animals is blurred (Bradshaw 2009; Swart and Keulartz 2011). In aiming for coexistence and applying management strategies of alignment like disciplining wild animals, this sensitive line between domestic and wild animals is put under pressure. Cassidy and Mills (2012, 506) clearly describe what is at stake when wild animals enter so-called human spaces by stating "the confusion over what to call an animal that roams across spaces humans imagine to be limited to themselves". Additionally, processes of power between individual humans and animals or groups, also depend on the place of a relation (Urbanik 2012). Finally, in this debate about the ambiguous use of 'wild' animals, it might also be questioned if 'wild' has solely become a strategic site in environmental politics (Whatmore and Thorne 1998). Whatmore and Thorne (1998) argue that the 'wild' is in fact surrounding us and not restricted and spatially confined to such strategic places as wildlife sanctuaries, nature reserves and other forms of areal separation. However, the designation 'wild' is still necessary to wildlife management practices since it requires characterizations of wildlife such as species distribution, species density and species abundance in the service of achieving its goals, for instance, the protection of the wild animals in question.

Concluding, as discussed by Lulka (2004), not only wild animals and humans move, boundaries move as well. Both cases have involved a variety of boundaries, and both revealed that, in practice these boundaries are dynamic and become blurred. To manage human-wildlife conflicts and aim for coexistence detailed knowledge about what and how boundaries are drawn in these local practices is needed, since, as the cases have shown, examining the boundaries drawn in policy directives alone does not suffice. In fact, the cases indicated that in managing human-wildlife interactions flexible, dynamic boundaries are a prerequisite to come to solutions. Additionally, in managing human-wildlife conflicts as relational endeavours, particularly strategies of alignment are essential. But, for humans and wild animals to coexist, both strategies of confinement and alignment seem to be required.

References

Baruch-Mordo, S. 2007. *Black bear-human conflicts in Colorado: Spatiotemporal patterns and predictors*. MSc Thesis, Department of Fish, Wildlife, and Conservation Biology, Colorado State University.

Baruch-Mordo, S., S.W. Breck, K.R. Wilson, and J. Broderick. 2009. A tool box half full: How social science can help solve human-wildlife conflict. *Human Dimensions of Wildlife* 14(3): 219–223.

Baruch-Mordo, S., S.W. Breck, K.R. Wilson, and J. Broderick. 2011. The carrot or the stick? Evaluation of education and enforcement as management tools for human-wildlife conflicts. *PLoS ONE* 6(1): e15681. doi:10.1371/journal.pone.0015681.

Bear, C. 2011. Being Angelica? Exploring individual animal geographies. *Area* 43: 297–304.

Bear, C., and S. Eden. 2011. Thinking like a fish? Engaging with non-human difference through recreational angling. *Environment and Planning D: Society and Space* 29: 336–352.

Bradshaw, G.A. 2009. *Elephants on the edge: What animals teach us about humanity*. New Haven, CT: Yale University Press.

Buller, H. 2013. Animal geographies I. *Progress in Human Geography* 38(2): 308–318.

Cassidy, A., and B. Mills. 2012. "Fox tots attack shock": Urban foxes, mass media and boundary-breaching. *Environmental Communication* 6(4): 494–511.

Center of the American West. 2001. Growth data sheet. http://www.centerwest.org/futures/archive/data_sheet.html.

Colorado State Demography Office. 2014. Population totals for Colorado and sub-state regions. http://www.colorado.gov.

Colorado Wildlife Commission Policy. 2009. Management of wildlife populations. June 11, 2009, 1–3.

Donaldson, S., and W. Kymlicka. 2011. *Zoopolis: A political theory of animal rights*. Oxford: Oxford University Press.

Francis, R.A., J. Lorimer, and M. Raco. 2011. Urban ecosystems as "natural" homes for biogeographical boundary crossings. *Transactions of the Institute of British Geographers* 37: 183–190.

Groot Bruinderink, G., D.R. Lammertsma, H. Baveco, R.M.A. Wegman, A.J. Griffioen, and G. J. Spek. 1999. *Aantallen wilde zwijnen in het Veluws Bos/heidegebied op basis van het natuurlijk voedselaanbod*. IBN rapport: Wageningen. (In Dutch).

Hinchliffe, S., and S. Whatmore. 2006. Living cities: Towards a politics of conviviality. *Science as Culture* 15(2): 123–138.

Hinchliffe, S., M.B. Kearnes, M. Degens, and S. Whatmore. 2005. Urban wild things: A cosmopolitical experiment. *Environment and Planning D: Society and Space* 23: 643–658.

Ingold, T. 2005. Epilogue: Towards a politics of dwelling. *Conservation and Society* 3(2): 501–508.

Johnston, C. 2008. Beyond the clearing: Towards a dwelt animal geography. *Progress in Human Geography* 32(5): 633–649.

Jones, R. 2009. Categories, borders and boundaries. *Progress in Human Geography* 33: 174–189.

Keulartz, J. 2009. Boundary work in ecological restoration. *Environmental Philosophy* 6(1): 35–55.

Knight, J. 2000. *Natural enemies: People-wildlife conflicts in anthropological perspective*. London: Routledge.

Latimer, J. 2013. Being alongside: Rethinking relations amongst different kinds. *Theory, Culture & Society* 30(7/8): 77–104.

Latour, B. 2004. *Politics of nature: How to bring the sciences into democracy*. Cambridge, MA: Harvard University Press.

Lorimer, H. 2005. Cultural geography: The busyness of being "more-than-representational". *Progress in Human Geography* 29: 83–94.

Lorimer, H. 2006. Herding memories of humans and animals. *Environment and Planning D: Society and Space* 24: 497–518.
Lorimer, J. 2008. Living roofs and brownfield wildlife: Towards a fluid biogeography of UK nature conservation. *Environment and Planning A* 40: 2042–2060.
Lulka, D. 2004. Stabilizing the herd: Fixing the identity of nonhumans. *Environment and Planning D: Society and Space* 22: 439–463.
Lulka, D. 2009. The residual humanism of hybridity: Retaining a sense of the earth. *Transactions of the Institute of British Geographers* 34(3): 378–393.
Marvin, G. 2000. The problem of foxes: Legitimate and illegitimate killing in the English countryside. In *Natural enemies: People-wildlife conflicts in anthropological perspective*, ed. J. Knight, 189–211. London: Routledge.
Milton, K. 2000. Ducks out of water: Nature conservation as boundary maintenance. In *Natural Enemies: People-wildlife conflicts in anthropological perspective*, ed. J. Knight, 229–246. London: Routledge.
Noske, B. 1997. *Beyond boundaries: Humans and animals*. Montreal, Canada: Black Rose Books.
Palmer, C. 1996. Taming the wild profusion of existing things? A study of Foucault, power, and human/animal relationships. *Environmental Ethics* 23: 339–358.
Philo, C. 1995. Animals, geography, and the city: Notes on inclusions and exclusions. *Environment and Planning D: Society and Space* 13: 655–681.
Philo, C., and C. Wilbert. 2000. Animal spaces, beastly places: An introduction. In *Animal spaces, beastly places: New geographies of human-animal relations*, ed. C. Philo, and C. Wilbert, 1–34. London: Routledge.
Provinciaal Bestuur van Gelderland. 2002. *Flora en Fauna in Gelderland Deel 2: Faunabeheer en schadebestrijding*. Arnhem (In Dutch).
Rinfret, S. 2009. Controlling animals: Power, foucault, and species management. *Society & Natural Resources: An International Journal* 22(6): 571–578.
Swart, J.A.A., and J. Keulartz. 2011. Wild animals in our backyard: A contextual approach to the intrinsic value of animals. *Acta Biotheoretica* 59: 185–200.
Thompson, C. 2002. When Elephants stand for competing philosophies of nature: Amboseli National Park, Kenya. In *Complexities: Social studies of knowledge practices*, ed. J. Law, and A. Mol, 166–190. Durham, UK: Duke University Press.
Thrift, N. 2008. *Non-Representational theory: Space, politics, affect*. London: Routledge.
Turnhout, E., C. Waterton, K. Neves, and M. Buizer. 2013. Rethinking biodiversity: From goods and services to "living with". *Conservation Letters* 6: 154–161.
Urbanik, J. 2012. *Placing animals: An introduction to the geography of human-animal relations*. Lanham, MD: Rowman & Littlefield Publishers Inc.
Vieira, M. 2011. *Black bear data analysis unit management plan, northern front range unit DAU B-3*, 34. Colorado: Colorado Divisions of Wildlife.
Von Drehle, D. 2013. America's pest problem: It's time to cull the herd. *Time Magazine*.
Whatmore, S., and L. Thorne. 1998. Wild(er)ness: Reconfiguring the geographies of wildlife. *Transactions of the Institute of British Geographers NS* 23: 435–454.
Wolch, J., and J. Emel. 1998. *Animal geographies: Place, politics, and identity in the nature-culture borderlands*. London: Verso.

Comment: Between Wild and Domesticated: Rethinking Categories and Boundaries in Response to Animal Agency

Sue Donaldson and Will Kymlicka

Abstract As the chapters is this section illustrate, we have to rethink our old categories of wild and domesticated animals. New relationships of mutual impact and hybrid management have been made necessary by relentless human expansion, anthropogenic climate change, and other ecological impacts. The animals involved in these new relations do not fit into the old dichotomy of independent wild animals untouched by humans on the one hand, or dependent domesticated animals under control of humans on the other hand. We need new ideas to help us understand the distinctive ethical challenges of these new relationships, with their mix of freedom and restriction, of independence and dependence, of self-willed agency and external control. The chapter authors of this section draw upon key concepts of animal ethics—care, flourishing, interests, intrinsic and instrumental value, capabilities, welfare, friendship—to negotiate human-animal entanglements. While broadly agreeing with their insights, we argue that their ethical approaches need to be integrated into a broader theory of interspecies justice which explicitly addresses issues of authority, responsibility and self-determination. The fact that humans inevitably affect and interact with ever more animals does not alter the fact that animals' lives are still theirs to lead, and that human management and intervention is legitimate only insofar as it respects animals as intentional agents. Our theorizing should begin by asking what kinds of lives animals want to live, what kinds of relationships, if any, they want to have with us, and whether our interactions with them bolster or inhibit their ability to lead such lives. We illustrate what such animal agency may mean using the case of the feral horses of Assateague Island.

S. Donaldson (✉) · W. Kymlicka
Queen's University, Kingston, Canada
e-mail: sld8@queensu.ca

B. Bovenkerk and J. Keulartz (eds.), *Animal Ethics in the Age of Humans*,
The International Library of Environmental, Agricultural and Food Ethics 23,
DOI 10.1007/978-3-319-44206-8_14

1 Introduction

The chapters in this section illustrate that our inherited dichotomy contrasting 'wild' and 'domesticated' animals is inadequate. Due to relentless human expansion, anthropogenic climate change, and other ecological impacts, an increasing number of animals fall into a liminal or contact zone, blending characteristics of wild and domesticated animals along various dimensions. Humans continue to move into areas of wild animal habitat, while some wild animals (the most adaptable ones) are moving into areas of human settlement. Moreover, some domesticated animals are being rewilded, and some wild animals are being captive bred in zoos and then released. The animals involved in these new relationships of mutual impact and hybrid management do not fit into the old dichotomy of independent wild animals untouched by humans, or dependent domesticated animals under control of humans. And as a result, many of the ideas that have governed wildlife management for the past century—ideas of conserving nature, respecting wildness, preserving species, or protecting biodiversity—no longer seem fully adequate. We need new ideas to help us understand the distinctive ethical challenges of these new relationships, with their mix of freedom and restriction, of independence and dependence, of self-willed agency and external control.

The authors in this section offer a number of thoughtful suggestions for how to think and act in an ethically responsible way regarding these complex new relationships. They draw upon key concepts of animal ethics—care, flourishing, interests, intrinsic and instrumental value, capabilities, welfare, friendship—to negotiate human-animal entanglements. While broadly agreeing with these insights, and with many of the more practical suggestions they give rise to, we will argue that these ethical approaches need to be integrated into a broader theory of interspecies justice. The language of justice is surprisingly absent in these chapters, even when describing what seem to be straightforward injustices, whether it is human instrumentalization of domesticated animals, or indifference to the harms we impose on wild animals.

Situating ethical issues of human-animal interaction in a broader framework of interspecies justice is necessary, in part to highlight their urgency. As Rawls famously said, justice is the "first virtue" of public institutions, in the sense that it sets the limits within which other goods and values can be pursued. While there are other goods than justice, those goods cannot be pursued using unjust means. So to name something as an injustice is to make clear that it is illegitimate, and not merely unfortunate or regrettable. The language of justice is important for another reason: it forces us to explicitly address issues of authority and responsibility. Much of the animal ethics literature takes as a given that humans always already have sovereign power over animals. Animals fall under our rightful authority and our jurisdiction, and the only issue is how to exercise our sovereign power (stewardship, management) more humanely or ethically. But whether animals fall under our authority is itself an issue of justice, and if we humans want to assert authority over animals or their territory, we need to show why we have a just claim to do so.

This commentary isn't the place to develop in depth an account of interspecies justice.[1] But the first step in developing such a theory is to ask what kinds of lives animals want to live, and then, as a corollary, what kinds of relationships, if any, they want to have with us. Animals have their own lives to lead—they are not instruments or resources for us to use in pursuing our lives and ideals—and so we need to start by asking what kinds of lives they want to lead, and whether our interactions with them bolster or inhibit their ability to lead such lives. Historically, the impact of human interactions has almost always been to erode this self-determining capability, and in many contexts respecting animals' agency requires us to simply get out of their way. But there are other cases where we are able, and indeed required, to interact in ways that facilitate animal agency.

Emphasizing the will or agency of animals is hardly unique in the animal ethics literature, and related ideas surface in these chapters. But even similar-sounding ideas may turn out to be quite different in motivation and implementation. Consider, for example, the term 'self-willedness'. Some authors use this term in ways that overlap with our ideas of animal agency, autonomy and self-determination (Keim 2014), but as Palmer notes, 'self-willedness' in environmental ethics is also often used to invoke two other ideas: (i) a condition in which animals escape human influence, intentionality or expectation; (ii) a condition in which animals express what are viewed as essential or natural species-specific capacities and behaviors (7). On our view, these further ideas are neither necessary nor sufficient to respect animal agency, and may indeed contradict a commitment to animal agency.

Our conception of animal agency is not (or not only) about an absence of human domination which allows so-called natural or species-specific behaviours to emerge. It is about the conditions under which humans and animals are free to exercise control, to make decisions, and to develop and emerge as individuals (and as groups of individuals). It is therefore not limited to knowable and predictable behaviors or species capacities. It is about unique and unpredictable interactions and responses, and about capabilities which emerge in relationships with other intentional agents (including humans), and/or in relationship to the environment (Lestel 2011).

Put another way, the extent to which animals wish to explore relations with humans or other animals (or their environments), and thereby develop new activities and relationships not pre-determined by their species biology, is precisely one of the issues to be determined by reference to their will. To insist that animals maintain some preordained level of human independence and/or species-typical behaviour does not respect their agency, but rather pre-empts and constrains their agency.[2]

---

[1]We develop our fuller account in Donaldson and Kymlicka (2011).

[2]That the term 'self-willedness' as used in environmental ethics is often at odds with animal agency is further illustrated by the fact that self-willedness is often applied indiscriminately to animals, plants and ecosystems, collapsing the idea of agency as subjectively experienced by intentional beings with the idea of 'actant' in actor network theory. To equate animals and plants in terms of their 'self-willing' is precisely to erase animals' capacity for intentional agency.

If we start from a commitment to animal agency, it will quickly become clear that animals are far from homogenous in the sorts of lives they want to lead, or in the sorts of relationships they want to have with us. And this is true not only for animals in the expanding liminal zone where the boundary between wild and domestic is blurred.[3] Acknowledgement of, and respect for, animal agency requires rethinking our approach to all animals, wherever they lie on the continuum of wildness or domestication. Whatever our relationship with animals—close or distant, intimate or stranger—our guiding concern should be to respect and uphold their ability to lead their own lives, rather than undermining the possibilities for them to do so.

2 Rethinking the Wild-Domestic Continuum Politically

What would this commitment to animal agency mean in practice? We will return to the complexities of animals in the liminal zone below. But to begin it is helpful to consider the category of domesticated animals, and to think about animal agency in that context. As noted by several authors, there are many competing definitions of domestication. Animals (including humans) can become domesticated through a variety of processes involving different degrees of purpose, mutuality, and intensity. For the purposes of our discussion, we will use the term in the 'control' process or 'suppressing' sense identified by Swart to describe a practice in which humans enclose and breed animals to serve human purposes (3). Domestication in this sense is by definition a process which instrumentalizes animals: violating their rights (through confinement, coerced breeding, premature death, etc.) and deliberately modifying them in order to make them more useful to, and dependent upon, humans.

The process of domestication in this control sense is pretty much the opposite of respecting animal agency, and any plausible conception of interspecies justice would need to prohibit any further controlled domestication of wild animals. But the more challenging question is how we should think about the agency of currently-existing domesticated animals. Some commentators have argued that, since domestication in the control sense violated animal agency, we should seek to de-domesticate or rewild the animals who resulted from this process, as exemplified in the efforts to rewild cattle, horses and red deer in Oostvaardersplassen (OVP, discussed by Swart). By withdrawing human support and care, and moving them into large unfenced territories, de-domesticating these animals enables them, over time, to become more 'self-willed'.

In fact, however, the OVP program of rewilding has nothing to do with respect for animal's intentional agency. The horses, cattle and red deer were given no

[3]We discuss liminal animals in Donaldson and Kymlicka (2011): Chap. 7. Swart uses the term "semi-wild" animals. Palmer refers to animals in the "contact zone".

choice about whether to be the subjects of this experiment, and the OVP was in no sense a response to the will of the animals themselves. It would be quite possible to create the circumstances that would allow individual animals to safely explore the option of a gradual and partial retreat from human society, and thereby initiate a process of rewilding. If such options were made available, it is almost certainly the case that some animals would choose gradual exit from human relationships, while others would choose closer and more intimate friendship and cooperation with humans.[4] But the OVP made no attempt to solicit or be responsive to the will of the animals themselves. On the contrary, they were simply expelled from human society in a unilateral decision by human planners. To describe this as respect for self-willedness involves a conflation of individuals and species—a conflation that Bovenkerk notes is pervasive in the literature. In this case, action to achieve greater self-willedness at a species level obscures disregard for the agency of individuals.

And so, rewilding, like domestication, can be an instrumentalizing and dominating process—one which violates the rights of individual animals—by deliberately leaving domesticated animals to fend for themselves in rewilding zones with the expectation that many will die but over time there will be re-adaptation at the species level. A domesticated animal left to her fate in the wild becomes a less self-determining agent, not more. Her lack of skills, experience and adaptation to this environment mean that her capabilities are diminished. She falls outside "the range of conditions that animals can still flourish under", in Swart's terms (12), regardless of the presence or absence of fences or of humans. This is typical of many rewilding projects, which view individual animals as mere tools for the achievement of a human desire for 'the wild'.[5]

The idea that forced rewilding somehow compensates or remedies forced domestication is deeply misguided. It misconceives the injustice of the domestication process, which was the imposition of human power to suppress individual animals' agency in order to pursue human ends. The OVP involves precisely the same exercise of human power at the expense of animal agency.

It is crucial therefore to distinguish the injustice of domestication as an instrumentalizing *process* from any moral evaluation of the individuals *resulting from* that process, wherever they fall along the wild-domestic continuum. The fact that someone's existence is the result of an unjust process doesn't alter her inherent value, or her claims to justice. We are not morally diminished beings because of the immoral actions of those who have instrumentalized us, or made us more dependent and vulnerable, or less 'wild'.

What this points to is that justice requires that we start from the claims of existing individuals. We need to start from who they are, in whatever circumstances of social incorporation and dependency they have inherited. And we must support their agency (whatever their level of dependency), without subsuming it to some aspiration about an ideal state of wildness or domestication that *we* are trying to

---

[4]We discuss such a process in Kymlicka and Donaldson (2014).

[5]For other examples, see Shelton (2004).

achieve. Treating individuals justly means that *they* control key dimensions of the nature and trajectory of their lives, and that we ask ourselves how we can enable these actually-existing animals to lead the lives they want.

Now let us turn to the other extreme of the domesticated-wild continuum—that is, animals who live, or try to live, remotely from humans in wilderness areas—and ask what sorts of lives they want to lead. Many of these 'truly wild' or 'wilderness' animals evade, insofar as they can, human management and control through avoidance, and withdrawal from areas of contact. They are typically unable or unwilling to adapt to human settlement or disturbance, and will languish or die if there is no available (and ecologically suitable) wilderness to retreat to. Some fashionable talk of 'The Anthropocene' implies that we can safely ignore or discount their wish to avoid human management as obsolete, as if there were no longer any possibility of defining territories that fall outside the property and sovereignty of human beings. It may well be true that anthropogenic climate change means that all parts of the earth are now affected by human activity. But again, the first requirement of justice is to ask what sorts of lives animals want to lead. And the fact that most of the planet is feeling the effects of human activity does not negate the desire or ability of many wild animals to live free from human management. Anthropogenic climate change doesn't mean that all human-animal relations are now unavoidably in the liminal zone of inevitable human contact, co-existence and/or management. Palmer notes the important difference between human 'effects' on wild animals and human 'constraints' on them (10). Many animals still live remotely from humans in relatively intact ecosystems, relatively free from human control—though not from human-caused effects—and they have a right to continue doing so.[6] While talk of the Anthropocene may originally have been intended to draw attention to the pernicious effects humans are having on wilderness, it has had the perverse effect of making continued human encroachment on, and management of, wild animal habitat seem inevitable. It thereby erases the agency of billions of non-human creatures who are actively seeking to live free of human management.[7]

[6]As Brandon Keim (2014) notes, the Chernobyl disaster provides an important lesson about the difference between human effects on animals and human domination of animals. A 30 K exclusion zone around the nuclear plant has been in place for 30 years now. Countless individual animals in the region suffered terribly from the original radiation release, but as radiation levels have dropped, wildlife populations and diversity have not just rebounded but now significantly exceed surrounding regions. As one researcher notes: "It's very likely that wildlife numbers at Chernobyl are much higher than they were before the accident. This doesn't mean radiation is good for wildlife, just that the effects of human habitation, including hunting, farming, and forestry, are a lot worse" (Jim Smith, University of Portsmouth, quoted in https://news.vice.com/article/chernobyls-exclusion-zone-is-now-a-thriving-wildlife-habitat). Even a devastating human spillover effect on an ecosystem, like radiation poisoning, pales in comparison to the constraining/controlling impact of human presence, numbers and direct destruction of animals and their habitats.

[7]See Crist (2013) for a compelling argument for rejecting the discourse of the Anthropocene on these grounds. See also Weurthner et al. (2014).

Justice requires acknowledging this agency, and granting these animals the rights to territory that are a precondition for their self-determination.[8]

For many wilderness animals, the best way for us to support their agency is to get out of the way. As we have argued elsewhere, this will require cessation of human colonization and despoliation of remaining wild(er) areas, and in some cases retrenchment from key habitat to restore necessary connectivity and migration corridors. It will require that humans drastically reduce our resource consumption, by living smarter and more sustainably, and by halting human population growth (Donaldson and Kymlicka 2011: Chap. 6). This doesn't mean that humans will cease having impacts in wild(er) areas, or that we should be prohibited from helping animals in the wild, or from being visitors or denizens in wild animal territories. But it does mean that these lands belong to animals, and they should be free to live there without the threat of human invasion and management. And to repeat, the reason why colonizing wilderness is unjust is not that it fails to respect the value of 'nature' or 'wildness'. Like Palmer, we are sceptical of the direct value of wildness as such (14). The problem with human colonization of wilderness is that it constrains, and thereby violates the right of wild animals to live self-determining lives on their territory.[9]

So we would argue that respect for animal agency/self-determination, as the first principle of interspecies justice, requires rethinking how we approach the familiar categories of domesticated and wilderness animals. However, the main focus of the chapters in this section is on cases that do not fit these traditional categories. The authors primarily consider cases where animals overlap space and interests with humans in ways that require co-existence strategies, even if animals' day-to-day lives do not involve close cooperation and interpersonal relationships with humans. How would a focus on supporting animal agency help to achieve justice for liminal animals in the contact zone?

[8]We realize that it is reductive to talk about animals' agency in these general terms, given the millions of different species and their unique situations. For some animals, the realm of meaningfully self-determined action may be minimal or non-existent. For others, it will be capacious. Whatever the case, the burden lies on humans who would manage animals' lives and animals' territories to demonstrate that they have looked for and been responsive to animal agency, and not simply to ignore or dismiss it. We should also note, contra Swart (13), that recognizing the importance of animals' agency, autonomy, or self-determination does not presuppose the idea of self-sufficiency. For all of us, autonomy is a relational construct involving various kinds of inter-dependencies.

[9]Work in deep ecology shares a similar view that humans must learn to see ourselves as visitors in wild animal habitat, without rights of colonization, but tends to defend this view in terms of the inherent value of nature, rather than the sovereignty rights of wild animals (eg., Wuerthner et al. 2014). This disagreement about underlying justification need not and should not preclude cooperation between animal right advocates and deep ecologists, but as we noted earlier, the value of nature must be pursued within the constraints of justice. Otherwise, the inevitable result is the instrumentalization of animals in pursuit of a human vision of nature, as in the OVP rewilding projects.

3 Rethinking the Contact Zone: The Feral Horses of Assateague Island

Unlike animals in the wilderness, who tend to avoid us, liminal animals can and do share territory with us, and so we must learn to co-exist. Unlike domesticated animals, it is typically neither possible nor desirable to enter into cooperative activities or intimate friendships with them. As Drenthen notes, attempts at forming closer relationships often lead to conflict and disaster for the animals involved.[10] We need to find a form of peaceful co-existence which manages the risks that humans (often unintentionally) impose on animals and the risks they impose on us. These forms of co-existence and risk management sometimes require restrictions on mobility—fences and barriers to protect them from us or us from them—as well as targeted interventions to deal with issues of disease or population pressure. The authors in this section offer a number of thoughtful reflections on these measures, but here again we would argue that the legitimacy of these restrictions and interventions can only be assessed within a broader theory of interspecies justice. A fence can mean very different things, depending on the broader political context.

To illustrate, we will briefly describe the case of the feral horses of Assateague Island,[11] which offers a helpful illustration of the politics of governing the contact zone, and the different meanings and purposes of boundaries, fences and interventions.[12] Assateague is a long (approx. 60 k) narrow barrier island off the east coast of the United States, bisected by the Maryland-Virginia state border. The northern part of the Island, in Maryland, is managed by the National Park Service. The southern section, in Virginia, is managed by the US Fish and Wildlife Service (USFWS). Assateague is home to herds of feral horses, rewilded descendants of domesticated horses who first appeared on the Island 300–400 years ago. The breed has adapted over time to the challenging coastal environment of limited vegetation and shelter, wild storms, temperature extremes, and mosquitos. The State boundary fence separates the equine population into two separate groups of approximately 120–150 individuals. Both herds share the same genetic inheritance and adaptation to a particular ecological niche. On both sides of the fence, population growth is controlled in order to prevent overgrazing and degradation of the ecosystem which supports the horses, and other wildlife (including deer, and many protected species

[10]For a similar argument, see Panagiotarakou (2014).

[11]We should note that the following discussion is based on an analysis of the policy statements of the agencies which manage the feral horses, not on first-hand knowledge. Information in this section is largely derived from the National Park Service website and reports: http://www.nps.gov/asis/learn/nature/horses.htm;
http://www.fws.gov/uploadedFiles/Appendix%20D_CHN%20Draft%20CCPEIS.pdf
http://www.nps.gov/asis/upload/feralhorsemanag.pdf
https://themustangproject.files.wordpress.com/2011/02/nps-feral-horse-management-at-assateague-island-national-seashore-2006.pdf.

[12]We are grateful to pattrice jones for alerting us to this case, and some of its implications, although she bears no responsibility for our interpretation of it.

of birds). Assateague Island is a popular destination for human visitors wishing to hike, camp, fish/hunt, kayak, and observe wildlife.

Thus, we can say that the Assateague equines are liminal or semi-wild animals sharing an overlapping space with humans, and under some level of management, stewardship or intervention by humans in this contact zone. Inherited ideas of 'wildness' are not particularly helpful for guiding our interspecies interactions in this instance. Given the horses' unique history, their island location with its strictly limited carrying capacity, and the presence of other resident animals also reliant on the Island ecology, human intervention seems to be necessary to prevent catastrophic decline. The animals wouldn't benefit from humans completely removing ourselves from the equation. So how do we ensure that our interventions enable rather than undermine these animals' agency, in a relationship of interspecies justice?

As it happens, while both herds are affected by human interventions, the policies adopted on the different sides of the State border have vastly different implications for animal agency, and thus help to illuminate the relationship between interventions and justice. Following convention, we will call the northern, Maryland equines "Assateague horses" and the Virginia herd "Chincoteague ponies". The Chincoteague ponies are owned by the local Volunteer Fire Company, which manages them in cooperation with the USFWS. These ponies are kept within large fenced areas to restrict them from wandering into roads and sensitive habit. This mobility restriction sometimes constrains their ability to shelter from storms, escape insects, and find forage and water during drought—which in turn makes it necessary for humans to provide the ponies with supplemental water and food when conditions reduce the availability of natural sources within their appointed areas. (Ponies sometimes escape their containment areas and are rounded up and returned.) Human visitors can view the ponies from a distance, but direct interaction is discouraged. The pony population is kept from increasing by auctioning off dozens of foals every year, with proceeds going to the Fire Company. In July, cowboys round up the horses, drive them into the ocean and force them to swim to nearby Chincoteague Island. They are penned near huge crowds of humans on hand to watch the pony swim and take part in the auction (events which are the biggest tourist draw to the area). Then the foals are forcibly removed from their families, and sent for 're-domestication' and an unknown fate on the mainland. Their parents swim back to Assateague.[13]

By comparison the Assateague horses seem to be subjected to many fewer interventions. They roam unrestricted over the northern Island, taking care of their own needs for food and shelter. They associate freely, form bands, and raise families—structures which are not torn asunder by humans. They tend to keep their distance from humans, and can be quite aggressive when approached. Horse-human

[13]The swim can be very stressful for the horses, especially young foals, and lactating mares who sometimes have to be treated for hypocalcemia. Horses also suffer cuts from oyster and clam shells. See https://www.avma.org/News/JAVMANews/Pages/081101f.aspx.

relations are managed through education (keep a distance; don't feed the horses; drive carefully) rather than internal fences. (There is a small state park where visitors can camp, and this is where human-horse conflicts tend to arise.) The horse population is regulated through the use of contraceptives which are administered remotely by dart, thereby avoiding any need for corralling or handling by humans. Every mare is allowed to give birth at least once, but is eventually put on a contraception program for the rest of her life. This policy has successfully limited the total population, while maintaining an adequate gene pool. Birth control has also led to better health and longer life for the mares.[14] The horses are not otherwise handled or treated by vets (unlike the Chincoteague ponies who are rounded up twice yearly for health checks and vaccination, in part to keep infectious diseases from transferring to the mainland horse population).[15]

On our view, the Maryland policy appears to be much more respectful of animal agency than the Virginia policy. We suspect that most of the authors in this section would agree, and the chapters offer various concepts to explain why. On Palmer's analysis, we might say the Assateague horses, compared to the Chincoteague ponies, retain a greater degree of "self-willedness", in which humans "[resist] the urge to control nature and [allow] it to find its own way" (Monbiot 2013, 9 cited in Palmer). In Swart's terms, their care falls further towards the "non-specific care" end of the spectrum: "care for the animal's natural environment, not specifically tailored to the individual traits of the wild animal but rather to ecosystem elements that will contribute to the survival, subsistence, and natural flourishing of the animal" (Swart p. 10). In Boonman-Berson's terms, human-horse relations for the northern herd are governed by a comparatively higher degree of "alignment" rather than "confinement" to serve human ends. In Drenthen's terms, the relationship comes closer to friendship conceived as a "combination of loving commitment and recognition of insurmountable otherness" best served "by maintaining a certain distance from one another" (p. 12). In Bovenkerk's terms, the Assateague horses have greater freedom, such as "control over one's own decisions, space to roam around, and being able to live according to one's own nature or carrying out one's natural behaviour" (9).

We find all of these descriptions suggestive but incomplete. As noted earlier we believe that the key right at stake here is agency or self-determination. The Assateague horses are more self-determining, and this is the crucial distinction between them and their relatives to the south.[16] Our goal in the remainder of this

---

[14]The contraceptive darting is probably stressful for the mares, and sometimes can cause temporary inflammation, but does not cause serious injury. We find it curious that the papers in this section make frequent reference to culling as a means of population control, but do not discuss contraception.

[15]The other vet intervention in the North is that Assateague horses are sometimes euthanized if they are suffering a painful decline/death.

[16]We don't wish to idealize the situation of the northern herd. They are more self-determining relative to the Chincoteague ponies, but this doesn't mean that human interactions are based on this value, or consistently achieve it.

commentary is to show how the concept of self-determination illuminates: (i) positive interventions versus dominating ones; and (ii) empowering boundaries versus captivity.

4 Positive Intervention Versus Domination

The idea that animals have an interest in agency is implicit in most of the chapters, but it needs to be made more explicit and applied more comprehensively. As noted earlier, we are concerned here with intentional agents, i.e. beings with a subjective welfare, whose lives can go better or worse from the perspective of the individual living the life. And by 'agency' we mean self-willed or initiated action which carries an expectation of efficacy. In other words, agency is realized in the relationship between individuals, other agents, and their environments. Our social and physical environment can support and enable our agency, or ignore, disrupt, thwart and suppress it. On any account of wellbeing, intentional agents have powerful interests in being able to realize their agency. This is the basis for many of our basic human rights including negative rights not to be harmed, killed, or held captive; and positive rights to mobility, free association, and self-determination. Our right to have our agency respected and supported is constrained by the comparable rights of other agents. But when our agency is unfairly thwarted (i.e., in a way which fails to recognize the full and equal recognition of our rights), we are dominated—subject to the unjustified power of another in an unequal hierarchical relationship.

Many animals have the capacity for intentional agency, although humans frequently ignore and suppress this in our relationships with them. They have the need and desire to exercise control in their lives—not just to make temporally localized decisions (what we have called "micro agency") about when to eat or to sleep, but also significant decisions about the general shape and structure of their lives ("macro agency") concerning where and how they live; who they mate with, live with and associate with; what sorts of activities they learn about, engage in, and pursue mastery of.[17] These larger ends are not given or determined by species biology, although species membership certainly sets key parameters of wellbeing. Like us, horses are not just members of a species; they are individuals with different capabilities, interests, personalities, and desires. And, like us, they need extensive freedom in order to explore, learn, and make choices—to shape their own identity in relationship to their social and physical environment. Unlike (some of) us, they do not rationally conceive a life plan in light of their interests and values, or rationally revise such a plan in light of experience. But they do explore, learn from experience, and make choices in light of their needs, desires, developing identity and understanding.

---

[17]We develop this notion of agency, and the micro/macro distinction, in Donaldson and Kymlicka (2016).

When we recognize that many animals have an interest in macro agency, and not just wellbeing defined as preference satisfaction or species-typical functioning,[18] then we can see with greater clarity what distinguishes the Assateague and Chincoteague herds. They belong to the same species and breed, and are equally adapted to cope with the physical environment that they share. And they are both subject to human intervention to some degree. If we take too broad a lens, we might say that all of the Island equines are in the same boat. Their ancestors were abandoned (or shipwrecked) by humans, and over generations underwent a process of (unassisted) de-domestication and adaptation which undoubtedly caused great suffering for many individuals. They are all subject to a human political regime which has divided the Island into two parts, bisected by a fence which keeps the two herds apart in perpetuity. Latterly, humans have become concerned about protecting the environment of the Island, with its rare plants, animals and shoreline habitat; and its recreational and aesthetic opportunities. This has resulted in population management policies to keep the equines from degrading the environment. And quite recently it has become clear that the Island is being affected by climate change, and this will bring yet another level of human management and concern which will affect the animals.

Similarly, if we take too narrow a lens, we might say that the day to day lives of the equines on either side of the fence aren't so different. They roam, graze, seek shelter from the sun or wind, look for fresh water and sleeping spots, enjoy (or tolerate, or reject) each other's company, engage in sex, gallop along the beach to escape mosquitos, or just for the thrill of it.

Neither of these perspectives captures the essential difference, which is that in some fundamental sense, the Assateague horses' lives belong to them in a way that the Chincoteague ponies' do not. What matters from this perspective is not *that* humans affect the equines' lives along some dimensions. It's *how and why* humans do so. From this perspective, the Chincoteague ponies are subject to extensive domination whereas the Assateague horses are not. And this domination is illuminated by thinking about how the ponies' macro agency is systematically suppressed on the Virginia side of the border.

The Chincoteague ponies are a resource in the crudest sense, owned and managed for the profit of the Fire Company. Mobility restrictions constrain the ponies' ability to take care of themselves, making them more dependent on the actions (and whims) of humans to make sure they don't run out of food and water, or lack for adequate shelter. They are subject to various kinds of coercive treatment (corralling, medical procedures, forced swims) undertaken primarily for human purposes and benefit. Young foals are kidnapped from their families, thereby severed from those who care for them most, and banished from the only world they know and have learned to negotiate and thrive in. They are sold (for approximately $1500) to mainlanders who are free to use these horses for their own ends (family pet, work

[18]Palmer and Bovenkerk discuss these preference-satisfaction and functionalist conceptions of wellbeing, none of which adequately capture the importance of macro agency.

horse, roadside attraction, investment, etc.). The removal of foals severs family relationships, group identities, and intergenerational learning. Through these practices of interference, coercion and destabilization, the ponies' ability to function as intentional agents is systematically undermined. Humans exert power over the ponies for their own ends, rather than interacting with them as full and equal agents in their own right.

5 Boundaries Versus Captivity

Macro agency, and its centrality to a theory of interspecies justice and non-domination, can help us analyze the issue of captivity raised by many of the papers in this section. Captivity is an inherently normative notion, identifying a condition in which an individual is deprived of certain basic rights or interests (freedom of movement, association, etc.). We are harmed, or wronged, by captivity, by definition. This contrasts with ideas of boundaries or restrictions, which are normatively neutral.[19] We are all restricted by our human nature from living unassisted under water, or in the air, or on Mars. This doesn't mean that we're captive on land/earth. Similarly, various restrictions (citizenship, field hockey club membership, chef school entrance requirements, militarized zones, private property, safety fences) may limit our mobility or options, but not in a way which necessarily renders us captive in normative terms. Restrictions don't render us captive (or unjustly excluded) as long as we still have many meaningful options open to us, and opportunities for exercising macro agency in writing the script of our own lives, and as long as restrictions on our agency don't have the intention or effect of subordinating our interests to the interests of others. What matters is not that all options are open to us, but that a meaningful range of options are open to us, and that we have the liberty and opportunity to explore them on the basis of equal consideration with others. The fact that we are restricted from some options does not mean we are captive or dominated.

Indeed, restrictions can work to enable our macro agency, by protecting us, and by providing secure boundaries within which we are free to act. This is a familiar theme in political theories of liberty. If the State restricts entrance to an area affected by a chemical spill, I can be relatively secure that I am safe to walk outside that zone without injury. If the State enforces traffic rules, I can feel relatively secure about driving. When the state enforces laws against trespass and stalking, it creates a zone of freedom and security.

Since restrictions aren't inherently dominating, the mere fact that animals are restricted (by external barriers like fences, buildings, or habitat conditions; or internal restrictions like states of interdependency) does not yet tell us whether they are captive or dominated. Some animal rights theorists believe that domesticated animals are necessarily captive if they live within enclosed spaces (bounded =

[19]See Streiffer (2014) for a helpful discussion.

captive), or because of their level of dependency on humans (dependent = captive). Similarly, wild animals are often viewed as captive by the mere presence of fences (physical restriction = captive), or the mere presence of human intervention in their lives (intervention = captive). But these chains of equivalence are all too quick. We need to analyze how restrictions in fact impinge on animal agency. Do they exclude the individual from a meaningful range of options for exercising macro agency and leading a flourishing life? Are they imposed in order to favour the interests of some individuals over others?

In the case of the Assateague horses in the north of the Island, it seems that the State fence operates (to a significant degree) to protect their agency, and that human intervention in their lives is (substantially) non-dominating. Their lives are physically bounded by their Island life, and by the fence along the Maryland/Virginia border. Their mobility and options are limited by these boundaries, but not in a way which denies them meaningful options or opportunities to exercise macro agency and live flourishing lives. Indeed, we could say that the boundary fence, far from rendering them captive, actually creates a zone of freedom for them in which they are protected from capture and harm, and from having their environment despoiled or settled. The issue of human intervention is more complicated. Enforced contraception is clearly a serious restriction on the horses' opportunities for raising family and extending their kin networks. However, each mare is able to have at least one foal. Moreover there are important trade-offs in terms of the affected horses' own interests (in terms of better health, and in not over-grazing their territory). Humans intervene in the lives of the Assateague horses, setting a boundary on their free movement and on their number of offspring. But these restrictions have limited impact on their options for leading flourishing lives. Moreover, they are imposed in a way that (at least to some degree) respects the horses' right to lead their own lives, and supports them in doing so. Theirs is a precarious freedom, especially in the face of climate change, and because it is freedom at the whim of a human regime which can remove the protections at any time if it decides that it no longer values the 'wildness' of the Assateague horses. But we could imagine a situation in which the existing policy was legally and institutionally entrenched as a right of the Assateague horses, protecting them from domination into the future.

The boundary fence which defines a zone of meaningful self-determination or macro agency for the Assateague horses is the very same fence that defines a zone of captive domination for the Chincoteague ponies. Their lives are not their own, but the property of the Fire Company which takes many of them from their families and auctions them for profit—an activity with many additional spin-off harms. Mobility restrictions within the Virginia portion of the Island further suppress the Chincoteague ponies' opportunities for developing and exercising macro agency, compounding the sense in which their lives are not their own.

The difference between the two halves of the Island is striking, yet it is not a difference we can detect simply by noting whether there are fences, or whether there is human intervention. It is a difference in agency and domination. It therefore nicely illustrates our fundamental claim that human interactions with animals, whether wild, liminal or domesticated, should be governed by the goal of supporting animals' self-determination/agency, and not, in the first instance, by values of preserving wildness, or species, or biodiversity for their own sake.[20]

References

Crist, E. 2013. On the poverty of our nomenclature. *Environmental Humanities* 3: 129–147.

Donaldson, S., and W. Kymlicka. 2011. *Zoopolis: A political theory of animal rights*. Oxford: Oxford University Press.

Donaldson, S. and W. Kymlicka 2016. 'Rethinking membership and participation in an inclusive democracy: Cognitive disability, children, animals. In *Disability and political theory*, ed. B. Arneil and N. Hirschmann. Cambridge: Cambridge University Press (forthcoming).

Keim, B. 2014. 'Earth is not a garden,' *AEON*, September 18, 2014. https://aeon.co/essays/giving-up-on-wilderness-means-a-barren-future-for-the-earth.

Kymlicka, W., and S. Donaldson. 2014. Animals and the frontiers of citizenship. *Oxford Journal of Legal Studies* 34(2): 200–219.

Lestel, D. 2011. What capabilities for the animal? *Biosemiotics* 4: 83–102.

Monbiot, G. 2013. *Feral: Rewilding the land, the sea and human life*. Toronto: Penguin.

Panagiotarakou, E. 2014. Right to place: A political theory of animal rights in harmony with environmental and ecological principles. *Ethics Forum* 9(3): 114–139.

Shelton, J-A. (2004). 'Killing animals that don't fit; Moral dimensions of habitat restoration,' *Between the species*, issue 4 (http://digitalcommons.calpoly.edu/bts/).

Streiffer, R. 2014. The confinement of animals used in laboratory research: Conceptual and ethical issues. In *The ethics of captivity*, ed. L. Gruen, 174–192. Oxford: Oxford University Press.

Wuerthner, G., E. Crist, and T. Butler. 2014. *Keeping the wild: Against the domestication of earth*. Washington, D.C.: Island Press.

[20]Our position is not that there aren't values in wildness and biodiversity that we should attend to. But justice must set the parameters for our pursuit and realization of them.

Part III
Between Freedom and Captivity

Towards an Animal Ethics for the Anthropocene

Jozef Keulartz

Abstract The complex problems of wildlife conservation during the current stage of the Anthropocene—the 'Great Acceleration'—are forcing us to develop an alternative to the traditional (utilitarian und deontological) approaches within animal ethics. I will put forward Martha Nussbaum's capability approach as a promising alternative to these traditional approaches, with the proviso that the current version of her list of basic animal capabilities will need to undergo some revision.

> A dog has the right to be a dog.
> Article 12 of the Constitution of the Republic of Užupis

1 Introduction

Today, the animal world is under increasing pressure, given the magnitude of anthropogenic environmental stress, especially from climate change and habitat fragmentation. *Climate change* facilitates the introduction, establishment, and spread of invasive species at the expense of native species. It also confronts native species with the alternative to move outside their historic ranges or to go extinct—to 'move it or lose it' (Minteer and Collins 2010). *Managed relocation* may be required when threatened species are not able to move on their own to other regions where conditions are more suitable. Relocated animals will have to go through a long process, from capture and captivity to transport and release to novel areas.

This chapter was first presented at the *Human Development and Capability Association* Conference in Washington, September 10–13, 2015, in a session with Martha and Rachel Nussbaum, Amy Linch and Jeremy Benedik-Keymer.

J. Keulartz (✉)
Radboud University Nijmegen, Nijmegen, The Netherlands
e-mail: J.Keulartz@science.ru.nl

B. Bovenkerk and J. Keulartz (eds.), *Animal Ethics in the Age of Humans*,
The International Library of Environmental, Agricultural and Food Ethics 23,
DOI 10.1007/978-3-319-44206-8_15

Habitat fragmentation leads to the ongoing conversion of what were originally continuous populations to so-called 'metapopulations': collections of subpopulations, that are spread geographically over patches of habitat. Because these patches are usually small and because the movement of the animals between these patches is restricted for lack of connectivity, local extinctions of subpopulations are a common event. This situation asks for *metapopulation management*. Central to this type of management is the exchange of animals between in situ populations (in the wild) and ex situ populations (in captivity). On the one hand, captive populations can be used for restocking in areas with declining populations or for reintroduction in areas where populations have gone extinct; on the other hand, the demographic and genetic viability of ex situ populations can be boosted by supplying genetic founders from wildlife populations.

As a result of these global environmental changes, the distinction between in situ and ex situ conservation is gradually breaking down. Instead of a stark contrast between the wild and the walled, we now encounter a continuum of environments more or less impacted by human activity. Animals are becoming increasingly dependent on care in conditions of temporary or permanent captivity. In this situation, we need to develop an alternative to the existing leading accounts of animal ethics: the utilitarian (animal welfare) approach and the deontological (animal rights) approach. The main problem with these traditional approaches is that they offer no guidelines at all for a morally sound management of wild animals in captivity.

Here, I will put forward Martha Nussbaum's capabilities approach as a promising alternative to the traditional animal welfare and animal rights approaches. According to this approach, the welfare of animals must be measured against the possibilities an environment offers animals to actually display their basic natural capabilities. If appropriate care is given, animals can flourish in less natural and more human environments.

To answer the question what actually is involved in appropriate care, i.e. care that allows animals to employ their natural capabilities, I will make a distinction between three types of care: care for the environment (habitat) of animals, species-specific care, and individual care. I will argue that the moral adequacy of various human-animal practices depends on whether they succeed in achieving a balanced mix of these types of care. I will show how the balance will gradually shift if we move across the continuum between wild and captive contexts: in protected nature areas the weight is on care for the habitat of animals, in zoos the weight is on species-specific care, while performing animals will predominantly depend on individual care (Keulartz and Swart 2012).

But first, I will briefly discuss the pros and cons of Nussbaum's capabilities approach vis-à-vis utilitarianism and the rights theory. Although I am convinced that Nussbaum's approach offers a favorable alternative to these approaches, I nonetheless feel that there are some contradicting treads running through her account. I will therefore sometimes have to think 'with Nussbaum against Nussbaum'.

2 The Pros and Cons of Nussbaum's Capabilities Approach

According to Palmer (2010) the three most important approaches to animal ethics—the utilitarian approach, the rights approach, and the capabilities approach, are 'capacity oriented'. All these approaches argue that animals that possess certain keystone capacities, such as sentience, self-consciousness, or rationality, are morally considerable and are therefore entitled to respectful treatment. Palmer points out that, in the case of utilitarianism and the rights theory, this capacity orientation takes the form of a 'class system'. The members of a group of animals with similar keystone capacities deserve equal moral consideration, regardless of their species (Palmer 2010, pp. 45–46). For Tom Regan, who is well known for his animal rights theory, this class includes all mentally normal adult mammals over the age of roughly 1 year, that can be considered 'experiencing subjects-of-a-life', while utilitarian philosopher Peter Singer assumes an even broader class of animals that are entitled to equal moral consideration, namely all sentient beings with a capacity for suffering.[1]

The problem with this view on the classification of animals manifest in utilitarianism and rights theories is that it is too general to provide for sufficient guidance in shaping morally sound human-animal relationships. Rights theories are even less able to give such guidance because they are almost exclusively concerned with negative duties, i.e., duties not to act, and to abstain from certain kinds of actions, such as harming, killing, interfering, infringing on liberty et cetera. Most work on animal rights are of little or no help in dealing with captive animals because it only tells us what we should *not* to do to animals (Palmer 2014, p. 707).

Although the capabilities approach is also capacity oriented, it differs from the utilitarian approach and the rights approach in one very important respect. Nussbaum rejects the view, taken by both these approaches, that species membership itself is of no ethical and political significance at all. Following James Rachels, Nussbaum calls this view 'moral individualism'.[2] The capabilities approach, by contrast, does in fact attach moral significance to species membership as such (Nussbaum 2006, p. 362/3). It is based on a species-specific norm of flourishing, that tells us what the appropriate benchmark is for judging whether a member of a species has decent opportunities for flourishing, and that commits us to bring members of that species up to that norm (idem, 365). What is important for

[1]"No matter what the nature of the being, the principle of equality requires that its suffering be counted equally with the like suffering—insofar as rough comparisons can be made—of any other being" (Singer 1975, p. 8).

[2]According to Rachels, "moral individualism is a thesis about the justification of judgments concerning how individuals may be treated. The basic idea is that how an individual may be treated is determined, not by considering his group memberships, but by considering his own particular characteristics" (Rachels 1990, p. 173).

Nussbaum is the fulfillment of species-specific capabilities, where each living being flourishes as the sort of being it is.

The main advantage of Nussbaum's species-specific norm compared to moral individualism and the associated class system is that it can provide more and better guidance to professionals and practitioners working in various human-animal practices. The capabilities approach has more concrete guidelines to offer than utilitarianism and rights theories. And whereas rights theories stress negative duties and tell us above all what we should not do to animals, the capabilities approach has a strong affirmative character. According to Nussbaum, we have a positive duty to support the capabilities of all morally considerable beings, up to some minimum threshold level specific to each species. Utilitarian approaches are less negative than rights theories, but they are usually also less affirmative than the capabilities approach. This also applies to Singer's preference utilitarianism. By focusing on the passive state of preference satisfaction, Nussbaum explains, "utilitarianism shows a deficient regard for agency. Contentment is not the only thing that matters in human life; active striving matters, too" (idem, 73; 283). The capabilities approach places great emphasis on activity and flourishing. "Its basic goal is to address the need for a rich plurality of life activities" (idem, 346).

However, the considerable conceptual gains that Nussbaum is able to achieve through the introduction of the species-specific norm of flourishing in animal ethics are at least partly being undone by the way she compiles a catalogue of innate or 'basic' capabilities relevant to animal species. On the one hand, Nussbaum's account of animal capabilities seems to be distinctly pluralist. The capabilities approach is attentive to the fact that each species has a different form of life, and is capable of recognizing a wide ranges of types of animal dignity, and of the corresponding needs for flourishing (idem, 327). But on the other hand, Nussbaum suggests a one-fits-all approach, that has a distinct anthropocentric character as it applies the same human yardstick to all animal species.[3] Although she fully acknowledges that species-specific entitlements of animals are based upon their various characteristic forms of life and flourishing, she nonetheless wants to use the existing list of human core capabilities "to map out, in a highly tentative and general way, some basic political principles that can guide law and public policy in dealing with animals" (idem, 392).[4]

---

[3]In *Women and Human Development*, Nussbaum argues that the central capabilities "are held to have value in themselves, in making the life that includes them *fully human*" (emphasis added) (2000, p. 74).

[4]In her review of Steven Wise's book *Rattling the Cage*, Nussbaum points to an important difference in the ethical evaluation that is involved in preparing capabilities lists: "With the human capabilities, we are evaluating ourselves. If we get it wrong, we are the ones who take the consequences. With animals, we are again the ones performing the evaluation—and there is great danger that we will get it wrong" (Nussbaum 2001, p. 1542/3).

3 The Predator Problem

What seems most problematic, when applied to animals, is the *Other Species* capability, i.e. the capability or entitlement to be able to live with concern for and in relation to animals, plants, and the world of nature (Cripps 2010, p. 8). This capability, Nussbaum suggests, "calls for the gradual formation of an interdependent world in which all species will enjoy cooperative and mutually supportive relations with one another. Nature is not that way and never has been. So it calls, in a very general way, for the gradual supplanting of the natural by the just" (2006, p. 399). This clearly implies that we should stop all predators from hunting and eating prey.[5]

Like animal ethicists of all colors, Nussbaum is also struggling with the so-called 'predator problem'. This problem arises from the need to reconcile two seemingly conflicting views. On the one hand, most animal ethicists believe that, if animals truly have a right to life, then we ought to protect them from being killed by their predators. But, on the other hand, they generally agree that there is a basic presumption against interference with animals in the wild (Cowen 2003).

Peter Singer, for instance, fears that interfering with nature to save animals from predation would cause more harm than good.[6] However, as a matter of principle, Singer believes that "if, in some way, we could be reasonably certain that interfering with wildlife in a particular way would, in the long run, greatly reduce the amount of killing and suffering in the animal world, it would, I think, be right to interfere" (Singer 1973).[7]

Tom Regan is also opposed to interference with nature to protect prey animals. Although wild animals can certainly harm one another, Regan argues, they cannot violate one another's rights since, in contrast to human predators, nonhuman predators are not moral agents, but only moral patients. "Animals are not moral agents and so can have none of the same duties moral agents have, including the duty to respect the rights of other animals" (Regan 1983, p. 357).

Interestingly enough, in the preface of the second edition of *The Case for Animal Rights*, Regan has put forward an idea that might shed new light on the predation problem. In response to Carl Cohen's critical question, why we should save a

---

[5]This demand calls to mind the utopian conditions after the Second Coming of Christ, where 'The wolf shall dwell with the lamb' (Wissenburg 2011).

[6]Saving wild animals "on any large scale would have disastrous ecological consequences", according to his fellow-consequentialist Aaron Simmons (2009, p. 26).

[7]White's Professor of Moral Philosophy McMahan (2010) has embraced "the heretical conclusion that we have reason to desire the extinction of all carnivorous species". He is actually in favor of selecting carnivorous species for extinction and herbivorous species for survival, and would also support using genetic modification to gradually turn carnivorous species into herbivorous ones.

human life from predators but not an animal's life when doing so would be equally within our power (Cohen 1997, p. 95), Regan asks us to imagine two cases: one in which a wild animal is threatened by a predator and another case in which the predator is threatening a human child. He argues that the difference between the two cases is that wild animals possess a certain 'competence' and are capable of 'using their natural abilities' to survive on their own in the wild, whereas human children do not. We honor this competence of wild animals by just letting them be, even if their lives are threatened by predators (Regan 1983, pp. xxxvi–xxxviii).

This recourse to the notion of competence with regard to prey animals could open the way for a less ambivalent solution to the predator problem than we have seen to date. This solution would in fact be perfectly in line with the capabilities approach, that centers on the idea that a creature's well-being is dependent on its opportunities to realize some basic natural abilities. But this avenue is blocked as long as we include the *Other Species* capability, that calls for cooperative and mutually supportive relations between species, in the list of central capabilities. As Nussbaum has argued, however, this list is "open-ended and humble" and "can always be contested and remade" (Nussbaum 2000, p. 77). It is, therefore, always possible for items to be both added to or deleted from the list. In the next section, we will show that, without a revision of the current version of the list, the capabilities approach runs into some substantial trouble when it comes to the management of wilderness areas.

4 Wilderness Areas and the Care for Habitat

Like most animal ethicists, Nussbaum attaches moral weight to the possibility for animals to enjoy sovereignty. She endorses "the idea that species autonomy is part of the good for nonhuman animals" (Nussbaum 2006, p. 375). At first glance, she seems to endorse the view that animals can pursue their own flourishing best when left to their own devices, and that we have no positive duties to support their welfare, providing them with food, shelter and healthcare. Such a 'benevolent despotism' of humans over animals might even be perceived as morally repugnant, because part of what it is to flourish for animals "is to settle certain very important matters on its own, without human intervention, even of a benevolent sort" (idem, 373).

On closer inspection, however, Nussbaum does not fully accept the view that we have no positive duties towards animals in the wild, although she admits that there is "much truth" in this view. The reason is that in today's world it is hardly the case anymore for animals to live sovereign and autonomous lives, unaffected by human interference. The environments on which animals depend for their survival are being increasingly disturbed or destroyed by human activity, and their opportunities for nutrition, shelter, and free movement are in constant decline. In this situation, where the possibilities for flourishing are severely limited, Nussbaum believes that we should not leave the animals to their fate but should take measures to preserve "at least some part of the creature's original habitat" (idem, 376).

Since wild animals are currently under increasing pressure from human activity, Nussbaum clearly assumes that we have a positive duty of care for their habitat. However, living in the early stages of what has been called the planet's 'sixth mass extinction', care for the habitat of wild animals cannot rely merely on preservation and protection. The rate and magnitude of the anthropogenically driven 'defaunation', asks for more offensive and interventionist strategies such as recreation, restoration and rewilding (Seddon et al. 2014).

On both sides of the Atlantic, such strategies are directed at the enlargement of nature areas, their connection into coherent and comprehensive networks that allow the exchange of individuals between populations, and, last but not least, the (re)introduction of keystone species i.e. those species whose impact on their environment is disproportionately large relative to their numerical abundance—their return is vital for the restoration of the evolutionary and ecological potential that was lost with their removal (Keulartz 2016).

In North America, most emphasis is on the restoration of large carnivores, because of their role in the top-down regulation of ecosystems. Large predators occupy the highest trophic level and create impacts that ripple downward along the trophic ladder. They activate trophic cascades that are essential to the preservation of biodiversity and the maintenance of ecosystem integrity. A famous case in point concerns the reintroduction of the grey wolf in Yellowstone National Park in 1995. With the return of the wolf the elk herd, one of the world's largest elk herds, declined 40 % in 5 years. The wolves prevented elk from overbrowsing willow and aspen near rivers and streams, and this gave rise to a substantial rebound of the beaver, itself a keystone species that may increase species diversity et cetera.[8]

Most animal ethicists will oppose the (re)introduction of predators. Although they are not in favor of interfering with nature on behalf of prey animals, they are even less inclined to intervene at their expense. Rights theorists, for instance, will argue that by (re)introducing predators, humans will become implicated in the killing of prey animals, and because, unlike predators, humans are moral agents, we have to do with a rights violation in this case (Milburn 2015).[9]

Nussbaum is also opposed to the introduction of 'natural predators' to control animal populations. She prefers any non-violent method of population control to such a violent method. The 'painless predation' of animals through human hunting,

[8]Recently, some doubts have been raised regarding this success story (Mech 2012).

[9]According to Dale Jamieson, moral evaluation is clearly in order when "predation is in some way affected by human agency, either because we have structured the encounter or because the predator is under our direct or indirect control" (Jamieson 2008, p. 186/7).

she argues, may be an alternative to "other deaths that elks would die, such as starving or being torn apart by wolves" (Nussbaum 2006, p. 394). In an interview with Carla Faralli, Nussbaum puts it this way:

> Sometimes people think that they have done a great good thing if they make hunting illegal and then, when the deer are reproducing too rapidly and can't find enough to eat, they introduce wolves to tear the deer apart. Actually, I am sure that for the deer the hunter's gun is better than the wolves' jaws, more sudden and less excruciating (Nussbaum and Faralli 2007, p. 158)

However, "painless predation" is a misleading term for human hunting. Hunters will in fact have to shoot far more prey animals than the numbers that predators would be able to kill. This is because prey populations are controlled not only top-down by predation but also bottom-up by food availability. Such a bottom-up approach, in which the system is regulated by energy moving upward from lower to higher trophic levels, i.e., from plants to herbivores to carnivores, is prevalent in Europe. Here, most emphasis is on the restoration of wild large herbivores and the introduction of naturalistic grazing, a policy that has been initiated by the management of the Oostvaardersplassen, a Dutch polder situated 5 m below sea level and just half an hour from Amsterdam. Apart from red deer, roe deer, wild boar, and wisent, European rewilders also use proxies for large ungulates that went extinct, such as the aurochs and the tarpan (the European wild horse) (Klaver et al. 2002).

As a consequence of the resource-driven bottom-up approach, the population size of large herbivores in the Oostvaardersplassen and similar reserves is regulated by limited food availability. Consequently, it is not allowed in these reserves to prevent starvation, either by proactive culling or by supplementary feeding.[10] Only reactive culling to prevent unnecessary and prolonged suffering of moribund animals is allowed.

It is important to realize that the top-down regulation by predation and the bottom-up regulation by food availability are forces working together simultaneously (Miller et al. 2001). The relative influence of these forces varies among ecosystems and depends on a range of variables, such as rainfall patterns, soil fertility, and especially the body size and diversity of predators and prey in the system, as can be illustrated by the graph below about the Serengeti (Hopcraft et al. 2009, p. 122). Here, large herbivores, such as elephant, tend to be regulated by food abundance (dashed blue line), whereas smaller herbivores, such as wildebeest, are regulated by food quality (dotted green line). Only the smallest herbivores, such as oribi, are mainly predator regulated (solid red line).

[10]Varner (1995) has termed this type of culling 'therapeutic hunting'.

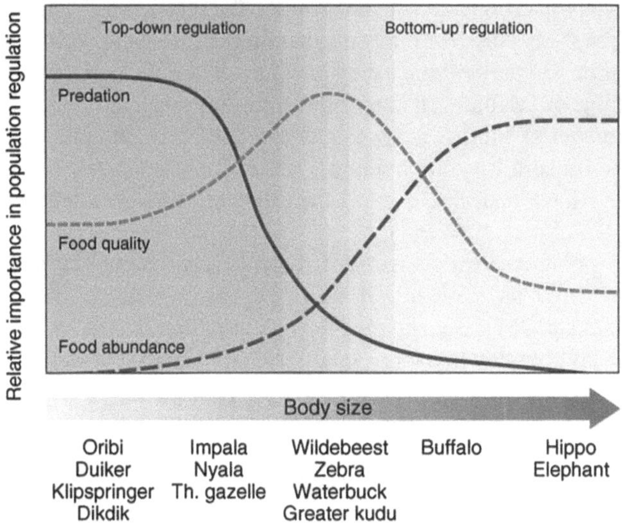

In short, populations are in varying combinations regulated top-down by predation and bottom-up by food availability.[11] Because hunters have to substitute for both top-down and bottom-up control of prey populations in order to prevent starvation, they generally have to shoot very large numbers. For instance, in the Veluwe, with 90,000 ha one of the Netherlands' largest nature reserves, 60 % of the red deer and 80 % of the wild boar are shot annually. It is estimated that you need 1,400 wolves during 8 months to realize such a large-scale cull.

This type of 'fauna management' has a disastrous impact on the natural mechanisms that animals in the wild have developed to cope with periodic food shortages. Red deer, for instance, are able to resort to a strategy of what has been called 'hidden hibernation'; their heart rate goes down to 30 beats per minute, and their energy use decreases to 13 percent of the annual average (Arnold et al. 2004). Another coping mechanism concerns the reduced fertility of female animals due to severe weight loss. They skip giving birth for one and sometimes even 2 years, which will offer them the opportunity to regain strength. A perverse effect of large-scale culling is the artificial increase of the reproduction rate of the animals. Red deer in the Veluwe produce up to three times more calves than in the Oostvaardersplassen, where they are only subject to reactive instead of proactive culling.

[11]In higher latitudes with ecosystems with only one major predator and a few prey species, such as tundra, desert, boreal and temperate woodlands, bottom-up control of prey is, in fact, dominant.

So, it is highly questionable whether "for the deer the hunter's gun is better than the wolves' jaws", as Nussbaum has claimed in the interview with Carla Faralli—human predation seems far from being a worthwhile substitute for natural predation. Abolishing or inhibiting natural predation, is problematic, if only because a fairly large number of animals are both predator and prey. But, more importantly, it is problematic because it would definitely make an end to species sovereignty and autonomy, and have a negative impact on the flourishing, not only of predator species but also of prey species.

Nussbaum points out the danger "of romanticizing nature, or suggesting that things are in order as they are, if only we humans should stop interfering" (Nussbaum 2006, p. 367). But she runs the risk of falling into the other extreme, by demonizing nature. Following John Stuart Mill in his essay 'Nature', she portrays predators as vicious criminals, merciless executioners and great monsters, inflicting painful torture and gruesome death on other vulnerable and defenseless creatures. This picture is deceptive because the hunter's gun is seldom more sudden and excruciating than the wolves' jaws; quick kills are rare, and many animals suffer prolonged, painful deaths when hunters severely injure but fail to kill them.[12] On the other hand, a prey animal that is being chased by a predator will produce endorphin, a chemical that is kin to opiates, and that acts as an analgesic, a natural pain-killer.

But, above all, Nussbaum's picture is a caricature of both predators and prey, who have evolved over many millennia in interaction with each other. During this tight evolutionary relationship, prey species have developed a stunning array of mechanisms to cope with predators for every stage of their struggle: the avoidance of detection by the predator (such as camouflage, refuge use, nocturnality), the avoidance of attack once detected (mimicking animals with strong defenses, signaling to the predator that pursuit is not worthwhile), the avoidance of capture once attacked (fleeing, bluffing strength), and the avoidance of consumption once captured (playing death or 'thanatosis', sacrificing body parts or 'autotomy').

In a certain sense, one can say that predators and prey fit together like puzzle pieces. Without predators, prey animal's possibilities for flourishing will be diminished, because all the amazing capabilities they have gained over evolutionary time to cope with predators might be rendered meaningless. All in all, we can safely conclude that it is counterproductive to extent the *Other Species* capability to the animal kingdom. Instead of working to ensure that all species will enjoy

[12]A British study of deer hunting found that 11 % of deer killed by hunters died only after being shot two or more times and that some wounded deer suffered for more than 15 min before dying (Bradshaw and Bateson 2000).

cooperative and mutually supportive relations, we should respect the natural capabilities of animals, be they predator or prey, without romanticizing or demonizing their agonistic interactions.[13]

5 Zoos and Species-Specific Care

Given the global processes mentioned in our introduction, traditional in situ (place-based) conservation methods seem no longer sufficient to save threatened species (Sandler 2012). The magnitude of anthropogenic environmental stress from bioinvasion, habitat fragmentation, nitrogen deposition, biodiversity loss, and, above all, climate change, makes it unavoidable to replace the hands-off approach that has guided mainstream species conservation until recently by a more proactive and interventionist strategy. However, this new strategy has led to manifold conflicts between wildlife conservationists and animal protectionists (Minteer and Collins 2013).

Animal ethicists generally question the intentions of conservationists. For instance, Dale Jamieson is convinced that the motivation for reintroduction and rewilding programs "centres on the satisfaction of human preferences rather than on concerns about animal welfare or the maintenance of ecological values" (Jamieson 2008, p. 194). He shares the view of Bekoff (2000) that these programs are nothing but humans' attempts to 'redecorate' nature. Conservationists, for their part, are strongly opposed to animal rights proponents in particular. According to Michael Hutchins, "it is time to face up to the fact that animal rights and conservation are inherently incompatible and that one cannot be an animal rights proponent and a conservationist simultaneously" (Hutchins 2008, p. 816).

As Hutchins rightly points out, animal rights proponents have fought vehemently against virtually every form of wildlife research or management. They oppose programs to control or eliminate destructive invasive species that involve techniques such as hunting and trapping or make use of pesticides such as piscicides, chemical substances which are poisonous to fish (Keulartz and Van der Weele 2008). They also oppose managed relocation, because of the chronic stress that relocated animals will experience at all stages of the process, from capture and

---

[13]David Schlosberg has also rejected Nussbaum's idea that we should protect prey animals from predators. He contends that she has a too narrow view of what it means to flourish as a prey animal: "we need to understand and accept that part of the flourishing of animals is to be the protein for other life forms… To be food for others is the essence of functioning for some beings" (Schlosberg 2007, p. 151). Elizabeth Cripps has rightly remarked that this specific argument is only convincing, so long as it concerns the species as a whole. "It may be part of the functioning of the species that it is food for another species; but to say that it is part of the functioning of that particular gazelle to be so overlooks precisely the concern for the capacity of individual animal lives to go better or worse that Nussbaum wants to recognize" (Cripps 2010, p. 10).

captivity to transport and release to novel areas. And, last but not least, they oppose ex situ conservation through zoos and aquaria.

As a response to the ongoing decline in effectiveness of in situ conservation and the accompanying loss of biodiversity, zoos began to turn their attention to the conservation of endangered species and wildlife in the 1970s and 1980s. 'Captivity for Conservation' became a crucial slogan for the modern zoo. A major milestone in this development was the Convention on Biodiversity which was signed at the Earth Summit in Rio de Janeiro in 1992. In the wake of the Earth Summit the first World Zoo Conservation Strategy was launched in 1993. Its conclusion explicitly stated that, at a time when species, habitats and ecosystems worldwide are threatened with extinction, modern zoos must commit to the conservation of species and wildlife.

Most animal rights proponents consider infringing an individual's right to freedom for the sake of the preservation of the species as morally wrong. For Regan any type of captivity or manipulation of a sentient animal is morally unacceptable, irrespective of the possibly beneficial consequences for the protection of rare or endangered species. The rights view's answer to the question whether zoos are morally defensible, "not surprisingly, is No, they are not" (Regan 1995, p. 46). Utilitarian Peter Singer, on the other hand, seems to accept some reductions in animal welfare when the survival of entire populations or species is at stake. He feels, however, that most zoos today fail to live up to their conservation mission. They tend to confine animals for our amusement in ways that are contrary to their interests. Even if these zoos do occasionally preserve an endangered species, "what is the point of preserving animals if they are having miserable lives?"[14] His fellow-utilitarian Jamieson is even more skeptical about zoos than Singer; he considers the benefit of species conservation "not significant enough to overcome the presumption against depriving an animal of its liberty" (Jamieson 1995, p. 60).

The problem with the rights approach and the utilitarian approach is that they say little or nothing to guide us in the case of animals that depend on care in conditions of temporary or permanent captivity. As we have seen, these approaches hold the view that species membership in itself is of no moral relevance; they do not recognize significant differences within the broad group of animals that belong to the class that deserve equal moral consideration because they possess some core capacities, such as sentience or being an experiencing subject-of-a-life. Nussbaum's capabilities approach, by contrast, does attach moral significance to the species to which a creature belongs. Her species-specific account of flourishing commits care takers to support the capabilities of captive animals, and enables us to judge whether they have ample and appropriate options and opportunities to perform species-specific activities.

Compared to utilitarians and rights proponents, Nussbaum has a rather positive opinion about zoos. With respect to endangered species, she argues that we should not only try to preserve at least some part of their original habitat, but that we

[14]http://www.mkhumanists.org.uk/node/73.

should also make "intelligent and careful" use of zoos. "Many animals will do better in an imaginative and well-maintained zoo than in the wild, at least in present conditions of threat and scarcity" (Nussbaum 2006, p. 376). In the interview with Carla Faralli, Nussbaum expresses her belief "that zoos have a good role to play, both in educating the public about animals and in developing breeding programs. The new type of zoo is a place not for painful captivity but for wide-ranging movement, and these zoos are sites of important research" (Nussbaum and Faralli 2007, p. 160).

There are, however, limitations to the extent to which species-specific activities can be carried out in zoos. It may be true that there is a strong tendency toward the 'naturalization' of zoos, but this process runs up against limits. Zoos cannot include the reproduction of natural contingencies. Some forms of predatory behavior, such as chasing and killing prey, cannot realistically be simulated in captivity. Likewise, in the absence of predators, some forms of prey behavior, such as vigilance, may not be exhibited at appropriate levels in captivity.

In short, captivity usually deprives wild animals of the necessity and opportunity to pursue the tasks of survival, such as finding food and avoiding enemies. Heini Hediger, know as the 'father of zoo biology', considered this lack of occupation of the captive animal as "one of the most urgent problems in the biology of zoological gardens" (Hediger 1950, p. 158). Quite a lot of studies show that animals prefer to work for their food, rather than to be fed ad libitum; they will most often voluntarily work for their food, even if the same food is available free (Neuringer 1969; Anderson and Chamove 1984; Laule and Desmond 1998). Especially mammals are unsuited to an existence in which no effort on their part is required to meet their basic needs (Kreger et al. 1998).

A solution to this problem is 'substitution.' Animals, particularly mammals, are flexible enough to modify their behavior to suit a wide range of situations and to substitute one form of action for another depending on the facilities available. In *Frontiers of Justice*, Nussbaum mentions an example of substitution. Modern zoos have to face the problem of allowing the capabilities of predatory animals like tigers to be exercised without actually harming or killing prey animals. The Bronx Zoo has found an answer to this question—instead of giving a tiger a tender gazelle to crunch on, it gives the tiger a large ball on a rope, whose resistance and weight symbolize the gazelle. "Wherever predatory animals are living under direct support and control," Nussbaum concludes, "these solutions seem the most ethically sound" (Nussbaum 2006, p. 370).[15]

The introduction of such novel objects—'toys and treats'—is just one option for providing 'environmental enrichment,' by which the expression and development of species-specific behaviors and abilities can be achieved, usually to the benefit of animal welfare. Feeding enrichment is also an important option. Claxton (2011) mentions three strategies in feeding enrichment: increasing the number of daily

[15]Nussbaum notes with some regret that "we do not have the option of giving the tiger in the wild a nice ball on a string to play with" (Nussbaum 2006, p. 379).

feeding sessions; making the feeding schedule less predictable; and making the food less easy to obtain, either by hiding it or by adding a level of complexity to the food manipulation process, such as providing live prey or full carcass meals (see also Ross 2002).

Another major enrichment strategy concerns the improvement of enclosure design. Most zoos suffer from a severe lack of space; the space for all the zoo animals in the world could easily fit within New York's 212.7 km^2 borough of Brooklyn (Conway 2011, p. 4). To address this problem, zoos have a number of options. They can reduce the number of species they maintain that are not threatened and specialize in species that are. They can also replace big charismatic mammals with smaller species, particularly amphibians, invertebrates and some species of fish, which occupy less space, are relatively inexpensive to keep, have a high birth rate and are easy to reintroduce (Keulartz 2015; in this volume).

A recent and also very promising strategy to tackle the problem of limited space concerns the creation of walkways between enclosures that allow animals greater freedom of movement. Building a network of trails, in particular top tree trails, gives animals the opportunity to rotate between various interconnected display and off-display areas. Animals may spend mornings in one area and afternoons in another. The concept of such animal rotation displays is based on Hediger's theory of territory, a wild animal's living space made up of a variety of special areas, such as dens, basking sites, and foraging areas, interconnected by regular pathways (Coe 2004).

Still other methods of environmental enrichment for captive animals include sensory stimulation in the form of auditory, olfactory and visual cues, and social enrichment, which provides animals the opportunity to interact with other animals, either conspecifics (same species) or contraspecifics (other species). A good example of the latter are mixed-species exhibits. They provide an interactive and dynamic experience for the animals, visitors and zoo staff. In mixed-species enclosures activity levels are typically higher, with more play behaviors, and this generally has a positive effect on both the physical and the mental health of the animals (Veasey and Hammer 2010, p. 151). Moreover, interspecific interactions can contribute to successful reintroductions. Animals from mixed-species exhibits are more likely to cope with the complexity of their natural habitat after introduction. Even 'negative' interspecific interactions, e.g. controlled nonlethal exposure to predators, may be beneficial in survival postrelease (idem, 153).

Finally, a special case of environmental enrichment concerns what Hediger has called 'occupational therapy'. "The captive animal," Hediger suggested, "must be given a new interest in life, an adequate substitute for the chief occupation of freedom… This substitute can take the form of biologically suitable training and assumes the importance of occupational therapy" (Hediger 1950, p. 158).

Animal training is an important component of animal enrichment programs as it facilitates exercise and mental stimulation. Training can also provide the animals with the motivation, skill, and confidence they need to use the enrichment devices they have been offered in the most successful way. Through training by positive reinforcement, animals can be taught to participate in daily health and husbandry care, without sedation or restraint. By voluntarily participating in their treatment during dental work, blood draws, urine collection, stethoscope examinations, artificial inseminations et cetera, they will obtain benefits in the form of preventive and curative health care. Also, the stress that is usually related to these procedures can be significantly reduced by teaching animals to voluntarily participate in their daily health and husbandry care.

Furthermore, training is also relevant for the relationship between the animal and its care giver, which is of critical importance for many species in captivity. The well-being of zoo animals is greatly enhanced if there is a trusting, amiable relationship with their keepers. The trust between animals and keepers can be reinforced by training animals to follow commands and teaching them simple routines using positive reinforcement. "Building a trusting and caring relationship, teaching the different procedures on cue and establishing strong reinforcement histories with our animals, enabling them to behave in a secure manner and provide the ability to anticipate what is happening to them, whether good or bad" (Brando 2010, p. 783).

Although training undoubtedly improves the welfare of animals and also makes routine examination and veterinary treatment less stressful, zoos are reluctant to use training, as Trevor Poole has noticed, "because they fear to be accused of not being serious or of turning into circuses" (Poole 1998, p. 91). For years animal training was motivated by the entertainment industry and has been associated only with animal shows.

6 Performing Animals and Individual Care

The circus has witnessed a transition from the use of punishment and other coercive methods to the use of methods that focus on positive reinforcement, which provides animals greater choices and greater control over their lives. The circus has also increasingly moved away from showing costumed animals displaying human-like characters toward showcasing them performing more naturalistic behaviors. Both developments are captured in the Code of Conduct for Animals by the European Circus Association (ESA 2007): "All animal training must be based on operant conditioning and the use of positive reinforcement and repetition of desired behaviors. Training should showcase individual animals' natural behaviors and athletics." These developments didn't prevent campaigns against the use of wild animals in circus acts to be successful in achieving bans almost worldwide. Nussbaum (2006, p. 392) seems to be quite positive about this development, although her capability approach does not preclude acceptance and even appreciation of animal performance training. Again, we need to think 'with Nussbaum against Nussbaum.'

When we turn to performing animals, the balance of care will shift once more, from species-specific care to individual care. Nussbaum argues that both types of care are important for animals to flourish. Her perspective is individualistic in making the living creature—not the group or species—the basic subject of justice, but at the same time she rejects the view that all moral relevance lies in the capacities of the individual. The capabilities approach should give a species-specific account of basic capabilities without losing sight of the capacities and personality of the individual animal. A respectful consideration of the species norm of flourishing should go hand in hand with a respectful attention to the capacities of the individual.

To gain insight in the prospects and problems of individual care for performing animals, including circus animals, it is useful to have a look at the work of two philosophers who have been theoretically and practically engaged in animal training: Donna Haraway and Vicky Hearne (Keulartz and Swart 2012).

Donna Haraway, a path-breaking feminist philosopher, best known for her 1991 essay *The Cyborg Manifesto*, has more recently been focusing her attention on human-animal relationships. In 2003, she published *The Companion Species Manifesto*, and in 2008 her book *When Species Meet* came out. What holds this book together is Haraway's relationship with her Australian shepherd Cayenne, with which she does agility work.

Haraway is less interested in Jeremy Bentham's famous question "Can animals suffer?" than in the questions: "Can animals play? Or work?" (Haraway 2008, p. 22)[16] To better understand and nurture our responsibilities for animals Haraway feels that Karl Marx's category of labor is more helpful than the Enlightenment's category of rights. According to Haraway, training, working with animals, does not imply submission and oppression; quite the contrary, it requires two-way interaction. Training relies on teamwork; trainer and trainee have to listen to one another if an act or exercise is going to succeed. Here, Haraway refers to the notion of 'isopraxis' used by the French ethologist Jean-Claude Barrey to describe the interaction between horse and rider, where the questions "Who influences and who is influenced?" can receive no clear answer (idem, 229).

Haraway is influenced by the work of Vicky Hearne, an animal trainer, philosopher, and poet, who died in 2001 (see Haraway 2003, pp. 48–54). Hearne, who also put much emphasis on mutuality and reciprocity in the working relationships between humans and animals, fiercely opposed the animal rights discourse. In an essay—'What's Wrong with Animal Rights'—that was included in the 1992 edition of *The Best American Essays Series*, edited by Susan Sontag, Hearne expressed her fear that animal rights activists, by branding training of any sort as torture, could deprive animals of the satisfaction they experience from work. She disagrees with the definition of 'happiness' as synonym for 'pleasure' and as antonym for 'suffering.' To better understand 'animal happiness' we should trade in Bentham for Aristotle, who considered happiness as activity in accordance with the highest excellence. Happiness is about a sense of personal achievement, like the satisfaction felt by a good woodcarver or a dancer or a poet or an accomplished dressage horse. "This happiness, like the artist's, must come from something within

---

[16]Remember that *Play* is one of the central capabilities on Nussbaum's list.

the animal, something trainers call talent, and so cannot be imposed on the animal" (Hearne 1994, p. 204).

Because Donna Haraway and Vicky Hearne primarily draw on experiences gained from training dogs and horses respectively, one might get the wrong impression that their work is only relevant with regard to domestic animals, not wild animals. However, Haraway explicitly challenges the domestic-wild dichotomy, while Hearne has written extensively and with admiration about trainers who work with wild animals. An example is the essay 'Wittgenstein's Lion' about Hubert Wells, well-known for his work with lions for Hollywood movies such as 'Out of Africa'. Another example is her essay 'Can an Ape Tell a Joke' about the acts of Bobby Berosini with his five Orangutans, an essay that was also included in the 1994 edition of *The Best American Essays Series.*

Hearne was only too well aware that wild animals were under increasing pressure due to ongoing habitat fragmentation and destruction. As a consequence, she writes,

> it looks as though we no longer have the option of simply leaving nature alone, looks as though something more radical is necessary in the way of a transformed relationship with nature than has yet been suggested. Wild-animal training is certainly not a solution to all problems, but the knowledge trainers have may contain clues to imaginative and enlightened ways to take up the burden of responsibility towards animals (idem, 192)

From the work of Haraway and Hearne, it can be concluded that responsibility for animals in working relations should be considered—to use Haraway's phrase—as 'response-ability' (2008, p. 88), the ability to listen to animals and to meet their needs. Following Despret (2004), Haraway argues that whether a specific setting is morally acceptable or culpable depends on the animal's possibilities of 'resistance' when its wishes fall on deaf ears and it is forced to compliance and docility. Marine-mammal veterinarian James McBain has formulated this requirement as follows: "We should work with animals as if gates and doors weren't there; as if they could leave any moment they wanted. If they then decide to stay and work with you, then you can say you have a good bond and trust and is the animal truly interested in being with you" (cited in Brando 2010, p. 783; cf. Brando 2012, p. 394).

On the one hand, it is obvious that the possibilities for care for the environment in a performance setting are almost absent, while the possibilities for species-specific care are fairly limited. The possibilities for individual care and attention, on the other hand, seem relatively large. One should of course, on a case-by-case basis, try to assess whether the happiness that results from the labor of training and performance can compensate for the lack of care for the environment and the deficit of species-specific care.

Now, Nussbaum seems to be somewhat ambivalent with respect to animal training. On the one hand she is convinced that most domestic animals, such as dogs and horses, have entitlements to 'suitable education', Nussbaum's name for training. These animals profit from some training and discipline. Indeed, if they are given the appropriate training they "are capable of fine feats of athletic excellence"

(Nussbaum 2006, p. 377). Nussbaum also seems to suggest that zoo animals are entitled to some forms of training, since she laments that "one of the greatest defects of most zoos has been their boringness, which constitutes a cruel assault on animals' opportunities for flourishing" (idem, 397). But, on the other hand, Nussbaum seems to be opposed to the use of wild animals in circus acts, as we have seen previously.

However, compared to a zoo environment, a circus environment is likely to address the cognitive abilities of the animals more often, as it may provide more and varied stimuli. That is at least the opinion of Hans Hopster and Ingrid de Jong, the authors of a recent report, commissioned by the Dutch government, that describes the scientific literature and expert views concerning the keeping, training and performance of sea lions in traveling circuses and the consequences for their health and welfare. One of their findings is that according to most experts, sea lion behavior strongly suggests that they are happy to be trained and like to perform. "They express anticipation to performances by looking for the trainer, by waiting at the gate of their accommodation, by willingly entering performance areas and taking their positions (station), by pre-empting clues for behaviors, by walking back and forth, swaying, stretching out to see where the trainers are, and other expressions of eagerness to participate" (Hopster and De Groot 2014, p. 20). Another finding is that there is a lot of trust between sea lions and their trainers, due to the fact that circus trainers literally live with their animals 24 h a day. "It is therefore highly likely that circus sea lions, besides being economically very valuable for their owners, also have emotional value to them. Within the constraints of the circus, sea lions probably get therefore the best possible care" (idem.).

It is of course questionable whether the benefits of training and performance for sea lions in traveling circuses outweigh the major hazards to their welfare—the Dutch report mentions pool dimensions, space, social conditions, food and water quality, and exposure to UV light as the major potential causes of poor welfare.

Nussbaum might add that the sea lions' expressions of happiness do not necessarily reflect the truth about their well-being. Their preferences for training and performance could be adaptive preferences—Nussbaum also speaks of 'submissive' or 'fear-induced' preferences (Nussbaum 2006, 344).[17] Adaptive preferences are formed under bad or unjust background conditions, as a generally unconscious and unintended adaptation to our actual situation which compares unfavourable with the desired situation.[18] People under circumstances of poverty, deprivation and injustice can adjust their wants to their possibilities and thus still consider themselves to

[17]As Rosa Terlazzo has rightly remarked, Nussbaum's capabilities approach "is centrally motivated by the problem of adaptive preferences" (Terlazzo 2014, p. 185).

[18]The term 'adaptive preference' was introduced by Jon Elster in his book *Sour Grapes* from 1983. Elster refers to La Fontaine's (originally Aesop's) fable of the Fox and the Grapes to explain what he means by preference adaptation. In the fable, after realizing that it cannot reach some grapes hanging high on the vine, a hungry fox turns away from them and declares them too sour for its tastes.

be happy and well-off—the Happy Slave being the paradigmatic example of this predicament.

We have, however, reason to be very cautious with applying the concept of adaptive preferences to animals. As Nussbaum herself has indicated, "the interpretation of animals' preferences is fraught with obscurity and difficulty" (idem, 343). Recall Nussbaum's example of the tiger in a zoo that appears satisfied when given a large ball on a rope whose resistance and weight symbolize the gazelle. The question is of course whether the tiger is really satisfied or whether he has merely adapted his preferences to his options. The latter is the case if one agrees with Daniel Crescenzo. that killing is part of exercising predatory instinct, and that the opportunity to fully exercise predatory instinct is a central capacity in its own right for predators and may thus be essential for predator flourishing. According to Crescenzo, "A large ball on a rope allows for some exercise of the predatory instinct, but the simulation of killing it provides is quite limited" (Crescenzo 2012, p. 195).

It is obviously very difficult, if not impossible, to distinguish the adaptive from the non-adaptive preferences of animals. More importantly, adapting one's preferences to ones situation can be a wise thing to do, a sign of resilience, i.e. the capacity to withstand shocks and surprises, and to maintain or regain psychological well-being in the face of changing and challenging circumstances. It seems at least a wise thing to do for wild animals that have become increasingly dependent on care in conditions of temporary or permanent captivity, as a result of the ongoing blurring of boundaries between in situ and ex situ conservation, and between the wild and the walled.

7 Concluding Remarks

Martha Nussbaum certainly deserves credit for breaking with the currently still dominant doctrine of 'moral individualism' that holds that species membership itself is of no ethical and political significance. With the introduction of the species-specific norm of flourishing in the animal ethics debate, Nussbaum has opened up promising avenues towards a morally sound guidance for professionals and practitioners working with wild animals in captivity, which is of imminent importance in the emerging Anthropocene era, the age of human dominion of the Earth, at least if we really want to stop or even reverse the defaunation process. However, to make this avenue accessible and passable, we should abandon Nussbaum's one-fits-all approach, which applies the same human yardstick to all animal species, and develop a truly species-specific account of central animal capabilities.

The task of determining which capabilities members of an animal species typically have is a complex one and calls for a broadening of Nussbaum's methodological perspective. Her conception of flourishing is thoroughly evaluative and ethical (Nussbaum 2006, p. 366). Nussbaum insists that we cannot simply "read off" species norms directly from observation of animals' characteristic ways of life (idem, 347 and 497). She suggests that we can come to recognize what animal flourishing entails through an Aristotelian-style "considered judgment" in conjunction with 'sympathetic imagination' (idem, 352–53). But, in order to create a really species-specific account of core capabilities, we cannot be content with moral evaluations only but we will have to combine them with empirical investigations from a wide range of animal sciences such as ethology, wildlife ecology, zoo biology, and conservation biology. Some of Nussbaum's statements seem to support such a plea for an empirically informed approach to the evaluation of what is essential to the flourishing of different species.[19]

References

Arnold, W., T. Ruf, S. Reimoser, F. Tataruch, K. Onderscheka, and F. Schober. 2004. Nocturnal hypometabolism as an overwintering strategy of red deer (*Cervus elaphus*). *American Journal of Physiology. Regulatory, Integrative and Comparative Physiology* 286: 174–181.

Anderson, J.R., and A.S. Chamove. 1984. Allowing captive primates to forage. *Standards in laboratory animal management, part 2*, 253–257. Hertfordshire, UK: The Universities Federation for Animal Welfare.

Bekoff, M. 2000. Redecorating nature: Reflections on science, holism, community, humility, reconciliation, spirit, compassion, and love. *Human Ecology Review* 7(1): 59–67.

Brando, S. 2010. Advances in husbandry training in marine mammal care programs. *International Journal of Comparative Psychology* 23: 777–791.

Brando, S. 2012. Animal learning and training. Implications for animal welfare. *Veterinary Clinics: Exotic Animal Practice* 15: 387–398.

Bradshaw, E.L., and P. Bateson. 2000. Welfare implications of culling red deer (*Cervus Elaphus*). *Animal Welfare* 9: 3–24.

Claxton, A.M. 2011. The potential of the human–animal relationship as an environmental enrichment for the welfare of zoo-housed animals. *Applied Animal Behaviour Science* 133: 1–10.

Cohen, C. 1997. Do animals have rights? *Ethics and Behavior* 7(2): 91–102.

Coe, J. 2004. Mixed species rotation exhibits. Jon Coe Design Pty Ltd. (originally prepared for ARAZPA 2004 Annual Conference).

Conway, W. 2011. Buying time for wild animals with zoos. *Zoo Biology* 30: 1–8.

Cowen, T. 2003. Policing nature. *Environmental Ethics* 25: 169–182.

Crescenzo, D.L. 2012. The problem of predator-prey relations and predator flourishing in Nussbaum's capabilities approach to justice. *Environmental Ethics* 34: 177–197.

Cripps, E. 2010. Saving the polar bear, saving the world: Can the capabilities approach do justice to humans, animals and ecosystems. *Res Publica* 16: 1–22.

Despret, V. 2004. The body we care for: Figures of anthropo-zoo-genesis. *Body & Society* (2–3): 111–134.

Elster, J. 1983. *Sour grapes: Studies in the subversion of rationality*. New York: Cambridge University Press.

[19]"It seems best for humans not to engage in too much second-guessing of animal capabilities, but try to observe what each creature actually considers important, on the basis of what it does… Part of respect for other species is a willingness to look and study, learning the internal rhythms of an animal community and the sense of value the way of expresses" (Nussbaum 2006, p. 371/2).

European Circus Association (ESA). 2007. Code of conduct for animals. http://moderngov.dover.gov.uk/Data/Scrutiny%20(Policy%20and%20Performance)%20Committee/20070625/Agenda/Agenda05_AppendixE.pdf. Accessed July 2016.

Hopcraft, J.G.C., H. Olff, and A.R.E. Sinclair. 2009. Herbivores, resources and risks: Alternating regulation along primary environmental gradients in savannas. *Trends in Ecology and Evolution* 25(2): 119–128.

Haraway, D. 2003. *The companion species manifesto. Dogs, people, and significant otherness.* Chicago: Prickly Paradigm Press.

Haraway, D. 2008. *When species meet.* Minneapolis, MN: University of Minnesota Press.

Hearne, V. 1992. 'What's wrong with animal rights of hounds, horses and jeffersonian happiness. In *The best American essays,* ed. S. Sontag. Boston, MA: Houghton Mifflin.

Hearne, V. 1994. *Animal happiness. A moving exploration of animals and their emotions.* New York: Skyhorse Publishing.

Hediger, H. 1950. *Wild animals in captivity.* London: Butterworths.

Hopster, H. and I.C. De Jong. 2014. *Welfare of sea lions in travelling circuses.* Wageningen UR Livestock Research. Lelystad, The Netherlands, Research Report 770.

Hutchins, M. 2008. Animal rights and conservation. *Conservation Biology* 22(4): 815–816.

Jamieson, D. 1995. Zoos revisited. In *Ethics on the Ark*, ed. B. Norton, M. Hutchins, E. Stevens, and T. Maple, 52–66. Washington: Smithsonian Institution Press.

Jamieson, D. 2008. The rights of animals and the demands of nature. *Environmental Values* 17: 181–199.

Keulartz, J. 2015. Captivity for conservation? Zoos at a crossroads. *Journal of Agricultural and Environmental Ethics* 28: 335–351.

Keulartz, J. 2016. Future directions for conservation. *Environmental Values* 25(4): 385–407.

Keulartz, J., and C. Van der Weele. 2008. Framing and reframing in invasion biology. *Configurations* 16(1): 93–115.

Keulartz, J., and J. Swart. 2012. Animal flourishing and capabilities in an era of global change. In *Ethical adaptation to climate change*, ed. A. Thompson, and J. Bendik-Keymer, 123–144. Cambridge, MA: The MIT Press.

Klaver, I., J. Keulartz, H. Van den Belt, and G. Gremmen. 2002. Born to be wild: A pluralistic ethics concerning introduced large herbivores in the Netherlands. *Environmental Ethics* 24(1): 3–23.

Kreger, M.D., M. Hutchins, and N. Fascione. 1998. Context, ethics, and environmental enrichment in zoos and aquariums. In *Second nature. Environmental enrichment for captive animals*, ed. D. Shepherdson et al., 59–82. Washington: Smithsonian Institution Press.

Laule, G., and T. Desmond. 1998. Positive reinforcement training as an enrichment strategy. In *Second nature. Environmental enrichment for captive animals*, ed. D. Shepherdson et al. pp. 302–313. Washington: Smithsonian Institution Press.

McMahan, J. 2010. The meat eaters. *The New York Times*, September 19, 2010.

Mech, L.D. 2012. Is science in danger of sanctifying the wolf? *Biological Conservation* 150: 143–149.

Milburn, J. 2015. Rabbits, stoats and the predator problem: Why a strong animal rights position need not call for human intervention to protect prey from predators. *Res Publica* 21(3): 273–289.

Miller, B., B. Dugelby, D. Foreman, C. Marinez del Rio, R. Noss, M. Philips, R. Reading, M. Soulé, et al. 2001. The importance of large carnivores to healthy ecosystems. *Endangered Species UPDATE* 18(5): 202–210.

Minteer, B.A., and J.P. Collins. 2010. Move it or lose it? The ecological ethics of relocating species under climate change. *Ecological Application* 20(7): 1801–1804.

Minteer, B.A., and J.P. Collins. 2013. Ecological ethics in captivity: Balancing values and responsibilities in zoo and aquarium research under rapid global change. *ILAR* 54(1): 41–51.

Neuringer, A. 1969. Animals respond for food in the presence of free food. *Science* 166: 399–401.

Nussbaum, M.C. 2000. *Women and human development: The capabilities approach.* Cambridge, UK: Cambridge University Press.

Nussbaum, M.C. 2001. Animal rights: The need for a theoretical basis. *Harvard Law Review* 114: 1506–1549.

Nussbaum, M.C. 2006. *Frontiers of justice. Disability, nationality, species membership*. Cambridge, MA: Harvard University Press.

Nussbaum, M.C., and C. Faralli. 2007. On the new frontiers of justice. A dialogue. *Ratio Juris* 20 (2): 145–161.

Palmer, C. 2010. *Animal ethics in context*. New York: Columbia University Press.

Palmer, C. 2014. The moral relevance of the distinction between domesticated and wild animals. In *The oxford handbook of animal ethics*, ed. T.L. Beauchamp, and R.G. Fey. New York: Oxford University Press.

Poole, T.B. 1998. Meeting a mammal's psychological needs. In *Second nature*, ed. D. J. Shepherdson, J.D. Mellen, and M. Hutchins, 83–97. Washington: Smithsonian Institute.

Rachels, J. 1990. *Created from animals: The moral implications of Darwinism*. New York: Oxford University Press.

Regan, T. 1983. *The case for animal rights*. Berkeley, CA: University of California Press.

Regan, T. 1995. Are zoos morally defensible. In *Ethics on the Ark*, ed. B. Norton, M. Hutchins, E. Stevens, and T. Maple, 38–51. Washington: Smithsonian Institute Press.

Ross, S.R. 2002. The effect of a simple feeding enrichment strategy on the behaviour of two Asian small-clawed otters (*Aonyx cinerea*). *Aquatic Mammals* 28(2): 113–120.

Sandler, R.L. 2012. *The ethics of species. An introduction*. Cambridge, UK: Cambridge University Press.

Schlosberg, D. 2007. *Defining environmental justice: Theories, movements, and nature*. New York: Oxford University Press.

Seddon, P.J., C.J. Griffiths, P.S. Soorae, and D.P. Armstrong. 2014. Reversing defaunation: Restoring species in a changing world. *Science* 345(6195): 406–412.

Simmons, A. 2009. Animals, predators, the right to life and the duty to save lives. *Ethics and the Environment* 14: 15–27.

Singer, P. 1973. Food for thought. www.nybooks.com. Accessed 1 July 2015.

Singer, P. 1975. *Animal liberation: A new ethics for our treatment of animals*. New York: Random House.

Terlazzo, R. 2014. The perfectionism of Nussbaum's adaptive preferences. *Journal of Global Ethics* 10(2): 183–198.

Varner, G. 1995. Can animal rights activists be environmentalists? In *People, penguins, and plastic trees*, 2d ed, ed. C. Pierce, and D. VanDeVeer, 254–273. Belmont, CA: Wadsworth Publishing Company.

Veasey, J., and G. Hammer. 2010. Managing captive mammals in mixed-species communities. In *Wild mammals in captivity. Principles and techniques for zoo management,* ed. D.G. Kleiman, K.V. Thompson, and C.K. Baer, 151–161 (2nd edn). Chicago: The University of Chicago Press.

Wissenburg, M. 2011. The lion and the lamb. Ecological implications of Martha Nussbaum's animal ethics. *Environmental Politics* 20: 391–409.

Animals, Freedom, and the Ethics of Veganism

Aaron Simmons

Abstract While moral arguments for vegetarianism have been explored in great depth, the arguments for veganism seem less clear. Although many animals used for milk and eggs are forced to live miserable lives on factory farms, it's possible to raise animals as food resources on farms where the animals are treated more humanely and never slaughtered. Under more humane conditions, do we harm animals to use them for food? I argue that, even under humane conditions, using animals for food typically harms animals by restricting their freedom. My argument raises an important question about the extent to which animals are harmed when their freedom is restricted. On one view, it's possible to restrict animals' freedom without causing them harm so long as we don't make them suffer. This view underestimates the value of freedom for animals. Even if animals aren't made to suffer, restricting their freedom can harm them insofar as it deprives them of freely pursuing their enjoyments in life. My argument has implications for not only the ethics of using animals as food resources but also the ethics of other human uses of animals that involve restricting animals' freedom, such as using animals in zoos and circuses. I examine these implications and also consider what we should do with farm animals if we cease using them for food.

1 Introduction

While the ethical questions surrounding vegetarianism have been explored in great depth by philosophers, there has not been much philosophical examination of the ethics of veganism. Veganism is a form of vegetarianism in which one abstains from consuming not only meat but also other animal products including milk and eggs. Many vegans also refuse to wear or use various other animal products in their lives, including products made from leather, fur, wool, and down feathers. Is it

A. Simmons (✉)
Marywood University, Scranton, PA, USA
e-mail: asimmons@marywood.edu

B. Bovenkerk and J. Keulartz (eds.), *Animal Ethics in the Age of Humans*,
The International Library of Environmental, Agricultural and Food Ethics 23,
DOI 10.1007/978-3-319-44206-8_16

wrong to use animals for their milk or eggs? Does morality requires us to adopt a vegan lifestyle?

According to one standard moral argument for vegetarianism, we ought to be vegetarian because it is wrong to kill animals for food, particularly when we are capable of living long healthy lives on a vegetarian diet. However, it might seem less clear why we morally ought to be vegan. On the one hand, many of the animals that we use for milk and eggs are raised on 'factory farms' where the animals are forced to live miserable lives and are ultimately slaughtered for food. However, it is possible to raise and use animals for milk and eggs on free-range farms where the animals are treated more humanely and allowed to live out their natural lives, never being slaughtered for meat.[1] Is it still wrong to use animals as resources for food and material, even under more humane conditions in which we do not ultimately kill the animals? More precisely, putting aside the environmental and health consequences that may come with raising and using animals for food, do we do anything wrong to the animals insofar as we use them as our resources?

Typically, when we do something wrong to another being, it is because we *harm* that being in some way that is morally unjustified. So let us ask, does using animals as our resources, even under more humane conditions, harm them in some way? This question of harm is the focus of my chapter. I argue that using animals as food and material resources, even under more humane conditions, still typically harms animals insofar as such practices *restrict animals' freedom* in various ways. My argument raises an important question about the extent to which animals are harmed when their freedom is restricted. According to one view, it is possible to restrict animals' freedom without harming them so long as we do not make them suffer or kill them. I argue that this view underestimates the value of freedom for animals. Even if animals are not killed or made to suffer, restricting their freedom can harm them insofar as they are deprived, to some degree, of freely pursuing their various enjoyments in life. Additionally, I contend that we underestimate the extent to which our confinement of animals may cause them to suffer in more subtle ways, such as through boredom and depression. My argument has important implications for the ethics of using animals as our resources in a variety of contexts, including not only our use of animals for food and material but also, for example, our use of animals in zoos and circuses.

I conclude my chapter by addressing what we should do with farm animals if we were to cease using them as food resources. I argue that it is in domesticated animals' best interests that we continue to take care of them, rather than setting them free to fend for themselves. I also briefly consider whether we ought to restrict domesticated animals' freedom to reproduce, including to the point of driving their species to extinction.

---

[1]To be clear, some free-range farms today still treat their animals in inhumane ways and the animals raised there are ultimately slaughtered for food. My point is just that we can imagine a free-range farm in which the animals are treated more humanely and allowed to live out their natural lives.

2 The Argument from Suffering and Death

According to a standard moral argument for vegetarianism, the practice of raising and killing animals for meat is unethical because it causes at least one of two significant harms to animals. First, most animals raised for meat are made to suffer in various horrible ways in industrialized factory farms. Second, even if animals are raised in more humane, less painful conditions, we harm the animals simply by prematurely ending their lives. Indeed, for many vegetarians, this is the most fundamental harm caused by the practice of raising animals for meat.

It may seem less clear how moral respect for animals requires us to adopt a vegan lifestyle since products such as milk and eggs can be obtained without killing animals. However, further investigation reveals that the same moral reasons for being vegetarian also support becoming vegan. Animals used to produce milk and eggs are commonly raised on factory farms and made to suffer in various ways. Cows are kept in small stalls, not allowed to freely move around. They are branded, tail-docked, or castrated without any painkillers. Cows suffer from health problems when they are forced into continuous birthing and lactation and given hormones to produce more milk than they naturally would. Baby calves can suffer if they are deprived of nutrients from their mothers' milk. Chickens used for eggs suffer when they are debeaked and crammed into small wire cages that restrict them from spreading their wings and moving freely. Mainstream dairy and egg industries are also closely linked with the meat industry. Many animals involved in the production of milk and eggs are ultimately slaughtered. This includes male cows and chickens, as well as female cows and chickens that outlive their usefulness in producing milk or eggs. So, support for the dairy and egg industries also typically supports the meat industry.

Opposition to human-inflicted animal suffering and death provides a strong argument in favor of becoming vegan. However, the argument has a limit. If using animals for their milk and eggs is wrong solely because the animals are made to suffer and killed in factory farms, then we do not wrong animals when we use them for food so long as we raise them in more humane conditions (e.g. keeping them in less restrictive spaces and refraining from physically injuring them) and we do not slaughter them when they outlive their usefulness to us. Assuming that it is possible to use animals for food under these improved conditions, the argument from suffering and death is a conditional argument for veganism. A similar conditional argument could be made in support of the use of animals in zoos and circuses: so long as the animals are not made to suffer or prematurely killed, we do nothing wrong to them when we use them for our entertainment. This sort of conditional argument stands in contrast to a more fundamental form of vegetarianism according to which no amount of improvements could justify the practice of killing animals for meat because it is wrong simply to kill animals for purposes of eating them.

Let us ask though whether there are any other ways in which we harm animals when we use them as resources for food (or use them in zoos and circuses) in ways that do not require killing them. In particular, are the animals harmed in some more basic way that is more difficult to avoid and, thus, in some way that might give us a more fundamental reason to be vegan?

3 The Argument from Freedom

Even if they are not killed or made to suffer in factory farm-like conditions, using animals as resources for food and material can also harm them in another way: *by depriving them of their freedom*. Under some definitions of freedom, it might seem that animals are not capable of being free. Freedom is sometimes equated with autonomy, which involves directing one's own life according to values or a conception of the good life that one has reached through reflection and reasoning. Arguably, most if not all animals are incapable of having this kind of freedom since they lack the capacity to form a conception of the good life through reflection and reason. However, although this is an important sense of freedom for highly self-aware beings, there is another, more basic sense of freedom that is applicable to many animals. Freedom in this basic sense involves being able to do as one wants without being subject to external constraints on one's actions. So long as animals are capable of having desires, they are capable of having this kind of freedom.

Using animals as our resources for food, including for milk and eggs, typically involves restricting their freedom to some degree. In more extreme cases, farm animals may be kept in small cages or stalls. But even if farm animals are not kept in small cages or stalls, we still restrict their freedom to some degree insofar as we control their movements and behavior, making them do things to satisfy human desires rather than allowing them to live as they please. For example, even under more humane conditions, dairy cows are artificially impregnated and milked for human consumption. Allowed to live as they please, the cows might not reproduce at the times we want them to do so and, when they do reproduce, they would feed their baby calves with the milk they produce.

Suppose I am correct that using animals as our food resources will usually entail restricting their freedom to some degree. How do we harm animals when we deprive them of freedom? One way in which confinement can harm animals is by causing them to suffer. Keeping animals in cages, stalls, or other confined spaces in which they lack adequate room to move around, exercise, and roam may cause them to feel distressed, frustrated, afraid, and sad. If confined animals are unable to properly move around and exercise, this could also result in painful physical ailments for the animals. The specific type of confinement can also be the cause of suffering. For example, wire cages or cold, cement floors can cause physical discomfort and injury. However, suppose that when we confine animals to use them for food, we take care to minimize their suffering, such as by giving them ample room to move around and exercise and, in general, providing them with comfortable environments that do not lead to physical discomfort and injury. Do we harm animals in any other ways when we deprive them of their freedom?

According to philosopher Cochrane (2012), the answer is no. On Cochrane's view, most if not all animals have no "intrinsic interest" in liberty—that is, liberty is not good in itself for animals. Rather, he contends, animals have only an instrumental interest in liberty, meaning that liberty is good for them only as a means to providing them with other goods, particularly the avoidance of suffering. Because animals have no intrinsic interest in liberty, Cochrane argues that they are not necessarily harmed when they have their freedom curtailed. He suggests that it is possible to keep and use animals in zoos, circuses, research laboratories, and farms without causing them harm so long as we eliminate the aspects of confinement which cause the animals to suffer.

By contrast, Cochrane contends that people have an intrinsic interest in freedom. They have an intrinsic interest in governing their own lives without interference from others. Freedom is good for persons independently of whether it makes them happier or helps them avoid suffering. Even if a person's life in confinement is overall pleasurable, she has an interest in not being controlled and manipulated by others. Likewise, a person has an interest in making her own decisions, even if her choices might make her life less pleasurable. Cochrane explains that people have an intrinsic interest in freedom in virtue of being "autonomous agents" who are capable of framing, revising, and pursuing their own conceptions of the good. This requires, among other things, being able to reason and reflect on one's desires and to change them according to one's values. Because animals are not autonomous agents in this sense, they cannot have this intrinsic interest in freedom (idem, 6).[2]

I believe that Cochrane's argument underestimates the extent to which confinement (i.e. having one's freedom restricted) can harm animals. Even if animals are not made to suffer in obvious ways, confinement can still harm them *insofar as it prevents them from freely pursuing their various desires and enjoyments in life*, including their desires to roam, explore, and play in their environments as well as their desires to interact with other animals. Two points deserve emphasis here. First, animals may suffer in less physical, more emotional ways when deprived of their enjoyments, such as by experiencing depression, frustration, listlessness, or boredom. Cochrane's view underestimates the extent to which confinement may cause animals to suffer in these more subtle ways.

But even if animals are not made to suffer from confinement, they are still harmed insofar as they are prevented from doing basic things that they desire or would enjoy doing, such as exploring their environments and interacting with others. Not all harms must be experienced as bad by the one who is harmed. Consider the harm of death—or more specifically, the permanent annihilation of one's consciousness. Epicureans contend that death cannot be bad for the one who dies because all harm consists in the experience of suffering and, although one may suffer in the process of dying, death itself cannot cause one to suffer. However, as other philosophers have pointed out, this view of death fails to appreciate that death

[2]Cochrane allows that some animals may qualify as autonomous agents and, therefore, have an intrinsic interest in liberty.

can be bad for the one who dies simply in virtue of the fact that it permanently deprives one of the various goods of conscious life, including the pursuit of various enjoyments in life. Similarly, confinement can be bad for animals insofar as it deprives them of certain basic goods—the fulfillment of various enjoyments they have in life—regardless of whether the animals suffer as a result of this deprivation.

It might be objected at this point that although it understandable how a being's death can be bad for it despite not causing it to suffer, it makes less sense to think that confinement can harm a being without making it suffer. When a being dies, there is a reason why it doesn't suffer: the conscious being no longer exists and is no longer capable of having any experiences. However, to confine animals does not annihilate their consciousness. So, if some particular instance of confinement is actually bad for animals, why wouldn't it cause them to suffer? In answer to this question, let me first emphasize, again, that we may not always know when animals are suffering in less obvious ways such as from boredom and depression. However, it is also the case that animals may not suffer because they are not aware that they are being deprived of goods. For example, cows that are being milked may not understand that they are being prevented from doing other things they would otherwise enjoy doing at that moment. The fact that they may not be aware that they are being deprived, to some degree, of living their lives as they please does not negate the fact that they are being deprived.

I do not wish to claim that we harm others whenever we deprive them of things that they desire. The concept of harm tends to be reserved for morally significant negative impacts that we make on an entity's welfare. Many instances of depriving others of things they want are not morally significant (and some deprivations may actually be good for a being). For instance, if I sit in someone's favorite seat on the bus, I prevent this person from satisfying a certain desire but, at least under normal circumstances, I do not thereby harm her. Similarly, if I deprive an animal of a treat that it would enjoy eating, I do not necessarily harm the animal. However, if I frequently or regularly prevent an animal from freely pursuing the things that it enjoys doing, then I am depriving the animal and negatively impacting its welfare in a way that is more cumulative and, thus, more morally significant.

My argument assumes that it is generally good for animals to satisfy their various desires and that it is generally bad for them to have their desires thwarted. For animals to satisfy their desires, generally speaking, is good as a means to their living more enjoyable lives. To frequently inhibit them from pursuing their desires is generally bad for animals insofar as it deprives them of living more enjoyable lives. There will be some exceptions to this general rule. Animals, like people, sometimes desire things which are not good for them. For example, an animal that wants to run across a busy road will likely get hit by a car and, therefore, it is better for the animal to not act on its desire. However, I see no special reason to doubt that, in general, it is good for animals to have opportunities to satisfy their various desires or enjoyments.

My argument implies that freedom, generally, is instrumentally good for animals: freedom is good for animals as a means to their living more enjoyable lives. It is unclear to me whether some animals also have an intrinsic interest in freedom—

that is, whether freedom is good for animals independent of whether it helps them obtain other goods such as enjoyment and the avoidance of suffering. However, I am not convinced that it matters ultimately whether animals have an intrinsic interest in freedom. One might think that it matters on the following grounds: if animals' interest in freedom is merely instrumental, then it would be possible to confine animals for human use without harming them so long as we can confine them without making them suffer, whereas if animals have an intrinsic interest in freedom, then to confine them would harm them independent of whether they are made to suffer. However, this reasoning is short-sighted. As I have argued, freedom is good for animals not only because animals may suffer when confined, but also because freedom allows animals to experience greater enjoyment in life. Even if it is possible to confine animals for human use in ways that eliminate their suffering (which is itself questionable), they are still harmed insofar as they are deprived, to some degree, of experiencing more enjoyable lives.

The fact that animals' interests in freedom may only be instrumental does not detract from the moral significance of their interests. Instrumental interests can be significant interests. Indeed, they can be among a being's most important interests. Arguably, the interest in continued life—for both people and animals—is only an instrumental interest. Life is good only as a means to various valuable conscious experiences. If a being permanently loses the capacity for conscious experience, yet remains alive, life is no longer good for that being. Nevertheless, life is among the most important interests for persons and animals. Additionally, even if freedom is only instrumentally good for animals, it could always or usually have this instrumental worth for them, such that animals are always or usually harmed by confinement. Similarly, although life has merely instrumental value for us, it is usually if not always the case that we would be harmed instrumentally by death—that is, being killed usually if not always robs us of having valuable experiences ever again.

In response to my argument, it might be objected that confining animals for human use can be done in ways that do not prevent animals from satisfying their desires or enjoyments. Indeed, Cochrane makes this objection in response to the idea that animals need freedom in order to satisfy their desires. I am unconvinced by this objection, however. Animals confined to be used as human resources are typically controlled, restricted, and made to do things to satisfy human desires rather than their own desires. It seems inevitable that, to some extent, they will be prevented from living as they please. Furthermore, we may not be able to know many of the specific things that animals enjoy doing (or would enjoy doing) when allowed to be free and explore. It seems reasonable to think that individual animals will typically know their own desires and enjoyments better than we will, since they are the only ones who actually experience their desires. The fact that they belong to different species than us, with different physiologies, and are unable to clearly tell us through language what their desires are further supports the view that the animals are typically in the best position to know what things they desire to do in any given moment. Animals are more likely to have their desires satisfied if, in general, we just 'let them be' and allow them to be in control of their lives, to pursue whatever it is that they may desire and enjoy.

There is also something confusing and perhaps contradictory in Cochrane's claim that animals can lack freedom but still have their desires satisfied. Freedom, in the basic sense that is applicable to animals, means the ability to satisfy one's desires absent of external constraints. If animals used as resources are able to fully satisfy their various desires, then it would seem that they do not lack freedom in the relevant sense. On the other hand, if the animals lack freedom in the relevant sense, then by definition, they are prevented from pursuing their desires to some extent. My suggestion is that confined animals used as resources typically do lack freedom to some extent, meaning that they are unable to pursue some of their desires, even if we make their environments more humane such that they are able to satisfy more of their desires.

To summarize the main points of my argument so far, I've suggested that using animals as our food resources usually involves, in various ways, restricting their freedom—that is, their ability to act on their desires without external constraints. To restrict animals' freedom harms animals by making them suffer in both obvious ways and more subtle ways such as depression and boredom. But regardless of whether animals are made to suffer, to restrict their freedom also harms them simply by preventing them from fulfilling their various desires or enjoyments in life. It follows that using animals as our food resources usually causes harm to them by depriving them of their freedom.

Supposing that my argument is correct, it remains an open question whether it is ethical to use animals as food resources. However, if we assume that it is generally wrong to intentionally cause harm to others, then we should conclude that it is usually wrong to use animals as our food resources. On this view—let us call it *the freedom-respecting view*—we should treat animals not as our resources but rather as free individuals, generally allowing them to control their own lives and live as they please. Note, however, that it could be ethical, on this view, to restrict animals' freedom when doing so is necessary to protect their welfare, such as preventing animals from running across busy roads.

I take this argument to be a central reason why we ought to adopt a vegan lifestyle, a lifestyle in which one abstains from using animals as food resources. It is, I would argue, a more fundamental reason for being vegan than the argument from suffering. For it suggests that even if we improve the conditions of farm animals, such that they are not made to suffer or killed when they outlive their usefulness, raising animals to use as food resources is still wrong. It is wrong insofar as it harms animals by restricting their freedom in various ways for human benefit, rather than allowing them the freedom to lead their own lives. This argument suggests that there is something fundamentally harmful about using animals as food resources. In this sense, it is similar to the argument that it is fundamentally harmful and wrong to kill animals for meat, even if they are raised and killed without being made to suffer. However, the harm caused to animals by restricting their freedom in the process of using them as food resources is, in general, significantly less than the harm caused to them when we kill them for meat, for death permanently prevents animals from pursuing all of their enjoyments in life, whereas the same is not true of confinement. It also should be pointed out that there may be some food products or materials from animals that we could obtain and use without harming animals by killing them, making them suffer, or restricting their freedom. For example, it seems possible to collect and use leftover animal manure for purposes of fertilizer without necessarily harming animals. Also, if hens lay eggs that do not hatch and the hens have no other use for them, it may be ethical for us to use these leftover, unwanted eggs.

In closing this section, let me address one final challenge to the idea that it is usually wrong for humans to use animals as their resources. In a recent article, Wayne (2013) defends the practice of using domesticated animals for their products or labor on the grounds that it is essential to incorporating them as participants in the human community, something which is beneficial for both animals and their human caretakers. Domesticated animals struggle to meet their own basic survival needs and, thus, they benefit from having relationships with human caretakers. In exchange, humans benefit from the products, labor, and affections that the animals can provide us. Wayne also questions how valuable freedom from human interference can be for domesticated animals given their dependence on humans for their survival and well-being. Her implication seems to be that domesticated animals are better off if included as participants in human communities, which includes providing food and labor to those communities, rather than being separated from them.

I am in agreement that domesticated animals, due to their dependent state, are generally better off being taken care of by human beings. However, it is unclear why we should think that the animals must serve as our resources, contributing their labor or products, as a condition of their receiving our caretaking. One might argue that beneficial relationships should be governed by a principle of fairness or reciprocity. However, especially since human beings are responsible for domesticating animals, I would propose that we owe it to the animals to take care of them without expecting any sort of beneficial return for ourselves. Assuming that we do take care of the animals independent of whether they give anything back to us, it is unclear how animals benefit when we use them for their products or labor, especially when it is not the case that they are voluntarily offering their services to us. To the contrary, I've suggested in my argument that using domesticated animals as our resources typically harms them to some degree, insofar as we prevent them to some degree from pursuing their own enjoyments or interests in life.

4 Farm Animals in a Vegan World

In a well-known article in environmental ethics, J. Baird Callicott suggests that the idea of "liberating" domesticated farm animals is meaningless and a practical impossibility. It is meaningless, he argues, because domesticated animals are not capable of living on their own. In contrast to wild animals being held captive, domesticated animals "have been bred to docility, tractability, stupidity, and dependency" (Callicott 1980, 331). Callicott further asks us to consider the practical consequences of liberating farm animals. If we simply release farm animals into the wild, many of them will starve or freeze to death. Some of these animals, he suggests, might be able to survive and recover some of their former wild selves, but then they would be competing for food and living space with other indigenous wildlife (idem, 331). This could be bad ecologically. Perhaps instead of setting farm animals loose into the wild, we could continue to house and feed them on farms without killing or using them for food. However, as these animals, free from natural predation, breed and increase their populations, more and more land use would be required to house the animals as well as to produce food for them (in addition to the plant food required to feed a vegan human population). This option also appears bad from an ecological standpoint. One final option would be to continue housing and feeding farm animals but not allow them to breed, ultimately driving them into extinction. Callicott suggests that this choice would be "ironic" since we'd be eliminating the very species that we are trying to protect and benefit by "liberating" them.

The idea of liberating domesticated farm animals is not meaningless, provided that we have a reasonable understanding of what "liberation" entails for farm animals. Liberation for farm animals need not entail setting farm animals free into the wild. As I understand it, liberation for farm animals means freeing farm animals from human exploitation, particularly in ways that are harmful to the animals. Nevertheless, Callicott's remarks raise an important question of what we should do with farm animals if we cease to use them as our food resources. Upon first glance, it does not seem ethical to set domesticated animals free to fend for themselves in the wild or on the streets in cities. Because they are domesticated, they are less able to fend for themselves, such as by finding food, making shelters, and defending themselves against predators. Many of these animals probably would not be able to survive on their own.

Rather than setting farm animals free to fend for themselves, perhaps we ought to continue to care for them due to their dependent state. Because human beings are responsible for domesticating these animal species, we incur a special moral responsibility to help them meet their basic needs. If we continue to take care of farm animals, they should be given ample space outdoors to move about, exercise, play, and socialize; their living spaces indoors should not be too restrictive; and we should not continue to treat them as our resources. This option might be impractical given the sheer amount of farm animals that would need to be housed and taken care of. Additionally, as Callicott points out, it also seems impractical and potentially disastrous ecologically to allow farm animals to continue breeding freely. This would leave us with two other options: either (1) we could prevent farm animals from breeding at all, which would ultimately lead to the extinction of these species or (2) we could significantly limit the breeding of farm animals, such that they would not be driven to extinction but also would not be allowed to become too populous. This latter option is currently the strategy used with cats and dogs in the United States: many animal welfare organizations emphasize the importance of spaying and neutering cats and dogs in attempt to control their population levels. We can imagine a similar option being used for farm animals like cows, pigs, and chickens.

Contrary to Callicott, I do not believe it would be "ironic" or self-contradictory to drive domesticated farm animal species into extinction by not allowing them to reproduce. As I understand it, the basic goal of liberating animals from human exploitation is to promote the welfare of existing individual animals or to protect them from human-caused harm. To allow farm animals to live out the rest of their lives under the protection of humans, but not allow them to reproduce, does not contradict this goal so long as we treat the existing individuals respectfully. Arguably, animals that have yet to come into existence do not have any interests or rights to come into existence. However, there may be other reasons to continue the existence of farm animal species. In particular, many people learn to empathize with and feel compassion for animals by interacting with domesticated animals and becoming familiar with their personalities and inner emotional lives. This can happen through the relationships we have with companion animals like cats and dogs or by spending time with farm animals on farm sanctuaries. Without domesticated animals, future human generations lose these important opportunities to develop their emotional and moral sensitivity to animals.

Still, the prospect of limiting animals' reproduction has its own moral problems. There are different means by which we can try to control animal reproduction, neutering being one of them. The practice of neutering animals is problematic since it requires inflicting some harm on animals. David Boonin details the harms imposed on cats and dogs when they are "fixed."

> [They] must be confined and taken to the vet, placed in unfamiliar surroundings, exposed to a frightening environment. Most animals who go through such procedures will surely experience a great deal of anxiety if not outright fear and terror. In addition, [they] must either be given a general anesthesia, which can cause a variety of adverse reactions and in some cases even death, or suffer a tremendous amount of physical pain during the procedure. The procedure itself exposes [them] to non-negligible risks of various infections and of complications that can arise from incomplete removal of the organs or from excessive bleeding. And when the procedure is over, [they] will suffer from a general disorientation as well as nausea and physical discomfort, lasting in some cases for several days (Boonin 2003).

Additionally, it may be thought that neutering animals harms them simply by restricting their reproductive freedom, frustrating their desires to have sex and reproduce. Indeed, it might seem contradictory to defend veganism by appeal to animals' interest in freedom but then also propose that we ought to restrict domesticated animals' freedom to reproduce. If it is wrong to use animals as food resources because doing so restricts their freedom, why isn't it also wrong to restrict animals' reproductive freedom?

Arguably, it is sometimes ethical to restrict another being's freedom when doing so is in the best interests of that being (e.g. preventing a child from playing in a busy street). However, to restrict the freedom of a farm animal to reproduce would not be done for the good of the animal; rather, it would be done to control the population level of the species. Nevertheless, controlling the reproduction of domesticated animals could be justified on the grounds that it is necessary to prevent animals from endangering the welfare of their offspring. If domesticated animals are allowed to freely reproduce, they will create more animals than we can take care of. As a result, many animals would be left to fend for themselves and could suffer and die from starvation or an inability to meet other basic needs. Now, animals are not morally responsible for their actions and they do not intend to endanger their offspring. Still, we are sometimes justified in preventing them from doing things that would harm or endanger the welfare of others, even if it requires that we cause some harm to them.

Although neutering animals deprives them of the freedom to have sex and reproduce, it is also questionable how much animals are truly harmed if they can no longer have sex or reproduce. Arguably, animals do not actually desire to reproduce because they lack the concept of reproduction. They desire to have sex for pleasure and reproduction is the unintended, unforeseen result of this. To neuter animals, then, initially thwarts their desire to have sex, but it does so by eliminating that desire. So, it is not as if the animals go on wanting something that we do not allow them to have. And, in some cases, if neutering animals eliminates the ability to reproduce but not the desire to have sex, then the animals are not prevented at all from having something they want, since it was never their desire to reproduce in the first place.

The most serious harms potentially inflicted by neutering seem to be those described by Boonin, including the physical pain that can result from the procedure or from complications that arise from the procedure. Whether it is ethical to neuter domesticated animals to control their population levels will depend, in part, on the likelihood and frequency of the harms to neutered animals, as well as how serious or significant those harms are in comparison to the harms that offspring would suffer if domesticated animals are free to reproduce and we are unable to take care of all of the offspring.

One final option for dealing with domesticated animals in a vegan world would be to painlessly kill all of the animals and simply end the moral problems associated with domesticated animals. However, I find that I cannot endorse this option. Arguably, to permanently end the life of another conscious being who possesses various enjoyments in life is one of the worst harms that we can inflict on another being. It is true that there are some cases in which a conscious being is better off being killed than continuing to live, but typically those cases involve beings whose lives are so irreversibly miserable as to not be worth living. I do not think it could be reasonably argued that all or even most domesticated animals' lives are so irreversibly miserable as to not be worth living.

In closing, let me add that although it may be ethical to neuter domesticated animals to control overpopulation problems, it does not follow that it is ethical to do the same to wild animals. Domesticated animals are dependent on us for their survival and health, and we have a special responsibility to take care of them. If stray domesticated animals suffer and die because there are too many of them to be taken care of and they cannot fend for themselves, we are partly responsible for their plight. So, we have a special responsibility to take steps to reduce the number of domesticated animals that must suffer. By contrast, wild animals are more capable of fending for themselves. Moreover, as I have argued elsewhere, to adopt a widespread policy of intervening in the wild to prevent wild animals from suffering and dying would likely cause ecological problems that would harm many more beings (see Simmons 2009).

5 Conclusion

If using animals as our resources for food and material somehow wrongs the animals, it must be because it harms them in some way. The most obvious and perhaps most serious harm to animals when we use them as resources consists in the suffering and death we inflict on animals raised on factory farms. However, even under more humane conditions, using animals as our resources can harm them insofar as we control them and restrict their freedom in order to satisfy our wants. To restrict animals' freedom can harm them not only by causing them to suffer, but also insofar as we deprive them of goods: the goods of living their lives as they please and freely pursuing their enjoyments in life. In the remainder of this chapter, I considered what we should do with farm animals if we were to cease using them as our resources. None of the options are morally unproblematic and it is unclear what the morally best option is. Setting farm animals free to fend for themselves avoids the potential harms imposed on animals by trying to control their reproduction, but does so at the cost of potentially sending animals to live miserable lives on their own in which they are incapable of fending for themselves. On the other hand, we could continue to take care of farm animals while limiting their abilities to reproduce, but this option may run into practical obstacles of providing enough land and resources to millions of farm animals for the duration of their natural lives. Additionally, it may require causing physical harm to animals through a mass neutering effort, harms that cannot be justified on the grounds that it is in those animals' best interests.

References

Boonin, D. 2003. Robbing PETA to spay paul: do animal rights include reproductive rights? *Between the Species* 13(3): 1–8.

Callicott, J.B. 1980. Animal liberation: a triangular affair. *Environmental Ethics* 2: 311–338.

Cochrane, A. 2012. *Animal rights without liberation*. New York: Columbia University Press.

Simmons, A. 2009. Animals, predators, the right to life, and the duty to save lives. *Ethics and the Environment* 14(1): 15–27.

Wayne, K. 2013. Permissible use and interdependence: against principled veganism. *Journal of Applied Philosophy* 30(2): 160–175.

Should Captive Primates Have Reproductive Rights?

T.J. Kasperbauer

Abstract Reproduction of animals in captivity is tightly controlled. In stark contrast, regulating the reproductive behavior of human beings is generally seen as impermissible. Why is this? This chapter discusses reproductive control of primates in light of two main reasons that human beings are granted reproductive rights: the importance of autonomy and human interest in procreation. I argue that captive primates pose a challenge to the use of birth control in captivity, because they too have important interests in reproductive autonomy and procreation. If we are to be consistent, it seems we must grant reproductive rights to some captive primates, perhaps at least the great apes. However, I further argue that there is room to limit reproduction within the framework of reproductive rights. Even in the case of human beings, it is widely accepted that limiting reproduction is permissible in certain circumstances. In captivity, unlimited reproduction would quickly lead to overpopulation, causing significant negative welfare for members of the group. In such cases, I suggest, the use of birth control is permissible. To further develop my account, I review the potential welfare costs to using contraception. I also discuss why it might be important to grant reproductive rights in the context of routine contraception, particularly in an era in which anthropogenic influences on native habitats require long-term survival in captivity.

1 Introduction

In 2001, 87 primate species were on some form of contraception, involving over 40 different methods of contraception (Porton and Dematteo 2005). All indications suggest that contraceptive use in captivity has only increased since 2001, and will continue to increase. For example, some experts have argued, in response to recent controversies over the use of culling, that zoos should invest even more heavily in

T.J. Kasperbauer (✉)
Department of Food and Resource Economics, University of Copenhagen, Copenhagen, Denmark
e-mail: tjkasperbauer@gmail.com

B. Bovenkerk and J. Keulartz (eds.), *Animal Ethics in the Age of Humans*, The International Library of Environmental, Agricultural and Food Ethics 23, DOI 10.1007/978-3-319-44206-8_17

contraception (e.g., Kaufman 2012). This has been echoed even by those who support culling in zoos (Carter and Kagan 2010).

In this chapter I look at the use of contraception and other forms of birth control specifically for captive primates. In particular, I explore whether captive primates have a right not to have their reproduction interfered with. I am not aiming to provide a comprehensive analysis of reproductive rights for captive primates. Rather, I focus on whether captive primates have a moral right not to be prevented from conceiving, gestating, and giving birth to offspring—otherwise known as a right to procreate. I am considering cases where primates are able to reproduce and conceive, but there is a human interest in restricting this ability. Thus I will not address the question of whether primates should be assisted in reproducing.

Keeping primates in captivity, for example in zoos, is a practice that gives rise to moral controversy. I am not here taking sides in this debate. Rather, I assume that care must be provided to currently existing groups of primates, and I am asking if, as part of their care, they should be allowed to reproduce. Hockings et al. (2015) argue that in the Anthropocene—which is characterized by increased human influence on animals' native habitats—we must learn to think about the lives of primates within a range of human-built environments. I assume that the 'anthropogenic continuum,' as they call it, must also include captive environments. The continued destruction of primates' native habitats, due to human activity, requires reflection on what we owe to primates in captivity.

Primates present a particularly interesting challenge to the widespread use of birth control because (1) there is good evidence that contraception causes welfare problems for primates, and (2) some primate species, particularly the great apes, possess capacities often cited in defense of procreative rights for human beings. There are also new initiatives aimed at placing tighter controls on reproduction for some primate species. To take one example, the European Association of Zoos and Aquariums (EAZA) has decided to prevent interbreeding between the four subspecies of chimpanzees, in order to preserve their genetic distinctness (similar to what has been done for years with Sumatran and Bornean orangutans). This decision came in response to analyses showing diverse genetic subspecies in the wild, but significant intermixing in captivity (Carlsen and de Jongh 2015; Hvilsom et al. 2013). Preventing interbreeding in captivity will take significant effort, and will require a sharp change in the management of some captive chimpanzee groups. This case raises the difficult question of whether the goal of genetic distinctness can justify placing extensive limitations on chimpanzee reproduction.

I argue that there are good reasons to grant procreative rights to primates, but that there are countervailing factors that permit the restriction of reproduction. I further argue, however, that these restrictions are compatible with granting reproductive rights to primates. To develop these arguments, I begin by sketching some of the most basic reasons to think that human beings have a right to procreate. I then apply these considerations to captive primates. In order to further investigate whether restricting reproduction violates reproductive rights, I review the potential welfare costs to using contraception. I conclude by discussing why it might be important to grant reproductive rights in the context of routine contraception.

One additional note on the conception of rights discussed here: I am not addressing the question of whether animals are legitimate recipients of rights, nor the nature of rights for animals, nor the adequacy of rights theories generally. My goal here is justificatory: I am investigating the reasons underlying the right to reproduce for human beings, and critically examining the application of these reasons to captive primates. I use the terminology of rights in order to evaluate how we ought to treat captive primates and determine whether we are justified in interfering with their reproductive capacities. This should not be taken to mean anything particularly controversial. Assigning reproductive rights to primates is a way of saying that reproduction is particularly important for them and thus might require special protection. As my analysis illustrates, however, reproductive rights are not incompatible with consequentialist concerns; on the contrary, examining reproductive rights requires a detailed assessment of welfare impacts beyond just the individuals whose reproduction might be restricted.

2 Reproductive Rights for Human Beings

One of the best-known defenses of the right to procreate comes from Dworkin (1993). He grounds procreative rights in human autonomy, and the right generally for autonomous individuals "to make important decisions defining their own lives for themselves" (idem, 222). Dworkin offers two broad reasons to think that the decisions of autonomous individuals should be respected. The first is that "each person generally knows what is in his own best interests better than anyone else" (idem, 223). He ultimately rejects this as inadequate, however, because some people probably do not know what is in their best interest (e.g., those with severe mental disabilities). The second reason he offers is that "Autonomy encourages and protects people's general capacity to lead their lives out of a distinctive sense of their own character, a sense of what is important to and for them" (idem, 224). Regardless of whether our interests are misguided, Dworkin thinks it is important that we are allowed to pursue the interests that we take as particularly central in our lives. Reproduction is one such interest.

The idea that the right to reproduce should be protected because it is an important human interest has also been defended by Robertson (1994). He argues, "Procreative liberty should enjoy presumptive primacy when conflicts about its exercise arise because control over whether one reproduces or not is central to personal identity, to dignity, and to the meaning of one's life" (Robertson 1994, 24). For Robertson, satisfying strong sexual desires is important, but more central is satisfying the human interest in having offspring. Having the choice to contribute to the next generation through one's children is, according to Robertson, one of the most fundamental human desires.

The importance of autonomy and strong human interests in having children are widely taken to provide good reasons to protect decisions to procreate. However, there is significant dispute about what exactly follows from these considerations with respect to a 'right to procreate.' For instance, some have argued that a right to procreate does not entail that we have the right to *unlimited* procreation (Conly 2005; Quickley 2010). Preventing someone from having *any* children violates the right to procreate, one might argue, but preventing the twentieth child, for example, does not. At the very least, we might think that governments are more justified in coercing people to limit their procreation than they are in preventing procreation at all.

We can also distinguish different components of the right to procreate, some of which may not be necessary to express one's right to procreate. For instance, consider the role of sex in producing offspring. It is of course possible to use reproductive technology to produce offspring without engaging in sex. So hypothetically, a government could use some form of technology to provide every citizen with offspring, while also preventing them from having sex, and still be legitimately protecting the right to procreate. This seems consistent with Robertson's account, for instance. Or consider the formulation found in the UN's Declaration of Human Rights, which states that all human beings have a "right to decide whether or when to have children." This right could conceivably be met through means other than sexual reproduction.

Rearing offspring might also be separated from the right to procreate. There is significant debate about whether humans have a right to rear their own offspring or to rear offspring generally.[1] Dworkin and Robertson's accounts do not seem to entail that rearing offspring is included in the right to procreate. O'Neill (2002) argues that procreation must involve the intention to give one's children a decent quality of life. Engaging in sex without such an intention thus is not actually exercising one's right to procreate, and so does not deserve to be protected. But this of course does not mean that you personally must provide a decent life for your offspring. Perhaps you would better fulfill your duty by letting someone else take over.

These concerns notwithstanding, we can identify a basic core to the right to procreate: non-interference in reproductive activities. Dworkin and Robertson's accounts claim that our strong interest in procreation—which includes both an interest in sex and in the resulting creation of offspring—should be protected. One reason for this is just that these interests are central to our lives and give our lives value. Another reason, offered by Dworkin, comes from the value of autonomy: it is important to allow people to control significant aspects of their lives.

Some of the complications just mentioned will be returned to below, but I set these aside for now in order to apply Dworkin and Robertson's considerations to primates. In the next section I examine whether the above considerations for human beings—autonomy and a strong interest in procreation—have similar implications for the procreative rights of captive primates. I also examine arguments put forth by Donaldson and Kymlicka (2011) for granting reproductive rights to domesticated animals, since many of the reasons they cite also apply to captive animals generally.

[1]For a review, see Eijkholt (2009).

3 Reasons to Grant Procreative Rights to Captive Primates

It is difficult to know whether captive primates have an interest in creating offspring in Robertson's sense—namely, in making a contribution to future generations. They do have a strong interest in sex, and many species show concern and affection for offspring, but it's not clear if they actively consider the importance of offspring to their lives. As a result, it's not clear how we might ground a right to procreate solely in a primate interest in procreation.

A similar obstacle pertains to autonomy. One view of autonomy (Velleman 2006) holds that an autonomous agent is one who is capable of determining future behavior based on past commitments. Primates are plausibly autonomous on this definition in that they have some sense of their past behavior, and are cognitively capable of future planning. But it is unlikely that they are autonomous with respect to their procreative commitments. And in captivity their autonomy is obviously limited in that human beings already limit much of their behavior. They require the support of humans to do the things they want to do.

Moreover, it is not clear whether many primates take an interest in autonomy as such. Some have argued that autonomy matters less for animals because they lack 'higher-order interests regarding autonomy and self-respect' (Ladwig 2015, 185; also see Cochrane 2012). Consider Dworkin's argument, as outlined above, which is based on people living their lives in their own distinctive way. Though it is not stated explicitly, this seems to assume that people value individuality in some sense; they also want to live on their own terms. If primates do not take a similar interest, then perhaps their autonomy accordingly has less moral weight.

These skeptical concerns about the right to procreate have been tempered recently by Donaldson and Kymlicka's (2011) arguments for granting reproductive rights to domesticated animals. They argue that a full fledged sense of autonomy is not needed in order to ground certain important rights. It is enough that animals have strong interests, that we can discern these interests from their behavior, and that at least in some cases granting them increased control over their lives can contribute to their flourishing. Donaldson and Kymlicka further argue that domesticated animals should be granted certain rights by virtue of their membership in human communities. Domesticated animals, like dogs and cats, have been brought into human communities and are dependent on humans for their survival. Their ability to survive outside of these communities has been fundamentally altered. As a result, Donaldson and Kymlicka (idem, 77–89) argue that domesticated animals should be considered citizens of human communities, and receive certain benefits and responsibilities akin to those granted to other citizens in those communities.

One of nine rights Donaldson and Kymlicka attribute to domesticated animal citizens is the right to sex and reproduction (idem, 123). It is precisely because the lives of domesticated animals are under tight human control that some important animal interests—like sex—should be protected. As they say, "In recognizing domesticated animals as citizens, to the extent that autonomous control over their sexual and reproductive lives is possible for such animals, then we should seek to restore it" (idem, 146). Domesticated animals may not have an interest in contributing to future generations, nor an interest in autonomy as such, but for Donaldson and Kymlicka this just provides additional reasons to protect their reproductive behavior. Domesticated animals do, for the most part, have an interest in sex, and their well-being is plausibly improved by being allowed to satisfy that interest (see e.g., Palmer et al. 2012). This gives us reasons to grant them some degree of control over their reproductive lives.

Similar considerations apply to captive primates. They are arguably undomesticated in the sense that they have not been selectively bred to be tamer, for instance, or to provide services to human beings. But their behaviors have been altered as a result of captivity, especially for species that have lived in captivity for multiple generations. Since they cannot leave captivity, perhaps they should be granted protections more congruent with their status as permanent members of human-built communities. They cannot participate in the lives of humans like some domesticated animals, but they possess many of the same dependencies. They also have a strong interest in sex, the satisfaction of which presumably has a large impact on their well-being.

One group of primates for which reproductive rights seem particularly important are the great apes. Gorillas, orangutans, chimpanzees, and bonobos are often understood to be more similar to human beings in their cognitive capacities than to the other primates. They are more likely to have an interest in autonomy, for instance (Arruda and Povinelli 2015; Beauchamp and Webber 2014). Given their intelligence, they are presumably more affected by not being able to make their own reproductive decisions (e.g., choosing their own mates). They also arguably have a greater capacity to appreciate the significance of reproduction. For instance, some have suggested that great apes understand the link between pregnancy and offspring, and might also understand the role played by their own sexual activity, though this hasn't been rigorously studied (Fouts and Mills 1997; Patterson 1987). Most populations of captive great apes are also permanent members of captivity, because of past human actions, and so cannot ever return to the wild. We may thus have a duty to restore as much autonomy as we can to apes, including control over reproduction.

In summary, Dworkin and Robertson's arguments for procreative rights do not neatly apply to all primates, though they seem to apply easier to great apes. This may not really matter though, because even a thinner sense of autonomy and interest in reproduction seem sufficient to ground procreative rights. Primates do have strong sexual desires, and they are capable of acting autonomously, even if they do not take a robust interest in bearing offspring or autonomy as such. On Donaldson and Kymlicka's account, these interests are enough to grant reproductive rights to captive primates.

4 Problems with Granting Procreative Rights to Captive Primates

There are many problems with granting procreative rights to captive primates, some of which flow from deficiencies in the considerations just mentioned, and others that stem from the challenges posed by captivity. Let's begin by discussing some issues that Donaldson and Kymlicka themselves raise with attributing reproductive rights to captive animals. They do not directly address zoo animals except for in a single footnote, where they say, 'restrictions on their sexual and reproductive choices must be justified in terms of the interests of the individual who is being restricted' (2011, 283). Ultimately, however, they think that the institution of captivity is impermissible, and because captivity entails total control over an animal's life, there is little room to talk about the implications of autonomy. It is only insofar as captivity must exist that Donaldson and Kymlicka think that decisions to control reproduction are justified. Nonetheless it is significant that they seem to think that this can be done while taking into account the interests of individual animals.

This can be better understood within the context of Donaldson and Kymlicka's account more broadly, according to which animal citizenship entails both rights and responsibilities. Along with reproductive rights come responsibilities as animal citizens to, in their words, "exercise their rights in ways that do not impose unfair or unreasonable costs on others, and that do not create unsustainable burdens on the scheme of cooperation" (idem, 146). One might argue that free breeding in captivity places unreasonable burdens on zoos, for instance, in a way that is unsustainable. If all zoos allowed primates to breed freely, an overpopulation problem would quickly arise. And since captive primate populations cannot self-regulate, we as their caretakers are permitted to control their breeding. This could be seen as just part of the animal citizenship package. So according to Donaldson and Kymlicka's account, captive animals do possess reproductive rights (for similar reasons as domesticated animals), but because captive populations are not capable of self-regulation, humans must impose some degree of restrictions on reproduction.

This point must be pushed further, however, since the issue might otherwise just seem like a problem with the institution of captivity. If zoos, for instance, do not have the capacity for primates to breed more freely, then perhaps zoos should not exist. It is not the fault of the animals if zoos cannot sustain a breeding population.

In response to this criticism of captivity, one could argue that restricting breeding is also in the *animals'* interest; it's not just zoos that would carry a burden. There are no predators in captivity or extra space to spread to when space becomes limited. Even if a zoo could expand significantly, there may nonetheless be costs to other animals in a group by allowing individuals to breed more freely. Competition for resources is inevitable in any context where there is prolific breeding. Captive primate citizens thus must also cooperate with each other. But assuming that they cannot adequately respond to population pressures on their own, it is up to human caretakers to do so in a way that balances the interests of all the individuals in a population—not just those looking to reproduce. Though Donaldson and Kymlicka do not argue for any of these conclusions, these are the sorts of concerns that seem to flow from their citizenship model, as applied to captive primates.

Another issue related to the balancing of interests is the balancing of rights. There is a good chance that unrestrained breeding will put other rights at risk. Even in the human case, the right to procreate can be outweighed by other rights, like the right not to be harmed or a right to a minimally decent life. These are of course potential outcomes when overpopulation occurs. In such cases, we may be justified in controlling reproduction in order to avoid causing harm to other members of a group.

In zoos, these population issues are sometimes dealt with through culling, or killing an animal that is considered surplus. Certain captive primate species are routinely culled, though the practice is highly controversial (Plowman et al. 2005). The right to reproduce may in any case go hand-in-hand with culling; adopting one practice necessitates the other. One might think this is in fact a good thing: through culling, zoos could create sustainable populations while also granting reproductive rights. But this may yet be objectionable because it infringes on the right not to be killed. Some zoos cull animals at a very young age, depriving them of a life at all. This, one might think, is not a tradeoff that is justified by the right to procreate.

The well-being of offspring raises an additional problem with free breeding. As mentioned above, on some accounts (e.g. O'Neill 2002) procreation involves the intention to give one's children a decent quality of life. This is particularly relevant for captive primates, who in some cases many not be willing or able to provide a good life for their offspring. Primates' decision to have sex might not be a decision to do anything but that. Their right to procreate may thus not be worth protecting because it does not include an intention to take care of potential offspring. Relatedly, in the human case it is often claimed that reproduction should be limited if the children cannot expect a good life, whether that is directly a result of parental decisions or not (e.g., Conly 2005). This too may limit the importance of procreative rights for certain groups of primates.

To summarize the discussion thus far, even if we grant captive primates the right to procreate, we still seem justified in limiting their reproduction if (1) the population is not self-regulating, (2) unrestrained reproduction is unsustainable or puts significant burdens on others, (3) the rights of others are infringed, or (4) the well-being of individuals, particularly offspring, is significantly reduced by free breeding. These considerations are not unique to captive primates, however, but are also frequently mentioned in the context of human reproductive rights. So it's not that primates necessarily lack reproductive rights. Rather, their reproductive rights can be trumped by other important concerns.

This general line of reasoning is also frequently applied to human reproductive rights. It is often argued that procreative rights must be sensitive to cost-benefit analysis. Benetar (2010), for instance, argues that limiting reproduction is permissible if doing so produces significant benefits with no other risks. Humans are typically allowed to reproduce freely because this is rarely if ever the case. If human overpopulation reached a critical level (as one might argue it has, though I will not discuss the issue here), limiting reproduction would be more justified. For instance, if human reproduction had consequences that were more similar to what we see in captivity, there would be similarly good grounds for placing restrictions on humans' reproductive rights. It is only because of the unique circumstances of captivity that the reproductive rights of primates must be curtailed in a very different way than for human beings.

We should take seriously the idea that procreative rights must be sensitive to cost-benefit analysis. This cuts both ways though. Limiting primate reproduction may produce greater risks than benefits, in which case it would seem unjustified. I have considered reasons to restrict the right to procreate mainly based on the costs of free breeding. But we must also consider the costs involved in limiting reproduction. I do this in the next section by reviewing the welfare costs of contraception for primates, in terms of both physical and psychological health.

5 Welfare Costs of Using Contraception on Primates

Contraception does seem to have significant psychological and physiological effects on captive animals. A number of analyses have found that contraception is not reversible for many species, or causes significant delays in reproduction (Williams and Hoffman 2009). Callitrichids, chimpanzees, and lemurs, for instance, are known to have problems reproducing after being on contraception. There are also cases where contraception that was thought to be reversible resulted in permanent sterilization (e.g., Goeldi monkeys; Ballou et al. 2010, 223). These are not directly welfare effects, but they do suggest that there may be underlying welfare issues with contraception.

Asa and Porton (2010) review a range of behavioral effects contraceptives have on females from a variety of species, including mood changes, depression, lethargy, and docility. These all likely involve a negative impact on welfare. Asa and Porton recommend caution in choosing to use contraceptives, because while "the ramifications can be multifaceted," they admit that "there is limited information on what impact contraception has on individual and social behavior" (2010, 476).

The situation is somewhat different for primates, however, where more is known about the effects of contraception. The American Zoological Association's (AZA) Wildlife Contraception Center (2015), for instance, maintains a large database on the different forms of contraception used in AZA accredited zoos. They make recommendations for which form of contraceptive to use, based on the needs of the facility as well as the species and individuals who will be on the contraceptive. Many of their recommendations are also followed by zoos outside of North America (including members of EAZA).

For primates, the Wildlife Contraception Center recommends the use of MGA (melengestrol acetate) implants and Depo-Provera (medroxyprogesterone acetate) injections, which are used only on females, and GnRH agonists, which can be given to either males or females. For great apes, there is also the option to use human birth control pills. All of these are commonly used for captive primates. MGA implants and human birth control pills are arguably the most common, while Depo-Provera injections are typically reserved for exceptional cases. Although GnRH can be used with males, it is more common to use GnRH to control unwanted behaviors than to limit reproduction. Vasectomy is more common in order to prevent males from producing offspring (Silber et al. 2013). Another option, which is not specifically recommended by the Wildlife Contraception Center but is widely used, are IUDs.

According to the Wildlife Contraception Center, all of these options carry a degree of risk. MGA implants involve a surgical procedure that requires social separation afterward. IUDs require a similar process. MGAs, Depo-Provera, and birth control pills all can cause excessive weight gain. Depo-Provera can also cause mood swings and increased aggression, and long-term use has been linked to reduced bone density. GnRH agonists do not seem to carry significant health risks, but do induce hormonal changes that can affect social behavior. Human birth control pills can also produce various side effects that can be detrimental in certain contexts (like reducing the effectiveness of antibiotics).

So it seems quite clear that contraceptives can have negative welfare effects on captive primates. Numerous other studies have come to similar conclusions (e.g., Bourry et al. 2005; Porton and Dematteo 2005). Welling's (2013) review of the social and psychological effects of hormonal contraceptives (like human birth control pills) argues that these effects can be observed across many primate species, including human beings. There is evidence that hormonal contraceptives have significant effects on female's social preferences, including their choice of mates and their ability to form relationships with others. There is also evidence that hormonal contraceptives increase agitation and aggression while reducing affiliative behaviors and, in some cases, reducing interest in sex.

There are numerous other consequences of contraception that are harder to measure but are nonetheless important to consider. One that is important for several species, but particularly for chimpanzees and bonobos, is the significant reduction in the size of females' anogenital swellings. These can sometimes be quite large (e.g. the size of a basketball) but when on contraception are significantly diminished (Bourry et al. 2005; Porton and Dematteo 2005). This presumably has extensive effects on group behavior. These swellings are, by all accounts, enormously influential on not just sexual behavior but social behavior as a whole. In bonobos, for example, anogenital swellings play a large role in non-reproductive sexual behavior, between both males and females, and works to reduce aggression and solidify social bonds (Furuichi 2011; Hohmann and Fruth 2000; Ryu et al. 2015; Savage-Rumbaugh and Wilkerson 1978). When swellings are severely reduced in size they cannot have the same impact, thus depriving bonobos (and other primate species) one of the most important aspects of their social lives.

Another consideration is how frequently interventions are required in order to properly control reproduction. Human birth control pills, for instance, typically require daily doses. Other methods require routine monitoring. MGA implants, for example, must be replaced every 2 years and require a surgical procedure. Though this might seem relatively minimal, additional interventions are often required due to errors, inefficacy, and changes in breeding programs. Chimpanzees, for instance, are known to remove IUDs (Bourry et al. 2005; Gould and Johnson-Ward 2000). Porton and Dematteo (2005, 123) note that MGA implants have efficacy problems, with 9–54 % of implants falling out or being removed, depending on the species. Properly addressing these issues requires repeated human interventions that presumably have negative welfare impacts but are hard to measure.

One final consideration is the consequences of spending an entire life on contraception. This too is hard to measure, but some have argued that we can see significant welfare consequences from forcing captive animals to spend all of their adult lives in a non-reproductive state (Glatston 1998; Penfold et al. 2014). Some of these are external to the individual; for instance, life might be very boring in groups where no reproduction is permitted, and thus there are no young offspring and enormous gender imbalances. Other welfare consequences, like those mentioned above, pertain to individuals and presumably aggregate over a lifetime. Moreover, Penfold et al. (2014) suggest that negative welfare effects are exacerbated in individuals who have never produced any offspring.

So what conclusions can we draw from the welfare effects of contraception? Though it is evident that there are numerous drawbacks to using contraception, most experts seem to agree that captive primates can be placed on contraception for significant periods of time with relative safety. Bolton et al. (2012), for instance, argue that there are no serious health risks from common forms of contraception, based on data collected from gorillas, bonobos, and chimpanzees. There are significant risks, as I have outlined, but some of the risks are manageable (e.g. weight gain), and overall it's not clear if the risks outweigh the benefits of being able to precisely control population growth. There are also negative welfare effects of free breeding that must be considered (e.g., those caused by pregnancy and giving birth). The pervasive use of contraception suggests that institutions with captive primates have concluded that the costs of unrestricted reproduction are much greater than any welfare costs caused by contraception.

This creates a somewhat complicated situation. If we take reproductive rights seriously, the welfare impacts of contraception must be taken into consideration before restricting reproduction. Routine contraception is not unproblematic. Its widespread use might just reflect reluctance on the part of captive facilities to bear any costs from growing populations, in which case primates' reproductive rights are being restricted unjustifiably. In the next section, I will outline some final challenges with attributing reproductive rights in the context of routine contraception.

6 Meaningful Reproductive Rights?

A skeptical take on my reasoning thus far is that there is no meaningful sense in which captive primates have reproductive rights. The same might be said for domesticated animals. Granting some control over their reproductive behavior, one might think, does not make up for the severe limitations placed on the rest of their lives. And as I have argued, there are in fact good reasons to place significant constraints on reproduction. For many captive primates then, reproductive rights might be granted only in principle, but not in practice. So what is gained from saying that captive primates have reproductive rights?

Reproductive rights are useful in reminding us why we cannot restrict reproductive behavior for *any* reason at all. The basic line of reasoning stemming from Donaldson and Kymlicka's account above seems sensible: We can restrict reproduction when it is in the interests of the individuals being restricted, and there are in fact occasions in captivity where that is the case. The danger comes in taking this line of reasoning to justify any form of restriction on reproduction. For example, it would be illegitimate to conclude from this that it always permissible to restrict reproductive behavior if doing so improves the overall sustainability of the group or the species. This is incompatible with reproductive rights. There are other conditions that must be met first, including benefiting the individual under contraception, preventing significant harm to others, preventing the violation of other rights, or preventing the group from becoming *un*sustainable.

A significant challenge, of course, is identifying when exactly these conditions are met. Zoos, for instance, likely lack the capacity to precisely weigh the reproductive rights of individuals against all the potential factors that would dictate limiting reproduction. Contraception, for instance, is typically adopted in accordance with a breeding plan or in order to prevent overwhelming financial and other burdens from multiple new offspring. Making the decision to use contraception or other forms of birth control may thus not ever really consider the rights of individuals.

The hope is that getting clear on primate reproductive rights can provide a corrective to this. There should be a presumption in favor of allowing primates to reproduce unhindered. As argued above, their autonomy and interest in sex seem sufficiently important to grant them control over their reproductive lives. This is bolstered by the fact that we are responsible for their captivity. Because we have placed primates in captivity, and made them dependent on us, we owe it to them to provide autonomy where we can. So although restrictions are permissible, and can be defended on ethical grounds, captive facilities must actually consider whether the relevant conditions have been met. It may turn out that most cases of routine contraception do in fact meet some of the conditions identified above, but this must be demonstrated rather than assumed. Reproductive rights should be part of the equation, even if the above conditions cannot be determined with perfect precision.

One may further object, however, that reproductive rights are still not really doing any work here. If the outcome is the same—routine contraception can continue—then why insist that reproductive rights should be taken into consideration? In response, I can envision certain concrete changes in the ways captive facilities use contraceptives. For instance, zoos could perhaps be more selective in which individuals to place on contraception. They could allow all individuals to procreate once, for instance, assuming that doing so does not cause significant harm to other group members or put the group's sustainability at risk. Great apes could perhaps be treated differently from other primates in this regard. Ape offspring of course entail huge financial investments for captive facilities, but it may be worth finding a way to accept these costs for apes, even if not for other primate species.

Captive facilities may also need to reconsider how they invest their resources in controlling reproduction. For instance, presumably some groups are tightly controlled merely because the institutions that manage them refuse to increase enclosure sizes or invest the resources required for new offspring. There is no easy fix for this; captive institutions face difficult resource tradeoffs when choosing to accommodate offspring. But reproductive rights seem sufficiently important to be prioritized in at least some cases. Limiting reproduction cannot be justified solely on the grounds that it is the cheapest option.

There are also difficult questions that captive institutions must face regarding how to deal with surplus animals. As mentioned above, granting reproductive rights might necessitate the use of culling. Primates present a particularly challenging case here: there are good reasons both to allow them to reproduce freely and to allow them to live long lives, but these will often come into conflict. Those who think it is wrong to kill primates may view birth control as the lesser of two evils. But the birth control methods required to adequately avoid surplus would place severe limits on reproductive behavior.

Some might argue that these difficulties indicate that the practice of keeping primates in captivity should be discontinued entirely. Why keep primates in captivity if they must have such severe restrictions placed on their reproductive lives? If we cannot properly care for them in captivity, which includes their reproductive behavior, then perhaps we should begin phasing out primate breeding programs.

I think there could be grounds for such a proposal. But there are many difficulties that must be sorted through as well. As mentioned above, Hockings et al. (2015) argue that primate conservation in the Anthropocene requires that we consider the lives of primates in a range of human-built environments. Captivity may be the only remaining sanctuary for certain primate species. If reintroduction is unlikely, reproductive rights could be particularly instrumental in providing a good life in captivity far into the future. Future conservation projects will also likely involve zoos. In such cases, reproductive rights will be important as zoos consider how to responsibly go about preserving the genetic health of primate populations. Spelling out the reasons to grant reproductive rights will have hopefully sharpened the importance for zoos and other captive institutions to address these sorts of issues.

Acknowledgments Many thanks to Sabrina Brando, Peter Sandøe, Clare Palmer, and David Wright for their comments on this chapter.

References

Arruda, C.T., and Povinelli, D.J. 2015. 'Chimps as secret agents,' *Synthese*. Advance online publication. doi:10.1007/s11229-015-0835-9.

Asa, C.S., and I.J. Porton. 2010. Contraception as a management tool for controlling surplus animals. In *Wild mammals in captivity*, 2nd ed, ed. D.G. Kleiman, et al., 469–482. Chicago: The University of Chicago Press.

AZA Wildlife Contraception Center 2015. Contraception methods. http://www.stlzoo.org/animals/ scienceresearch/contraceptioncenter/contraceptionrecommendatio/contraceptionmethods.
Ballou, J.D., C. Lees, L.J. Faust, S. Long, C. Lynch, et al. 2010. Demographic and genetic management of captive populations. In *Wild mammals in captivity*, 2nd ed, ed. D.G. Kleiman, et al., 219–252. Chicago: The University of Chicago Press.
Beauchamp, T.L., and V. Webber. 2014. Autonomy in chimpanzees. *Theoretical Medicine and Bioethics* 35(2): 117–132.
Benetar, D. 2010. The limits of reproductive freedom. In *Procreation and parenthood*, ed. D. Archard, and D. Benatar, 78–102. Oxford: Clarendon Press.
Bolton, R.L., N.J. Masters, P. Milham, and R.G. Lea. 2012. Environment and reproductive dysfunction in captive female great apes (Hominidae). *Veterinary Record* 170(26): 676–678.
Bourry, O., P. Peignot, and P. Rouquet. 2005. Contraception in the chimpanzee: 12-year experience at the CIRMF Primate Centre, Gabon. *Journal of Medical Primatology* 34(1): 25–34.
Carlsen, F., and De Jongh, T. 2015. *Strategy for management of chimpanzees (Pan troglodytes) in EAZA*.
Carter, S., and R. Kagan. 2010. Management of "surplus" animals. In *Wild mammals in captivity*, 2nd ed, ed. D.G. Kleiman, et al., 263–267. Chicago: The University of Chicago Press.
Cochrane, A. 2012. *Animal rights without liberation*. New York: Columbia University Press.
Conly, S. 2005. The right to procreation: Merits and limits. *American Philosophical Quarterly* 42 (2): 105–115.
Donaldson, S., and W. Kymlicka. 2011. *Zoopolis*. Oxford: Oxford University Press.
Dworkin, R. 1993. *Life's dominion*. New York: Knopf.
Eijkholt, M. 2009. Conceiving a new right to procreate. In *Reprogen-ethics and the Future of Gender*, ed. F. Simonstein, 37–52. Dordrecht: Springer.
Fouts, R.S., and S.T. Mills. 1997. *Next of kin*. New York: William Morrow.
Furuichi, T. 2011. Female contributions to the peaceful nature of bonobo society. *Evolutionary Anthropology* 20(4): 131–142.
Glatston, A.R. 1998. The control of zoo populations with special reference to primates. *Animal Welfare* 7(3): 269–281.
Gould, K.G., and J. Johnson-Ward. 2000. Use of intrauterine devices (IUDs) for the contraception in the common chimpanzee (Pan troglodytes). *Journal of Medical Primatology* 29(2): 63–69.
Hockings, K.J., M.R. McLennan, S. Carvalho, M. Ancrenaz, R. Bobe, R.W. Byrne, et al. 2015. Apes in the Anthropocene: Flexibility and survival. *Trends in Ecology & Evolution* 30(4): 215–222.
Hohmann, G., and B. Fruth. 2000. Use and function of genital contacts among female bonobos. *Animal Behaviour* 60(1): 107–120.
Hvilsom, C., P. Frandsen, C. Børstin, F. Carlsen, B. Sallé, B.T. Simonsen, and H.R. Siegismund. 2013. Understanding geographic origins and history of admixture among chimpanzees in European zoos, with implications for future breeding programmes. *Heredity* 110(6): 586–593.
Kaufman, L. 2012. When babies don't fit plan, question for zoos is, now what? *New York Times*, 2 August 2012.
Ladwig, B. 2015. Against wild sovereignty: An interest-based critique of Zoopolis. *The Journal of Political Philosophy* 23(3): 282–301.
O'Neill, O. 2002. *Autonomy and trust in bioethics*. Cambridge: Cambridge University Press.
Palmer, C., Corr, S., and P. Sandøe. 2012. 'Inconvenient desires: Should we routinely neuter companion animals?' *Anthrozoös* 25 (Suppl.), 153–172.
Patterson, P. 1987. *Koko's kitten*. New York: Scholastic Press.
Penfold, L.M., D. Powell, K. Traylor-Holzer, and C.S. Asa. 2014. "Use it or lose it": Characterization, implications, and mitigation of female infertility in captive wildlife. *Zoo Biology* 33(1): 20–28.
Plowman, A.B., N.R. Jordan, N. Anderson, E. Condon, and O. Fraser. 2005. Welfare implications of captive primate population management: Behavioural and psycho-social effects of

female-based contraception, oestrus and male removal in hamadryas baboons (Papio hamadryas). *Applied Animal Behaviour Science* 90(2): 155–165.

Porton, I.J., and K.E. Dematteo. 2005. Contraception in nonhuman primates. In *Wildlife Contraception: Issues, Methods and Application*, ed. C. Asa, and I. Porton, 119–148. Baltimore, MD: Johns Hopkins University Press.

Quickley, M. 2010. A right to reproduce? *Bioethics* 24(8): 403–411.

Robertson, J. 1994. *Children of choice*. Princeton: Princeton University Press.

Ryu, H., D.A. Hill, and T. Furuichi. 2015. Prolonged maximal sexual swelling in wild bonobos facilitates affiliative interactions between females. *Behaviour* 152(3–4): 285–311.

Savage-Rumbaugh, E.S., and B.J. Wilkerson. 1978. Socio-sexual behavior in Pan paniscus and Pan troglodytes: A comparative study. *Journal of Human Evolution* 7: 327–344.

Silber, S.J., N. Barbey, K. Lenahan, and D.Z. Silber. 2013. Applying clinically proven human techniques for contraception and fertility to endangered species and zoo animals. A review. *Journal of Zoo and Wildlife Medicine* 44(4 Suppl.): 111–122.

Velleman, D. 2006. *Self to self*. Cambridge: Cambridge University Press.

Welling, L.L.M. 2013. Psychobehavioral effects of hormonal contraceptive use. *Evolutionary Psychology* 11(3): 718–742.

Williams, S.E., and E.A. Hoffman. 2009. Minimizing genetic adaptation in captive breeding programs: A review. *Biological Conservation* 142: 2388–2400.

Wild Animals in Entertainment

Sabrina Brando

Abstract Wild animals in entertainment have long been popular. A wide variety of wild animals are held in captivity worldwide. Some are housed in modern zoos, sanctuaries, research facilities and wildlife centers, while others live their lives as actors in (traveling) circuses, in entertainment parks or on movie sets. Good animal welfare and quality of life matters first and foremost to the individual animal, but is also fundamental to meaningful and successful conservation, research and education programs. It is important to consider the animal's perspective and the species-specific requirements that are not always compatible with our entertainment goals. This chapter will focus specifically on the topic of entertainment and performing captive wild animals in zoos, marine parks and circuses worldwide and which conditions need to be fulfilled to warrant good animal welfare, i.e. thriving captive wild animals.

1 Introduction

This chapter flows from almost 25 years of working in and with animal facilities such as sanctuaries, zoos, marine parks, and wildlife centers worldwide. Many topics and discussions are important when considering and providing for good animal welfare. The mission statements of zoos and sanctuaries (and some circuses) around the world include education, conservation, research, and entertainment as primary goals and their aims of existence. Good animal welfare is and should be fundamental for achieving these mission statements and goals of zoos (Buchanan-Smith et al. 2001; Mellor et al. 2015). Understanding how wild animals are faring is one of the key responsibilities of modern zoo professionals and regulatory agencies today. This chapter will focus specifically on the topic of entertainment and performing captive wild animals in zoos, marine parks and circuses

S. Brando (✉)
AnimalConcepts, Lelystad, The Netherlands
e-mail: sbrando@animalconcepts.eu

B. Bovenkerk and J. Keulartz (eds.), *Animal Ethics in the Age of Humans*,
The International Library of Environmental, Agricultural and Food Ethics 23,
DOI 10.1007/978-3-319-44206-8_18

worldwide and which conditions need to be fulfilled to warrant good animal welfare, i.e. thriving captive wild animals.

My hope is that disciplines concerning animal welfare and ethics will continue to make meaningful changes for animals under human care through research, debate and practical applications.

A wide variety of captive wild animals, like birds of prey, sea lions, dolphins, parrots, but also bears, elephants and hippos are popular in shows and presentations. Some are used for photo opportunities and interaction programs in zoos, marine parks and circuses. Although some of these activities have an education or conservation function, the main focus is human entertainment, sometimes also referred to as 'edutainment'. Should wild animals perform in front of an audience for people to be engaged and educated, or be interacted with just for the sake of human entertainment without ulterior motives linked to conservation? Should they be trained to perform species-typical or circus-like acts? And what types of methods are acceptable when training wild animals? What about their living conditions? Can we ensure good welfare for performing wild animals?

In this chapter, *entertainment* is defined as the amusement or pleasure that people derive from watching performers, in this case watching wild animals. Entertainment can be in the form of animal shows and educational presentations, or outreach programs where animals are taken to meet the public—usually in a type of classroom or amphitheater to get a more up close and personal event. Other activities include photo opportunities, swim-with-dolphins or sea lion programs, tiger tug-of-war, trainer- or keeper-for-a-day.[1] It could also be seen as the engagement of the public to stay longer in front of an animal's exhibit through the use of environmental enrichment, exhibit design, and or animal training. Entertainment is predominantly viewed from the person's perspective, but professional and well-developed animal programs can offer entertaining opportunities for animals that provide for complexity, choice and control. Modern entertainment/edutainment programs should fulfill an essential role in educating and engaging the public. These programs can potentially engage the public to care about animals and nature in ways that they want to protect wildlife and come into action to conserve wild animal habitats. This can be achieved through meaningful programs and presentation in combination with an entertaining component to engage and empower the general public.

Furthermore, *performing* is defined as the engagement of an animal in an activity, and specifically animals carrying out species-appropriate behaviors and repertoires, such as playing, climbing or flying. I hold the opinion that performances should be seen as an education and conservation oriented presentation or activity. Presentations can come in the form of an appropriately themed or staged interpretation, showcasing animal capabilities and capacities. It can also be a display of different physical, behavioral and cognitive opportunities in a

[1]A tiger tug-of-war is a rope pulling interaction between a tiger and the visiting public at Busch Gardens Tampa Bay, Florida, USA.

species-specific and functional exhibit. This can be achieved through effective and dynamic environmental enrichment and behavioral programs that challenge animals physically and psychologically. Sometimes training can be used to ask animals to voluntarily participate in their care as well as in the aforementioned presentations. Modern-day performances, as I will argue, should exclude all activities in which animals are engaged or staged in scenarios that are not respectful to the individual or the species, and could jeopardize welfare as well as education and conservation efforts and messaging.

Wild animal species that can regularly be found in entertainment are too many to be named; those that follow are only some examples. Birds range from parrots and ibis to birds of prey, such as falcons, owls, hawks and large eagles. Primates include species such as macaques but also orangutans and chimpanzees. Small cat species like the serval and large cats such as lions, tigers and cougars. Animals in entertainment may also include bears, many ungulates species, including elephants, zebras and giraffes. Marine mammals like bottlenose dolphins, California sea lion, orcas and seals are frequently used too. Other mammals include coati, binturong, wallaby and llama. Snakes, like pythons, spiders, chameleons, lizards and larger insects such as walking sticks are also popular. Wild animals are also used in outreach programs that go beyond the confines of the zoo, such as TV performances.[2]

In this chapter, I will try to answer the following questions with respect to entertainment and performing captive wild animals: (1) Are performing wild animals morally acceptable? (2) Are performances only morally acceptable when there is a strong educational component or also for entertainment purposes alone? and (3) If morally acceptable, which conditions need to be fulfilled to warrant good animal welfare, i.e. thriving captive wild animals?

I will first discuss the criteria that are necessary to answer these questions. Next, I will successively examine the condition of wild animals in circuses and zoos.[3]

2 Criteria for Assessing the Condition of Performing Animals

I will distinguish three criteria necessary to assess the moral acceptability of animal entertainment practices: animal welfare, animal flourishing, and animal dignity.

---

[2]Domesticated species such as pigs, horses, dogs, cats, rats and cows, although commonly used and housed in zoos and circuses, will not be covered in this paper but require similar discussion and scrutiny.

[3]Marine parks, aquaria and zoological parks housing captive wild animals will all be referred to as zoos in this paper. Circuses will be kept in a separate category although some activities might be overlapping.

2.1 Animal Welfare

The assessment of animal welfare is an important yet often difficult task. The focus is on a holistic approach to animal welfare, considering physical and psychological perspectives. Animal welfare or animal wellbeing are used interchangeably, both referring to this holistic approach. Issues relevant to animals prospering in captive environments include the questions of whether they were taken from the wild (including when and how) or raised in captivity from birth; how they are housed and cared for; how much choice and control they have in and over their lives; how enriching their environment is; what their relationships with human caretakers and visitors are; and whether or not they lead fulfilling lives.

Relevant to animal welfare as well as to education and conservation goals are questions such as: What was the impact of the removal of animals from the wild on the population; what role(s) do the animals serve?; how are show and interactive animals housed in comparison to the same species in another role, like a display animal?; should we use these animals for our entertainment purposes; and how do we present them to the general public? When considering animal welfare many topics are relevant. We have to consider the animals' biology, anatomy, ecology, social life, how they sleep and rest, near species-specific adaptations, sensory systems and cognitive abilities. As animal welfare pertains to the individual, i.e. how the individual perceives, thinks and feels about the environment, social aspects and other events that happen, we have to try to get as close as possible to the individual's needs and preferences (without loosing sight of species requirements). In other words, how do animals cope (Broom 2001) as a minimum requirement (as coping is on the lower end of the scale when considering thriving) and are we considering welfare from a holistic perspective (Fraser 2008)? The practical and academic field of animal welfare is steadily growing. Methods used to gain insight in animal welfare, animal minds and affective states include: theory of demand (Cooper and Mason 2000), preference testing (Dawkins 1983), cognitive bias (Harding et al. 2004), and Qualitative Behavioral Assessment (Rutherford et al. 2012). Holistic approaches considering 'natural living', 'affective states' and 'basic health and functioning' by Fraser (2008) and the Extending the 'Five Domains' model for animal welfare assessment to incorporate positive welfare states by Mellor and Beausoleil (2015).

Another useful approach to understanding animals is one proposed by Gordon Burghardt, which is named Critical Anthropomorphism. In his 1985 paper, Burghardt argues we should transform sin to virtue. Use of Anthropomorphism, attributing human characteristics to nonhuman entities, is considered a grievous sin, while Anthropomorphism by Omission, unwittingly neglecting the animal's perspective and Umwelt is also considered a grievous sin. Burghardt proposes the use of 'Critical Anthropomorphism', using our stance as human beings to propose the occurrence of testable cognitive, emotional, and behavioral processes in other species informed by a rigorous incorporation of empirical knowledge and research findings on physiology, sensory and neural processes, ecology, behavior, sociality, and development. This is considered an 'Essential Virtue'. As an example he states that judging other species using markers such as screams, smiles, or even the speed of responding can be uncritically anthropomorphic and a hallmark of anthropocentrism.

A well-known concept in animal welfare science is "The Five Freedoms" which have long been used to direct animal welfare assessment and research. Since 2009 an expanded framework of animal welfare assessment criteria, established by the European project Welfare Quality® has been developed. Although designed with farm animals in mind, these 12 animal welfare assessment criteria can be used, or at least serve as guidelines to think about and assess captive wild animal welfare, both in zoos and circuses.

Good feeding	Absence of prolonged hunger
	Absence of prolonged thirst
Good housing/environment	Comfort around resting
	Thermal comfort
	Ease of movement
Good health	Absence of injuries
	Absence of disease
	Absence of pain induced by management procedures
Appropriate behaviour	Expression of social behaviours
	Expression of other behaviours
	Good human-animal relationship
	Absence of general fear

This framework has been expanded to 14 criteria to reflect positive emotions and experiences by Brando and Buchanan-Smith in a forth-coming paper. More information can be found on www.247animalwelfare.eu. Blokhuis et al. (2010) proposed that because animal welfare is a multidimensional concept, all criteria are important, so, for example, the ability to exhibit appropriate behavior does not compensate for poor health, and good health does not compensate for behavioral problems. In the case of many exotic species, it is unfortunate that there is little research-based information about the proper environments, the health and nutritional requirements and the behaviors and/or activities that are necessary and desired for good welfare and for a good quality of life. For some species there do not seem to be any husbandry manuals to guide their care.[4]

Some zoos, such as Chester Zoo in the UK, employ a specialized animal nutrition expert and many staff members participate in a wide variety of in situ and ex situ projects. They also have working relationships with universities and conservation programs. Another example is the extensive husbandry manual for elephants highlighting the development of care programs through research, monitoring programs and practical improvements and applications. A model for exemplary care

[4]However, many modern zoos are conducting research projects and continue to develop (extensive) husbandry manuals for species under their care, like the extensive Callitrichids husbandry manual counting close to 200 pages.

is the Dublin Zoo where elephants are housed in a family group, on deep sand for resting and sleeping in a social setting, space to walk and a pool with water canons to play, bath and swim. Extensive enrichment programs offer a wide variety of activities including foraging, play and problem solving. Positive reinforcement training is used to work with the elephants on a voluntary basis, providing preventive and proactive healthcare, and husbandry is managed, such as foot care and gating, through cooperative behaviors.

2.2 Animal Flourishing

According to Martha Nussbaum's capabilities approach, the welfare of animals must be measured against the possibilities an environment offers animals to actually display their basic natural capabilities. It is based on a species-specific norm of flourishing that tells us what the appropriate benchmark is for judging whether a member of a species has decent opportunities for flourishing, and that commits us to bring members of that species up to that norm. Nussbaum's capabilities approach does attach moral significance to the species to which a creature belongs. Her species-specific account of flourishing commits caretakers to support the capabilities of captive animals, and enables us to judge whether they have ample and appropriate options and opportunities to perform species-specific activities. If appropriate care is given, animals can flourish in less natural and more human environments, but capabilities will have to be species-specific (Keulartz in this Volume).

2.3 Animal Dignity

In *Ethics and Animals*, Lori Gruen worries that animals who are kept in captivity are denied '*wild dignity*'. Gruen bases her concept of wild dignity on something other than the Kantian notion of dignity, which is something all humans possess due to their rationality—which Kant describes as a second order capacity to reflect on our own desires and to judge the universality of our reasons. While (most) animals do not possess such second order thoughts, Gruen still thinks they can have dignity and that our making them look ridiculous or our portraying them as something different than what they are violates this dignity. Even though the animals in question themselves do not experience it as such, we do experience a bear dressed up in a tutu riding a bicycle as undignified. What could be the basis of this wild dignity?

One view seems to be that when animals are forced to behave as 'something other than they are' they are 'denied their animal dignity' (Gruen 2011, p. 153). In such an instance they are made to behave in a way that reduces our respect for them. The idea behind this seems to be that animals derive their dignity from their species-specific characteristics and behaviour. Although she agrees with the conclusion of this view, Gruen finds this basis for animal dignity problematic, because it assumes a static, essentialistic or naturalistic view of species-specific characteristics. After all, just because something is natural, this does not automatically make it good. In Gruen's view also learned capacities or behaviour could form the basis of dignity.

Instead of focusing solely on what is 'natural (functioning)' or 'innate', Gruen argues that animal dignity needs to be dynamic, i.e. adapted to the individual animal, and is only valuable when it is expressed and recognized as contributing to the well-being or flourishing of that individual. For example, animal training and artificial environmental enrichment might not be natural to a wild animal, but it might serve a meaningful function in restoring some of the individual's 'wild dignity' through e.g. giving more choice and control.

In the case of humans, dignity is something that is evoked in others and recognized by others and therefore dignity is a relational concept; it is 'a property of relations between human beings—between, so to speak, the dignifier and the dignified' (idem, 154). In the case of animals, the concept of 'wild dignity' is seen "within their own social networks as evaluated through ours". Wild animal dignity only comes into play when animals are taken out of their wild context and put into the human context. It appears, then, that for Gruen the basis of the attribution of wild animal dignity rests in our own attitude. When we try to change animals' species-specific behaviours and take away the control of their own lives we dominate them and thereby we violate their dignity, regardless of the question whether this is experienced as such by the animals in question. Therefore, when housing captive wild animals we need to consider and respect their needs and preferences, and what contributes to and interferes the least with their 'wild dignity'. This can concern behaviors that we might find off-putting and consider indecent from a human perspective, for example, behaviors such as masturbation or aggression that animals enjoy engaging in, or are important to maintaining social relationships. This concept also has relevance for how wild animals are portrayed to the public, housed, or used in the various activities offered by zoos and aquaria today as discussed earlier in this paper.

Gruen writes, "In contrast we dignify the wildness of other animals when we respect their behaviors as meaningful to them and recognize that their lives are theirs to live" (idem, 155). Although there will be restrictions to life in captivity, the concept of 'wild dignity' can serve to identify conflicts between human expectations versus animal needs and preferences, and propose ways to mitigate them.

3 Performing Wild Animals in Circuses

The provenance of animals housed in circuses is difficult to determine, and information is difficult to obtain. In her 1990 book Kiley-Worthington reported that 40 % of carnivores and 14 % of ungulates in 15 British circuses were unwanted zoo animals, whereas approximately 94 % of elephants were wild-caught, and animals can be procured through private breeders and suppliers (Kiley-Worthington 1989).

Lord Soulsby of Swaffham Prior FRCVS said on May 23rd 2006 in a debate on animal welfare, "We know so little about the physiology and the sentience of the exotic species that often used to be seen—and still are—in some traveling circuses" (in Radford 2007). The scientific related literature on circus animals is indeed scarce, and a modern comprehensive overview of all aspects of animal welfare in circuses is absent (Radford 2007; Brando 2015).

3.1 Animal Training

The Code of Conduct for Animals by the European Circus Association states, "All animal training must be based on operant conditioning and the use of positive reinforcement and repetition of desired behaviors. Training should showcase individual animals' natural behaviors and athletics."[5] Animal training can be an important pillar in the care of wild animals. Getting animals to participate voluntarily in their care and management can reduce negative stressors associated with these procedures. Animal training can also provide mental stimulation, and, if done with appropriate methods such as positive reinforcement and the building of good animal human-animal relationships, can function as environmental enrichment. Although animal training can be a positive addition to an animal care program, it is also time-consuming and requires a high level of skill on the part of the trainer. Animal caretakers and trainers need to regularly refresh and update theoretical and practical knowledge to stay up to date with the latest science and best practices in the animal training field.

Also, animals who are trained to participate in presentations and any other activities should have the option to opt out if they so desire, without negative consequence. Especially in small social groups it is important that animals who are involved in performances have the option not to participate. It often happens that because there are so few animals, all animals have to do all presentations, potentially all year round. This should be avoided by providing choices and other activities as well as active resting days. Professionally designed animal training programs can provide enriching experiences for animals but at the same time we have to be cognizant of the fact that these activities can also become boring (Wemselfelder 2005), when they become repetitive, and new learning is not taking place (Melfi 2013). There are still many aspects of animal training that need research and discussion, such as the different methods used, the intrinsic value of training and consequences, dependence on caretakers, anticipatory behaviors, human-animal interaction, and perceived choice and control.

However, animal training is only a small part in a complete and professional animal care program and is not a substitute for good environments, environmental enrichment and other important animal care components. The quality and quantity

[5]ESA Code of Conduct http://www.europeancircus.eu/?page_id=361.

of care of circus animals is often different from what similar species housed in a modern zoological park receive.

3.2 *Housing and Transport*

In circuses, due to transport requirements and small and substandard housing conditions, the opportunities to express species-natural behaviors are undoubtedly very challenging and thwarted. That tigers favor more room was demonstrated in a pilot study on exercise pen use in tigers by Nevill et al., who write, "However, several tigers were active up until the end of the activity period, indicating that exercise pens are an important form of enrichment and that access for periods longer than 40 min should be considered" (2006, p. 355).

The need for and benefits of species-specific social groups are difficult to achieve in an environment where social or small group housing is likely due to space restrictions. This can also have serious negative consequences for behavior, welfare and reproduction (Price and Stoinski 2007). Vidya et al. write, "The family group in both Asian and African elephants is one that is stable and a rich source of social interactions" (2005, p. 1200). With the way elephants are moved around, exchanged and housed in circuses, it is highly unlikely that facilitating qualitative social interactions is possible.

Life in a circus is a life often on the road, with little opportunity to express desired behaviors. Because they live on the road, animals are also deprived of opportunities that we do not consider very often, such as the desire for a view (to see what is happening), or control over the duration of an activity; we can imagine that these are important aspects for circus animals. Very few studies have looked at the effect of transport on the welfare of circus animals. There are a few studies, which only provide meager information and conclusions. For example, Dembiec et al. (2004) point out that if circus animals are going to be moved around a lot it is important to habituate the animals to these procedures in a positive manner. It should also be considered that animals would perhaps benefit from exercise before transport to reduce pacing, and have space to move around between performances and transport (Nevill and Friend 2003).

Animals are not moved just once, their lives revolve around being on the road, and the impact of continued exposure is poorly understood. As Teixeira et al. have observed, "most do not understand the importance of subclinical stress or the fact that the effect of successive stressors can be additive or accumulative" (2007, p. 1). Long-term effects of transport on circus animals are unknown, but they should be urgently investigated if wild animals in circuses are not banned. The suggestion by Clubb and Mason (2003) to not house wide-ranging species should also apply to circuses to minimize the adverse effects of captivity. Although animals get out of the wagons and can often stay in exercise pens (which are larger cages, connected to the wagon) once they have arrived on site, the majority of their time they are confined to small living quarters, considerably smaller than those of their zoo counterparts (Iossa et al. 2009), and larger animals like elephants are often chained. The complexity of the transport and holding areas for animals are severely impoverished, missing vegetation, environmental enrichment, hides and other choices in the environments. Space is very often limited and species like elephants can spend many hours shackled resulting in high levels of stereotypic behavior (Schmid 1995).

3.3 *Restraints to the Freedom to Express Natural Behavior*

Even if circus animals are exercised regularly and performing for the public, this is still a short time compared to waiting and transport. Being in sterile environments causes boredom that, in turn, can lead to negative welfare states (Wemelsfelder 2005). Furthermore, the aim is not solely to reduce negative welfare states such as pain, anxiety, fear or boredom, but to ensure positive welfare states such as being curious, exploring, relaxing and socially engaging. Animals should have a life worth living, a life in which they can thrive. Animals should be able to engage in species-specific behavior as outlined in the animal welfare assessment criteria.

The Animal Welfare Act of 2006 underlines the need to be able to exhibit normal behavior patterns. Apart from transport there are many circumstances and procedures strains in which animals cannot express natural behaviors. Picketing or single chaining is one such procedure strain that is often used for elephants. Picketing is the traditional way of restraining elephants by chaining one front and the diagonal rear leg to parallel cables or chains. Not only does this have an effect on stereotypic behavior (Friend and Parker 1999), it also hinders the elephant's ability to freely move around, to lay down comfortably, or engage in social contact beyond just being able to touch body parts in reach because these chains restrict the range of motion.

The life of animals in a circus is without a doubt distinctly different from their natural environment, not only with regards to ways of living but also to the behaviors animals might engage in. Their lives revolve mainly around the need for performances, for which a lot of training is necessary. This training is not always done with positive methods, like the use of positive reinforcement training with some control and choice over their environment. Kiley-Worthington (1989, 1990), did find that staff worked with the animals in a collaborative and positive manner, but also witnessed frustration and avoidance behaviors in animals, and refusal to perform certain acts. She also found that not all staff members were skilled animal trainers. Undercover videos and statements (many already older) of former circus staff show and describe negative interactions between handlers and animals, with beating, slapping, pushing and tying animals, but this is not to say that there are not circuses that have positive methods to achieve their goals.[6]

[6]https://vimeo.com/122718299.

Many species housed in circuses have rich mental lives and high cognitive abilities. For example, chimpanzees "know what conspecifics have and have not seen" (Hare et al. 2001, p. 139), elephants know when they need the help of another elephant (Plotnik et al. 2011), and sea lions can remember activities that they have not been engaged in for 10 years (Reichmuth Kastak and Schusterman 2002). The little time dedicated to training and the few enrichments the animals are receiving will not be enough to fulfill their cognitive and sentient needs, or their preferences and opportunities to engage with their environment and social companions.

Performances and routines can also result in negative behavioral repertoires as demonstrated in a study by Friend and Parker (1999), who showed that stereotyping in circus elephants increases prior to performance. Similarly, Krawczel et al. (2005) found a peak of stereotypic behavior in tigers in the hour before the performance. Being exposed to loud noises from transport or human crowds can pose problems and result in negative welfare states such as escape and huddle behavior, high vigilance or aggression, as has been found in primates and ungulates. One could mistakenly think that animals performing in circuses are habituated to all this commotion; we can refer back to the study mentioned earlier on accumulative stressors and the as yet unknown impact. The evidence for zoos, where animals have a generally higher quality environment, is that large crowds, particularly active ones, are often stressful (Hosey 2000), so in circuses, where large excited crowds are prevalent, we could expect the stressful effect to be worse if animals are not well habituated. One of the modifications to existing circuses could be the choice for seeking shelter, as proposed by Hosey: "The key to reconciling these apparently conflicting considerations is perhaps related to the amount of control that the animals have over their exposure to humans" (2000, p. 355).

We have many other species to consider, for which information in circuses is virtually non-existent, such as birds, other hoofstock, bears, hippos, small carnivores, rodents, and primates. Although the data are scarce, the papers and reports available as well as common sense would lead us to conclude that the welfare of circus animals is far from optimal. Many of their basic needs are not met, space is limited, social opportunities strained, natural behavioral opportunities few and boredom likely high. Iossa et al. write,

> Circuses have a limited ability to make improvements, such as increased space, environmental enrichment and appropriate social housing. Consequently, we argue that non-domesticated animals, suitable for circus life, should exhibit low space requirements, simple social structures, low cognitive function, non-specialist ecological requirements and an ability to be transported without adverse welfare effects. None of the most common species exhibited by circuses, such as elephants and large felids, currently meet these criteria. We conclude that the species of non-domesticated animals commonly kept in circuses appear the least suited to a circus life (2009, p. 129).

4 Wild Zoo Animals: The Case of Marine Mammals

Both the European Association of Zoos and Aquaria (EAZA) and the World Association of Zoos and Aquariums (WAZA) have clear guidelines and aims in their strategies to both ensure welfare, the ethical treatment and representation of species housed under the care of accredited facilities. I will focus on the Bottlenose dolphin (*Tursiops truncatus*) and the California sea lion (*Zalophus californianus*) as example species to consider the relevant guidelines aforementioned, as well as mention other marine mammal species.

How the various species live in the wild, including for example cognitive abilities, social life and feeding habits does not say anything per se on how life in captivity is perceived by the individual animal. One cannot conclude that because life in the wild is different from life in captivity that therefore welfare is negatively impaired, and that animals could not be experiencing positive welfare and thrive. At the same time it needs to be evident that our knowledge on social life, cognitive abilities, complexity of environment, and any other relevant information and research through observations in the wild as well as captive conditions guides the development of expansive and dynamic animal care programs relevant to the species. Knowledge on species-specific needs and preferences should also guide questions arising in animal ethics and the practical treatment of animals. Considering all the knowledge we might conclude that certain species can only be kept under certain conditions, or cannot be kept appropriately under human care.

4.1 Social Life

California sea lions are a highly gregarious and social species. They live in dynamic groups with certain males maintaining a harem part of the year. Unsuccessful males retreat during the mating season to bachelor beaches. Males congregate in all male groups outside the breeding season and tend to travel north. Females and juveniles disperse but stay close to the breeding grounds. Animals often rest close to each other, on the beach, rocks, or floating objects like docks and buoys. Sea lions in zoos and circuses are usually not kept in large groups and the social dynamics is fairly stable year round.

Bottlenose dolphins appear to form relatively permanent social groups and mother-calf bonds are long lasting, and although individuals have preferences who to associate with, bottlenose dolphin behavior is flexible and their society has a dynamic fission-fusion nature (Shane et al. 1986; Mann et al. 2000).

4.2 Environment

Marine mammals live in dynamic environments, with changing weather, currents, water depths and temperatures, a variety of substrates and vegetation. Limited information and guidance on marine mammal habitats can be found in official documents. The European Association for Aquatic Mammals (EAAM) provides guidelines for housing requirements for Bottlenose dolphins. It is stated that a pool area of at least 275 m^2 should have a minimum depth of 3.5 m (EAAM 2009, p. 13). However, considering that male and female dolphins can reach lengths of 3.8 and 3.7 m this seems largely inadequate.

Unfortunately pinniped or sea otter guidelines are not (publicly) available at the time of this writing. Marine mammal environments under human care are far behind with regards to other animals of similar body size, behavioral complexity and cognitive abilities. Great apes and elephants, as well as some polar bear exhibits have seen many changes in the last decade. Environmental complexities, through a variety of levels, substrates, and interactive features have been readily incorporated in these terrestrial mammals. Unfortunately many marine mammals are still housed in pools that are void of aforementioned complexity, choices and control. Is an important and interesting topic that needs further development, as the housing for example of polar bears in empty bear pits is not considered acceptable today. Similar housing for marine mammals like dolphins, sea lions, seals and otters seems to be an acceptable status quo within the zoo industry today.

There is a tension between explaining and showcasing how animals live in the wild and showing them in captivity as life in captivity will never be as life in the wild, nor are animals free. This tension can be lessened with efforts to come as close as possible to the functional complexity of the environment combined with choices and control for the animals.

4.3 Feeding

Sea lions spend much time foraging, in summer and early fall, at least, the majority of animals may leave hauling grounds early in the morning and return in late afternoon. Large groups feed within 15 miles of land. Farther offshore, up to 39 miles, sea lions usually occur singly or in small groups of 2–12 animals (Fiscus and Baines 1966). Sea lions can be seen chasing and hunting for prey, from small schools to individual fishes, between kelp beds, rocks and open seas. Polar bears can remain motionless for extended periods at a breathing hole awaiting seals to come for air or stalks prey during hunting (Stirling 1974).

Feeding in zoos and circuses revolve predominately round fish (or other species-specific food) being delivered from a caretakers hand, and most often contingent upon correct behavior. Foraging opportunities for sea lions and dolphins are very limited as most of the food is used in animal training sessions, shows and interactive programs.

4.4 Cognition and Communication

The complex cognitive abilities of marine mammals become visible in various individual as well as cooperative hunting strategies these animals engage in their flexible and dynamic habitats. Such hunting strategies are specific to certain groups and specific to certain locations ([7]Bottlenose dolphins: Gazda et al. 2005; Pinnipeds: Hanke et al. 2010). Culture has been studied in Bottlenose dolphins as they show tool-use and socially pass these traditions on to their offspring (Krützen et al. 2005). They form relationships and alliances that persist over decades (Connor et al. 2000) and they use signature whistles to identify and recognize each other (Janik and Sayigh 2013). They do not only recognize individual conspecifics but also themselves in mirrors (Reiss and Marino 2001). Language training experiments showed that they process semantic relations and understand syntactic information (Herman et al. 1993). Dolphins and other marine mammals use many different sensory modalities for communication and for perceiving the world, and have extensive cognitive abilities to process, analyze, file and retrieve information from memory (Reichmuth Kastak and Schusterman 2002). Understanding these sensory modalities and taking the animals' memory capabilities and other cognitive abilities into account will play an important role in the future treatment of marine mammals under human care.

In contrast to the many cognitive, social, physiological and psychological challenges the animals face in the wild; life in captivity can cause monotony and boredom (Wemelsfelder 2005). Many of the activities and choices over the environment have been removed in captivity, like where to go, whom to be with, when to eat, what to do, or sleep. The effects of captivity on marine mammals are still poorly understood. Enrichment and training can be mitigation options as they can hand back some amount of choice and control over the environment to the animals.

Up to now, most enrichment items in marine mammals are floating hard plastic objects that animals can push around, given during daytime and often under supervision. Delfour and Beyer (2012) researched the use of objects by bottlenose dolphins and found that only 50 % of the object elicited manipulative behaviors. Not every toy seems to be a successful enrichment device and not each behavioral change subsequent to the introduction of objects necessarily indicates an enrichment effect. Cognitive environmental enrichment has been proposed as one of the areas of development and improvement in marine mammal care programs (Clark 2013).

[7]The common bottlenose dolphin (*T. truncatus*) are the most commonly kept dolphin species in captivity; Pinnipeds include sea lions, seals, fur seals, and the walrus.

Advances in marine mammal captivity have been made over the past years (Brando 2010), but existing management and husbandry guidelines should be critically reviewed and analyzed and the lack of scientific literature on the topic should be tackled. New ideas should be developed. Some zoos find creative ways to increase positive welfare. The Kolmarden Zoo in Sweden for example provided sonar-activated water enrichment for dolphins. Animals in captivity need challenges, e.g. by working to get food, by solving problems, by using the limited space they have for swimming, exploring or playing. Challenges can be created through creating a more interactive environment, animals being able to control access to certain pools and control over indoor and outdoor areas to choose what species of fish to eat, which objects and toys to use or who to be with. The use of substrates in a creative way, like sponges or artificial vegetation, could encourage natural behavior without completely eliminating visibility.

4.5 Interactive Programs

Independent from questions of good husbandry the captivity of marine mammals is being criticized per se. As in the case of the great apes, some experts draw attention to animal rights arguments, asking for human-like rights for dolphins and whales because of their complex cognition, emotion and needs (White 2007; Declaration of Rights for Cetaceans 2010). A right to freedom would clearly ban captivity in zoos and aquaria. Humans could then exclusively encounter these fascinating animals in their natural habitat. But swimming-with-dolphin programs in the wild are also increasingly criticized. The presence of human swimmers and tourist boats change the animal's natural behavior. They spend much more time moving and much less time resting; they reduce social interaction with conspecifics and foraging. Serious health issues arise if swimmers feed the animals. Long-term effects for the animals' health and well-being, for reproduction and for the way they use their habitat are being expected (Christiansen et al. 2010; Samuels and Bejder 2004).

Problems associated with swimming-with-dolphins are seriously underestimated in captivity, too. They encompass risks of injuries and cross-infection (both for humans and animals). The lack of space leaves the animals no possibility to retreat, which can result in severe stress (Hunt et al. 2008; Brensing et al. 2005). Dolphins and other marine mammals are wild animals. Visitors, swimmers, patients and trainers need to be fully aware of this. The effects of swim-with programs are poorly understood as research is lacking. However, a study by Trone et al. (2005) found no detrimental effects on the behaviour of the dolphins they studied. More research is needed to understand the effects of swim programs and necessary parameters to ensure welfare. Many interactive programs with marine mammals consist of partly open educational presentation that features some biology and anatomy, and what to expect once entering the water and wading with the animals. The pre-or post activities might also include some information about threats that animals are facing in the wild and what people can do to help. Most of the interactive session is focused on people either in the water swimming, or wading, and animals can be requested to perform a variety of behaviors such as shaking pectoral fin, doing a bow on cue, doing a dorsal tow, or posed kissing or hugging a dolphin for a photo opportunity.

Many other animals who live their lives in sanctuaries due to illegal trade, poaching, or because they retire from a life working for humans, like former working elephants. A sanctuary is a safe haven where animals should be able to live out their lives. Many of these sanctuaries depend on outside funding to provide care to the animals and many rely on a variety of programs to sustain them like adopting an animal. Some of these activities involve animals engaging in direct or indirect contact with visitors and tourists, like many of the elephant sanctuaries in Thailand. Tourists pay to help the mahouts to bath and care for the animals. Although this is for a worthy cause that I recognize and applaud, removing elephants from abusive work environments, the elephants have to be under control and go through certain routines on a daily basis. Sometimes they are asked to perform certain behaviors like lifting the trunk, leg or let people sit or ride on their backs, therefore the same scrutiny should apply to these types of sanctuaries and programs.

4.6 Education and Conservation

Presentation styles vary from educational with naturalistic themed backgrounds and activities showcasing the animals' capabilities and natural adaptations like climbing, jumping, flying or swimming to highly orchestrated circus-acts like behaviors such as jumping through hoops, going down slides, riding on bicycles, to dressed up animals. Most circus-style shows feature princes and princesses, dream worlds, pirates and lost treasure islands or other themes along these lines. These often showcase the animals as actors in these stories, the "pirate" sea lion, the "naughty" parrot, the "sly" otter, or the "life-saving" dolphin. Lions and tigers are often asked to perform behaviors that instill awe and fear but in the context of the controlling human trainer, sending conflicting messages. Lions and tigers are indeed dangerous animals, but the likelihood of being attached or killed by them in the wild is extremely low and they could not be controlled in this way in the wild. Animals can be dressed up, carry or operate props like electric mopeds, or pretend vocalize as they are talking with the human voice-over.

Training dolphins or sea lions to jump through hoops, dance the lambada, wear sunglasses, or balance a ball on their nose can hardly be called natural, and luckily these behaviors are being faded out in many modern facilities. Although considered entertaining by most of the general public, very little to no educational information features in these types of shows and often lack a conservation message.

By contrast, educational presentations tend to cover general biology, anatomy, social life, cognitive abilities and conservation related topics. These presentations are geared to amaze the audience as well as educate people about the animals in their natural environment, informing them about potential threats to wild populations and empower the general public to helping conservation efforts. Miller et al. (2013) examined the short- and long-term effects of dolphin shows and interaction programs on visitors' conservation-related knowledge, attitude, and behaviour. They found that participants of both dolphin shows and interaction programs demonstrated a significant short-term increase in knowledge, attitudes, and behavioural intentions, and found sustained results over at least three months following the first survey. In this follow up survey participants also reported that they were engaging in more conservation-related behaviours. This study suggests that dolphin shows and interaction programs can be important contributors to a conservation education program in zoos and aquaria housing marine mammals. But if well designed, incorporating facts and interpretation, these types of presentations can be educational as well as entertaining (Visscher et al. 2009). Dolphins and sea lions can be asked to presents different body parts to highlight the adaptation to aquatic life, and exhale on a signal showing they are mammals just like humans. Hawks can be taught to fly between the tight space of two human standing close together to show their precision and agility, and parrots can do color discrimination. Small cats as well as owls can showcase their amazing hearing abilities through locating a hidden sound box, as if they are listening for a prey in hiding.

Educational presentations can be highly themed to entertain, engage and call into action the public. Themes can vary depending on topics such as 'life in the ocean, which can include types of fisheries and threats to the environment' and how people can eat more sustainably. Information on threatened fish stocks and the bycatch of marine mammals in large nets can move people to pay more attention to what fish they buy and, where relevant, from which company to buy dolphin-safe tuna (Brown 2005). Topics could also be feeding strategies, i.e. how do animals acquire food that in turn could highlight the animal's cognitive abilities, communication and sensory systems for feeding.

5 Discussion and Conclusions

This paper discussed the welfare aspects of performing animals in zoos and circuses. It highlighted the commitment of animal welfare and care organizations, the guidelines and strategies. It considered the species-specific needs, individual needs and preferences, training, education and conservation through science and practical applications. In light of this background I would like to return to the central questions to this paper.

(1) Are performing wild animals morally acceptable?

Circuses and zoos often argue that they are aiming for a similar goal: that seeing wild animals in person leads to a closer connection to animals, to wanting to save the species, that people can learn to be more kind to animals, and to create a better understanding of the natural world. If performance and performing, as defined in this paper, are acceptable definitions that pave the way to convey conservation messages, create more awareness about the threats to individuals and species, call people into action and create more empathy, then animal presentations, *of the kind defined earlier*, are morally acceptable.

We might argue that, accepting the outlined definition at the beginning of the paper and the research on visitor watching time, we are morally obliged to ensure people see animals performing species-specific and desired behaviors, to develop programs that engage and empower the public for long enough to connect with them in a meaningful manner. Performance and activities that do not fall under the stated definition are not morally acceptable, such as unnatural behaviors, ridicule, and disrespectful representation (e.g. dressing animals up).

Facilities must be able to provide for animals to lead fulfilling lives. Circuses cannot adequately provide for captive wild animals a thriving and flourishing life. The short time of training through human-animal interactions and activities does not outweigh the vast amount of time of confinement, inadequate housing, transport, social structure, lack of complexity, control and choices, among other concerns in the keeping of captive wild animals in circuses. Performing wild animals in circuses are therefore morally not acceptable.

(2) Are presentations morally acceptable for both educational and entertainment purposes?

If an engaging setting, themed exhibit or presentation facilitates learning and empowerment then performing wild animals in educational entertainment is morally acceptable. Learning is often easier and more fun when a combination of elements is used, such as sounds, sights, storytelling, awe and inspiration. Therefore educations through entertainment (often called edutainment) will likely hook people's attention, which in turn raises curiosity. Zoos should work with psychologists, sociologists, museum experts and other professionals to achieve high performance educational entertainment that catches young and old. Zoos should also do much more research in visitor experience and learning to verify whether indeed these programs are effective in achieving their goals of empathy for animals and awareness of nature conservation topics which result in positive actions for animals and the planet.

Zoos must develop presentations of a variety of kinds. Staged or not, the use of exhibits, enrichment and training, where wild animals are the star attraction and showcased for their amazing natural capacities and adaptations, will evoke respect for animals as individual and sentient beings. Critics of zoos argue that instead of looking at wild animals we look at something resembling but not really being that, something called 'Disneyization' by Beardsworth and Bryman (2001). The authors argue that these types of exhibits involving such elements may increase. Many shows with unrealistic animal representations exist worldwide, such as the 'pirate' sea lion mentioned earlier and the use of props to set a pirate island theme. This "locates animal presentation and animal performance in highly elaborate quasified settings, … and the probability exists that the exhibition of animals will become subordinate to the staging of elaborate quasifications of the 'wild'" (idem, 100).

Watching wild animals perform in television commercials can have detrimental consequences for conservation efforts, or can lead to the perception that these animals could be suitable pets. "The portrayal of chimpanzees in unnatural, human-like situations may have a negative effect on the public's understanding of their endangered status in the wild while making them appear as suitable pets" (Schroepfer et al. 2011, p. 1). Schroepfer and his colleagues conclude that their results "firmly support the hypothesis that use of entertainment chimpanzees in the popular media negatively distorts the public's perception and hinders chimpanzee conservation efforts" (idem.). Therefore serious questions need to be answered about the way wild animals are portrayed to the public. Although no other studies to my knowledge compare to this chimpanzee study it is plausible to expect similar results for other species in similar contexts.

How animals are portrayed to the general public should therefore be of concern. Animals should not be ridiculed, nor should they be shown in ways that are not species-typical. Some argue that it is not a problem to use animals as actors even when it makes them look silly or inadequate. They argue that the animal doesn't know that he or she is represented in this manner. Whether the animal knows or doesn't know is irrelevant: animals should be represented in a respectful manner that reflects the individual and the species. Animals should not be used as actors in non-species relevant themed presentations and shows. One could argue that especially *because* the animal is not aware of the representation we should take the utmost care not to misrepresent and be disrespectful. As the animals are not aware of the types of presentations they are used and cannot voice their own concern or objections it is important that we think about the effects. We have the moral duty to consider how animals are represented, whether the situation they are placed in is in their benefit or not.

The classic circus-format shows in zoos, marine parks and other entertainment parks should be ended and only presentations that are representative and respectful of the individual animal and species should be developed, this includes the educational programs and displays. Only if all criteria can be achieved can educational entertainment be morally acceptable.

(3) If morally acceptable, which conditions need to be fulfilled to warrant good animal welfare, i.e. thriving captive wild animals?

As outlined in the welfare section and species examples, there are many aspects that warrant close scrutiny to ensure a good quality of life. Animal welfare frameworks, lines of research, guidelines and strategies are many of the avenues that can inform and guide in the development of environments in which captive wild animals might thrive.

Animals should be able to perform within their exhibits and/or trained to enter a certain presentation area after which they can return again to the regular housing. Animals should have an open access policy at all times, meaning they have places to retreat to if they wish to do so. Animals should always have an option to opt out of any of the activities if they wish to do so. Real choice and control is perceived choice and control that is not contingent on reinforcement or other consequences.

For example, the option to earn or have access to free food, water, space and companions should always be there.

Putting forward the capabilities approach with a revised list to suit species-specific needs and wants is another valuable avenue to explore, and how it can be meaningfully adapted to further animal welfare and the ethical treatment of captive wild animals.

Performances and other entertainment activities are often short in time and only during times humans are active. However, as mentioned before, an animal's life should be considered from the animal's perspective, 24/7 across lifespan (cradle to grave). Although wild and domesticated animals can have many things in common, such as wanting to work to gain something or to play (Haraway 2008, p. 22), training will be only a small part of their day. As training and other activities will be a small part of their day and often dependent on the caretaker it is therefore important to give equal consideration to species and individual needs and preferences so overall welfare can be maintained at high levels. The development of semi-autonomous environments in which animals have choices and over which they have control is paramount.

Performing wild animals should be housed in environments that are species-specific and geared to the individual: they should also have complexity, choices and control for animals throughout a 24/7 approach (Brando and Buchanan-Smith, unpublished data).[8] Individual personalities, needs and preferences should be taken into consideration when developing animal care programs, as well as species-specific requirements.

6 Closing Remarks

Before I conclude I would like to highlight some issues (but this list is not exhaustive) that need urgent attention. If these practical issues are not solved or at least drastically mitigated, we can either continue to talk until the cows come home—as the expression goes, or watch many facilities worldwide fail or not achieve best practice standards throughout, for all species and individuals.

Between the ethical approaches, the welfare sciences, guidelines and strategies, and the practical application in the real world is a very long bridge that needs crossing. This bridge has many obstacles: lack of time, low staff numbers, limited budgets, contracted architects who do not listen and have no clue about animals but continue to build new parks, inadequate food supplies, old buildings, city and country laws, unmotivated personnel, staff that only works 1/3 of the day while animals are housed 24/7 across lifespan, dysfunctional breeding and TAG committees, lacking or incomplete husbandry guidelines, lack of empathy and understanding for animals, weather and geographical constrains, language barriers,

[8]www.247animalwelfare.eu.

contractors and builders, visitor expectation, cultural differences, opinion version evidence-based decision making, tradition and just-because-we-have-always-done-it-like-that attitude, irrelevant or low entry requirements or no schooling, to just name a few. Solving these problems would be a topic for another book but without solving many of these obstacles the efforts of thinking about or implementing valuable opportunities for animals will be suboptimal or in vain.

It is clear to me that there are some very good zoos and sanctuaries that are committed to the animals in their care, and committed to education, conservation and research. Animal wellbeing has to be first priority, and is the basis of and compatible with the achievement of successful and significant research, education and conservation programs, as highlighting in the mission statement of modern zoos today. Only if good animal welfare can be guaranteed, and predominantly positive welfare ensured over a 24/7 across lifespan, is the keeping of wild animals for entertainment morally acceptable.

References

Beardsworth, A., and A. Bryman. 2001. The wild animal in late modernity: The case of the Disneyization of zoos. *Tourist Studies* 1(1): 83–104.

Blokhuis, H.J., I. Veissier, M. Miele, and B. Jones. 2010. The Welfare Quality® project and beyond: Safeguarding farm animal well-being. *Acta Agriculturae Scandinavica Section A-Animal Science* 60: 129–140.

Brando, S.I.C.A. 2010. Advances in husbandry training in marine mammal care programs. *International Journal of Comparative Psychology* 23: 777–791.

Brando, S.I.C.A. 2015. Circus. In *Lexikon der Mensch-Tier-Beziehungen*, 431–435. Bielefeld, Germany: Publisher Transcript Verlag.

Brensing, K., K. Linke, M. Busch, and I. Matthes. 2005. Impact of different kinds of humans in swim with-the-dolphin-programs in two settings. *Anthrozoös* 18(4): 409–429.

Broom, D.M. 2001. Coping, stress and welfare. In *Coping with challenge: Welfare in animals including humans*, ed. D.M. Broom, 1–9. Berlin, Germany: Dahlem University Press.

Brown, J. 2005. An account of the dolphin-safe tuna issue in the UK. *Marine Policy* 29: 39–46.

Buchanan-Smith, H.M., J. Stronge, and M. Challis. 2001. The roles of zoos: Examples from Belfast Zoological Gardens. In *Biological collections and biodiversity*, ed. B.S. Rushton, et al., 163–170. Otley, West Yorkshire: Westbury Academic & Scientific Publishing.

Burghardt, G.M. 1985. Animal awareness: Current perceptions and historical perspective. *American Psychologist* 40(8): 905–919.

Christiansen, F., D. Lusseau, E. Stensland, and P. Berggren. 2010. Effects of tourist boats on the behavior of Indo-Pacific bottlenose dolphins off the south coast of Zanzibar. *Endangered Species Research* 11(1): 91–99.

Clark, F.E. 2013. Marine mammal cognition and captive care: A proposal for cognitive enrichment in zoos and aquariums. *Journal of Zoo and Aquarium Research* 1(1): 1–6.

Circopedia http://www.circopedia.org/Main_Page.

Clubb, R., and G.J. Mason. 2003. Animal welfare: Captivity effects on wide-ranging carnivores. *Nature* 425(6957): 473–474.

Connor, R.C., R.S. Wells, J. Mann, and A.J. Read. 2000. The bottlenose dolphin: Social relationships in a fission-fusion society. In *Cetacean societies Field studies of dolphins and whales*, ed. J. Mann, et al., 91–126. Chicago: University of Chicago Press.

Cooper, J.J., and G.J. Mason. 2000. Increasing costs of access to resources cause re-scheduling of behaviour in American mink (*Mustela vison*): Implications for the assessment of behavioural priorities. *Applied Animal Behaviour Science* 66(1): 135–151.

Dawkins, M.S. 1983. Battery hens name their price: Consumer demand theory and the measurement of ethological "needs". *Animal Behaviour* 31: 1195–1205.

Declaration of Rights for Cetaceans, Helsinki (2010). http://www.cetaceanrights.org/.

Delfour, F., and H. Beyer. 2012. Assessing the effectiveness of environmental enrichment in bottlenose dolphins (Tursiops truncates)'. *Zoo Biology* 31(2): 137–150.

Dembiec, D.P., R.J. Snider, and A.J. Zanella. 2004. The effects of transport stress on tiger physiology and behavior. *Zoo Biology* 23(4): 335–346.

EAAM. 2009. The European Association for Aquatic Mammals Standards and Guidelines for the management of bottlenose dolphins (Tursiops species) under human care. Published online by the EAAM. www.eaam.org.

EAZA. 2014. Guidelines on the use of animals in public demonstrations. http://www.eaza.net/assets/Uploads/Standards-and-policies/Animal-Demonstrations.pdf.

Entertainment. http://www.merriam-webster.com/dictionary/entertainment.

ESA Code of Conduct. http://www.europeancircus.eu/?page_id=361.

Fiscus, C.H., and G.A. Baines. 1966. Food and feeding behavior of the Steller and California sea lions. *Journal of Mammalogy* 47(2): 195–200.

Fraser, D. 2008. Understanding animal welfare. *Acta Veterinaria Scandinavica* 50: S1.

Friend, T.H., and M.L. Parker. 1999. The effect of penning versus picketing on stereotypic behavior of circus elephants. *Applied Animal Behavior Science* 64: 213–225.

Gazda, S.K., R.C. Connor, R.K. Edgar, and F. Cox. 2005. A division of labour with role specialization in group-hunting bottlenose dolphins (Tursiops truncatus) off Cedar Key, Florida. *Proceedings of the Royal Society B: Biological Sciences* 272(1559): 135–140.

Gruen, L. 2011. *Ethics and animals: An introduction*. Cambridge, NY: Cambridge University Press.

Hanke, W., M. Witte, L. Miersch, and M. Brede. 2010. Harbor seal vibrissa morphology suppresses vortex-induced vibrations. *Journal of Experimental Biology* 213: 2665–2672.

Haraway, D. 2008. *When species meet*. Minneapolis/London: University of Minnesota Press.

Harding, E.J., E.S. Paul, and M. Mendl. 2004. Animal behaviour: Cognitive bias and affective state. *Nature* 427: 312.

Hare, B., J. Call, and M. Tomasello. 2001. Do chimpanzees know what conspecifics know? *Animal Behavior* 61: 139–151.

Herman, L.M., S.H. Kuczaj II, and M.D. Holder. 1993. Responses to anomalous gestural sequences by a language-trained dolphin: Evidence for processing of semantic relations and syntactic information. *Journal of Experimental Psychology: General* 122(2): 184–194.

Hosey, G.R. 2000. Zoo animals and their human audiences: What is the visitor effect? *Animal Welfare* 9: 343–357.

Hunt, T.D., M.H. Ziccardi, F.M.D. Gulland, P.K. Yopchem, D.W. Hird, T. Rowleas, and J.A.K. Mazet. 2008. Health risks for marine mammal workers. *Disease of Aquatic Organisms* 81: 81–92.

Iossa, G., C.D. Soulsbury, and S. Harris. 2009. Are wild animals suited to a travelling circus life? *Animal Welfare* 18: 129–140.

Janik, V.M., and L.S. Sayigh. 2013. Communication in bottlenose dolphins: 50 years of signature whistle research. *Journal of Comparative Physiology* 199(6): 479–489.

Keulartz, J., and J. Swart. 2012. Animal flourishing and capabilities in an era of global change. In *Ethical adaptation to climate change*, ed. A. Thompson, and J. Bendik-Keymer, 123–144. Cambridge, MA: The MIT Press.

Kiley-Worthington, M. 1989. The training of circus animals. In *Proceedings of a Symposium organized by Universities Federation for Animal Welfare 26th–27th September*, 65–81.

Kiley-Worthington, M. 1990. *Animals in zoos and circuses: Chiron's world?*. Essex, UK: Little Eco-Farms Publishing.

Krawczel, P.D., T.H. Friend, and A. Windom. 2005. Stereotypic behavior of circus tigers: Effects of performance. *Applied Animal Behavior Science* 95: 189–198.

Krützen, M., J. Mann, M.R. Heithaus, R.C. Connor, L. Bejder, and W.B. Sherwin. 2005. Cultural transmission of tool use in bottlenose dolphins. *Proceedings of the National Academy of Sciences of the United States of America* 102(25): 8939–8943.

Mann, J., R. Connor, P. Tyack, and H. Whitehead (eds.). 2000. *Cetacean societies: Field studies of dolphins and whales*. Chicago: The University of Chicago Press.

Melfi, V. 2013. Is training zoo animals enriching? *Applied Animal Behaviour Science* 147: 299–305.

Mellor, D.J., and N.J. Beausoleil. 2015. Extending the "Five Domains" model for animal welfare assessment to incorporate positive welfare states. *Animal Welfare* 24(3): 241–253.

Miller, L.J., Zeigler-Hill, V., Mellen, J., Koeppel, J., Greer, T., and Kuczaj, S. (2013). Dolphin shows and interaction programs: benefits for conservation education?. *Zoo biology* 32(1): 45–53.

Mellor, D.J., S. Hunt, and M. Gusset (eds.). 2015. *Caring for wildlife: The world zoo and aquarium animal welfare strategy*. Gland: WAZA Executive Office.

Nevill, C.H., and T.H. Friend. 2003. The behavior of circus tigers during transport. *Applied Animal Behavior Science* 82: 329–337.

Nevill, C.H., and T.H. Friend. 2006. A preliminary study on the effects of limited access to an exercise pen on stereotypic pacing in circus tigers. *Applied Animal Behavior Science* 101: 355–361.

Nussbaum, M.C. 2006. *Frontiers of justice. Disability, nationality, species membership*. Cambridge, MA: Harvard University Press.

Plotnik, J.M., R. Lair, W. Suphachoksahakun, and F.B.M. De Waal. 2011. Elephants know when they need a helping trunk in a cooperative task. *PNAS* 108(12): 5116–5121.

Price, E.E., and T.S. Stoinski. 2007. Group size: Determinants in the wild and implications for the captive housing of wild mammals in zoos. *Applied Animal Behavior Science* 103: 255–264.

Radford, M. 2007. Wild animals in travelling circuses; The report of the chairman of the circus working group.

Reichmuth Kastak, C., and R.J. Schusterman. 2002. Long-term memory for concepts in a California sea lion (Zalophus californianus). *Animal Cognition* 5(4): 225–232.

Reiss, D., and L. Marino. 2001. Mirror self-recognition in the bottlenose dolphin: A case of cognitive convergence. *PNAS* 10: 5937–5942.

Rutherford, K.M., R.D. Donald, A.B. Lawrence, and F. Wemelsfelder. 2012. Qualitative behavioural assessment of emotionality in pigs. *Applied Animal Behaviour Science* 139(3–4): 218–224.

Samuels, A., and L. Bejder. 2004. Chronic interaction between humans and free-ranging bottlenose dolphins near Panama City Beach, Florida, USA. *Journal of Cetacean Research and Management* 6(1): 69–77.

Schmid, J. 1995. Keeping circus elephants temporarily in paddocks—the effects on their behavior. *Animal Welfare* 4: 87–101.

Schroepfer, K.K., A.G. Rosati, T. Chartrand, and B. Hare. 2011. Use of "entertainment" chimpanzees in commercials distorts public perception regarding their conservation status. *PLoS ONE* 6: 10.

Shane, S.H., R.S. Wells, and B. Würsig. 1986. Ecology, behaviour and social organization of the bottlenose dolphin: A review. *Marine Mammal Science* 2: 34–63.

Stirling, I. 1974. Midsummer observations on the behavior of wild polar bears (Ursus maritimus). *Canadian Journal of Zoology* 52(9): 1191–1198.

Teixeira, C.P., C. Schetini de Azevedo, M. Mendl, C.F. Cipreste, and R.J. Young. 2007. Revisiting translocation and reintroduction programs: The importance of considering stress. *Animal Behavior* 73: 1–13.

Trone, M., Kuczaj, S., and Solangi, M. 2005. Does participation in dolphin-human interactions programs affect bottlenose dolphin behaviour? *Applied Animal Behaviour Science* 93, 363–374.

Vidya, T.N.C., and R. Sukumar. 2005. Social and reproductive behavior in elephants. *Current Science* 89(10): 1200–1207.

Wemelsfelder, F. 2005. Animal boredom: Understanding the tedium of confined lives. In *Mental health and well-being in animals*, ed. F.D. McMillan, 79–93. Iowa, USA: Blackwell Publishing.

White, T. 2007. *In defense of dolphins: The new moral frontier*. Malden, MA: Blackwell Publishing.

Visscher, N.C., R. Snider, and G. Vander Stoep. 2009. Comparative analysis of knowledge gain between interpretive and fact-only presentations at an animal training session: An exploratory study. *Special Issue on Conservation in Zoos and Aquariums* 28(5): 488–495.

Comment: Caring for Captive Communities by Looking for Love and Loneliness, or Against an Overly Individualist Liberal Animal Ethics

Clemens Driessen

Abstract Animal ethics in its liberal, analytic style of academic writing can suffer from a form of excessive individualism that lacks a full view of life as experienced by many animals. A range of arguments against using and enclosing animals, or in favour of certain (pre)conditions of captivity, can be found to have a tendency to focus on generic and isolated individual organisms. In its most extreme form, this type of ethical thought sets up a truncated notion of the animal as separate from their conspecifics, limits animal interests to the desires of solipsistic individuals, fails to appreciate meaning that may emerge in human-animal relations, and renders invisible a range of concerns of animal ethics in view of the communal character of animal lives. Through a critical reading of the previous four chapters, this one will trace the extent to which reasoning in terms of welfare, freedom, capabilities or dignity may lead to granting attention and value to (many) animals as the idiosyncratic, relational, sociable beings which many of them are. Or can be, even in captivity, and even in the age of humans.

1 Introduction

This chapter reflects on the character and implications of excessive analytic individualism in animal ethics, as well as on potential alternative understandings of animal lives, through a reading of the four preceding chapters. It starts by discussing Aaron Simmons's "Animals, Freedom and the Ethics of Veganism" as bringing a liberal focus on particular understandings of harm and freedom. Next, it turns to TJ Kasperbauer's "Should Captive Primates Have Reproductive Rights?", as a largely exemplary case of the truncated view of animal life as implied by certain analytical analyses of liberal individualist animal ethics. Jozef Keulartz's chapter on animal capabilities offers an interesting alternative to the individualism of animals. However, in Keulartz's adapted capabilities approach, the view of the

C. Driessen (✉)
Wageningen University, Wageningen, The Netherlands
e-mail: clemens.driessen@wur.nl

B. Bovenkerk and J. Keulartz (eds.), *Animal Ethics in the Age of Humans*,
The International Library of Environmental, Agricultural and Food Ethics 23,
DOI 10.1007/978-3-319-44206-8_19

lives of animals is still limited to their species membership, as implying specific characteristics to be catered for. As opposed to more relational view that would open up to more truly relational life. With Sabrina Brando's chapter some prospects of an animal ethics that appreciates animal community are briefly explored.

There is a long history in animal (and environmental) ethics of accusations of individualism. But primarily as ecological illiteracy on how wild animals in nature figure as part of biotic communities (Callicott 1980). This is not the community I am referring to here. I am interested in how animals have meaningful relations with other animals in their living environment, implying a notion of community that is closer to the human sense—or even overlaps with human communities. What this chapter is concerned with is seeing animals as individuals with relations, as experiencing meaningful interactions with a range of others, some of whom can perhaps be human.

Of course the potential for meaningful intra-species and inter-species relations diverge wildly whether one is a snake or a sloth. But discussing whether and how to keep wild animals captive, I will argue, should involve an ongoing form of inquiry, performed together with enclosed animals who actively respond to interventions in the conditions but also in the communities in which they are kept.

The confines of this chapter do not allow me to elaborately engage with all the implications of an alternative take on animal ethics that would be more conducive to a view of (the endless variety of) nonhuman sociability. It is perhaps also a question to what extent this perspective needs to be formalized in a separate theory, rather than taken up in existing ones. It may be more a matter of acknowledging and promoting how this view already is actively practiced and experimented with in institutions that hold (wild) animals captive. But at least it may be clear that ethical theorizing around animals in captivity should actively acknowledge how animals, living in what is now officially the age of humans, potentially are not just suffering pain, boredom or humiliation, but also from loneliness[1] and lack of love.

2 Liberating Animals One by One

Aaron Simmons in his chapter argues it is not just harmful to animals to kill and eat them, but also to confine them. Whether using them for animal products or for entertainment in circuses and zoos, restricting freedom is presented as a case of harming an animal. Simmons describes this harm as depriving animals from "freely pursuing their various enjoyments in life" (p2). Besides physical pain that may be inflicted and the associated suffering which is generally invoked to argue against animal use, the harm produced here consists also of mental suffering due to

[1]We already commonly grant animals that they may be suffering from loneliness—think of the requirement of providing a single horse with a goat.

boredom and depression associated with captivity (p2). So also freedom is important to animals.

However, referring to Cochrane, Simmons distinguishes between freedom for animals and the human intrinsic interest in liberty, since humans are autonomous beings "capable of framing, revising, and pursuing their own conceptions of the good." We can rationally reflect on our desires and try to change these according to our values. Animals cannot. This means freedom is important for animals not in the way liberty is central to humans: as it allows us to develop a self, to learn, and become citizens. Animals therefore according to Simmons need only to be provided with (what I would call) a watered down version of human liberalism, only retaining more modest ('basic') notions of freedom and harm. Freedom in this 'basic sense' then gets defined as "the ability to satisfy one's desires absent of external constraints" (Simmons p8).

This may sound logical in a kind of materially obvious way: animals tend not to like being locked up. Liberating animals from confinement seems a *prima facie* preferable or even important thing to do. What is problematic though with this conception of the interests of animals, is how this argument espouses a rather atomistic idea of animal life and lacks any sense of (potential) sociability. Simmons discusses the interests of animals in what may strike as 'consumerist' terms of satisfying desires and individual preferences. Freedom is defined as in the service of the pursuit of happiness, for animals to "live their lives as they please" (p14). The role of ethics and politics is to provide conditions and guarantees for this.

The proposed (somehow very 'physical') ideal of freedom of individual subjects involves implicitly a notion of 'the other' as a constraint on satisfying one's desires. This reveals what could be labelled a traffic model of freedom and what it means to share space: others appear in sight only as being in the way of getting where you want to go, as (potential) obstacles to fulfilling one's personal desires.

This seemingly obvious notion of freedom, as pursuing the things that an animal desires and enjoys doing, ultimately pits everyone against everyone, humans and nonhumans, as caught up in a planetary zero-sum struggle over land and resources. Whereby, for those promoting the interests of animals, morality ideally functions to curtail the desires of individual humans in order to make sure individual animals do not unduly suffer. The human agent that Simmons addresses his argument to is the consumer of animal products, who needs to change his or her 'lifestyle'. Simmons has added an interest in freedom to the cow's list of having to suffer udder inflammation, chronic diarrhoea and hoof problems. And to him this interest is either a very strong one, or the desires of the lover of cheddar and brie rather trivial.

This type of radical rethinking of the ways in which animals are used, and exploring a range of arguments that would help us to see why it is wrong to use animals as 'mere resources', is of course commendable. My problem with Simmons' train of thought is not that animals are looked upon as individuals, but how in fact they are not: they only figure as generic specimens of their species, as solitary receptacles of either pleasure or suffering, not as individuals with meaningful experiences and idiosyncratic relations.

Gorillas that may come to 'love' another gorilla or at least actively choose to forge an intimate bond (cf Braverman 2013), or sea lions that grow attached to a particular trainer, testify to the possibilities of animals actively relating to others in ways that are central to their wellbeing and flourishing. Thereby the very idea that it is fully up to us to make these decisions, that we can survey the arguments, register the relevant interests, and weigh these according to coherent sets of principles, actually contributes to the ongoing marginalization of animals as subjective, agential beings (Driessen 2014). This style of ethical thought entrenches the idea of human superiority and the associated legitimation of domination, even if benevolent. Animals big or small are only passive moral patients delivered to either our cruelty or our generosity. They are outside of culture and politics, living in a separate zone, not partaking in meaning making or interpreting—thus we are happily ignoring our shared or at least overlapping biosemiotic environments.

What the mode of reasoning of Simmons misses, is an appreciation of a situation in which we (both a variety of animals and humans) are ideally embedded in a lively community that allows us to lead rich social lives. Others we encounter here do not figure as impediments to us getting what we want, or as mere resources to be used instrumentally, but as essential others for us to encounter, without whom there is little opportunity for self-realization. If we start from the idea of a multispecies community, then using others as mere resources is not solely an infringement on their freedom, but also may be inappropriate, offensive, and detrimental to the ideal of shared lives and sharing space in a more-than-human community.

Lots of questions in this view still remain. We could wonder whether perhaps we could still, together with Berger (2009), idealize meaningful human-animal relations on small scale traditional farms. Or whether this is no longer a mode of living together, to care for animals and eat them, that is meaningful to us (cf. also Burt 2005).

Simmons ends his chapter by contemplating what to do with all those farm animals after the world's consumers have collectively gone vegan overnight. Considering that death is a harm to animals, as it restricts their freedom, Simmons proposes to phase-out farm animals by maintaining the current population until they die of old age, diligently cared for by former farmers on spacious sanctuaries. Not letting them reproduce is not an undue infringement on animal freedom, as we would do so in the best interest of their offspring, which would otherwise grow too numerous to be properly taken care of. And Simmons argues animals were only interested in sex anyway: "it was never their desire to reproduce in the first place" (p12). This brings us to the chapter by Kasperbauer.

3 Sex as an Individual Animal Right

Kasperbauer in his chapter focuses on a particular aspect of the lives of animals in captivity: sex and reproduction. Adding deontological notions of rights and autonomy to the mix, he argues that, like humans, animals too "have important interests in reproductive autonomy and procreation", especially higher mammals such as great apes (p1). As did Simmons, he also believes there should be a limit though to this reproductive right, in the form of birth control. This is argued for in the name of the welfare of those individuals that will start to overpopulate the facilities in which they are held captive when procreation runs rampant. As in the case of Simmons, it is easy to agree with the core premise of the chapter by Kasperbauer. So captive animals not only have a strong interest in freedom, but also in having sex.

Does that make it just an academic concern in animal ethics to worry about the right reasons for treating animals in certain ways? Also in Kasperbauer's chapter, it becomes clear it does matter how to make these arguments—whether to see animals either solely as individuals with desires they would like to see satisfied, perhaps even with autonomy to be respected, or instead (in many cases) as primarily social beings that exist in meaningful relations to others.

In his formulations of the importance of the freedom to procreate, or rather control over whether or not to reproduce, Kasperbauer highlights how this—for humans—is "central to personal identity, to dignity, and to the meaning of one's life" (p5). However, this starting point to discuss it as an individual interest or right rather ignores the practical situation in which procreative decisions come up. Procreation, especially in mammals such as humans, obviously requires the meeting (and mating) of at least two individuals. At least potentially, also in a variety of nonhuman partnerings, the decision of who to mate with and with whom to seek to care for offspring is a complex interactive process of courtship and negotiation. Besides a matter of autonomy and freedom, in many human cultures it centrally involves notions less common in political and moral philosophy, such as love and commitment between life partners. One does not have to extend some bourgeois conservative ideal of the nuclear family as the cornerstone of society to the animal world in order to appreciate how animals may want to live their lives in close relation to (particular) others.

But could there for animals be a right to have sex? As an individual right, this seems rather unworldly. Perhaps a right to masturbate would be easily granted—a common sight in captive animals with opposable thumbs—even though it may disconcert unsuspecting human visitors. But beyond this rather meagre mode of sexual flourishing, in order to reap the fruits of a right to sexual intercourse, an animal will need to convince some significant other first. Or, infringe on some other animal's right not to have sex or procreate, at least not then and there, with that particular partner.

Kasperbauer does find it relevant to differentiate regarding what the process of mammalian procreation involves for different genders. Indeed, we can imagine their experience to be radically different of what it means to have sex and to procreate. Perhaps what many animals in captivity—from intensively farmed sows, to prize bulls, to bottleneck specimens in efforts of conservation breeding—would actively appreciate is a right not to procreate. Not to be the vessel of efforts to maintain the species or produce (more spectacular or otherwise genetically desired) individuals.

Indeed, what animals may be interested in general is a perhaps somehow shared form of reproductive autonomy rather than a guaranteed provision of sexual intercourse. In part, the ideal situation for many animals may be similar to the situation in 'the wild'. Although arguably the condition of confinement could be thought to increase the importance of a provision for animals to be protected from sexual violence. At least the spaces of confinement should be designed to cater for possibilities to flee the advances of overly eager conspecifics.

In the second part of his chapter, Kasperbauer does go into sex and procreation as a multi-individual process. The choice of mates is also part of 'procreative autonomy' as he envisages it. Thus it becomes clear that, at least potentially, the locus of decision making and of experiencing the outcomes of this fundamental interest in procreation are not just a solitary affair. Nevertheless, Kasperbauer's discussion of reproductive rights and interests assumes that for 'someone' having children figures in our moral considerations as an individual achievement and an expression of autonomy.[2]

Kasperbauer's chapter on animal sex thereby offers an interesting case of how, as soon as you start discussing the lives of animals in more detail, one has to leave behind the kind of truncated view of animal life presumed in blanket undifferentiated individualism. Even if one wants to put freedom central, the meaning of it for an individual is hard to understand without an idea of freedom as something achieved together with others, not just as means to private ends but as part of shared meaningful practices, which also for animals may be labelled to be of a 'cultural' nature (cf. Nimmo 2012).

As in the human case, animal behaviours and the meaning of animal cohabitation is not a static given, especially when they live in the vicinity or under the control of humans. As Bertrand Russell has summarized, animal behaviour may be seen to change in parallel to human cultures:

> In the seventeenth century, animals were ferocious, but under the influence of Rousseau they began to exemplify the cult of the Noble Savage which Peacock makes fun of in Sir Oran Haut-ton. Throughout the reign of Queen Victoria all apes were virtuous monogamists, but during the dissolute 'twenties' their morals underwent a disastrous deterioration (Russell 1959, p. 95/6).

Lots of animals in fact are found to be not much interested in procreation while in captivity. What they communicate with this lack of interest is of course unsure; From confinement induced lethargy (some animals seem to have problems to get

[2]Whereas most parents come to realize having children also involves a serious infringement of one's autonomy—as becoming a parent produces an ambivalent dynamic of self-realization and loss of individuality.

aroused behind bars[3]) to perhaps even an active decision not to seek to bring offspring into existence in the conditions in which they live? Why make tiger babies if you know they'll never grow up to be real tigers? Having sex seems at least for some captive wild animals to only be meaningful or worthwhile in a situation that suits the species experience of a particular ecological setting, and perhaps an associated 'cultural' life in which living in groups of a particular 'family' make-up (and seasonal absence or periodic transfers between these) are common practice.

But when Kasperbauer discusses the animal right to sex and reproduction, he does not mention other aspects of meaningfully relating to mates and or offspring. A limited utilitarian/rights focus seems to lead to only accord animals rather basic bodily pleasures, divorced from meaningful 'mental' social experiences.

While informed by a view of ethics that seeks to secure individual rights and wellbeing, Kasperbauer does discuss several reasons for granting animals the right to procreate, which include more relational ones:

> They do have a strong interest in sex, and many species show concern and affection for offspring, but it's not clear if they actively consider the importance of offspring to their lives (p...).

This of course is a very sweeping statement about giraffes and donkeys and dolphins and spiders, not to mention kangaroos and cockatoos. On the one hand Kasperbauer is willing to derive from observing behaviours that certain experiences or activities are meaningful to animals. At the same time he has provisos based on his criterion of them 'actively considering' this issue. With Donaldson and Kymlicka (2011), Kasperbauer argues that rights can be grounded without a full-fledged sense of autonomy, as long as interests of animals can be discerned from their behaviour and when for them having increased control over their own lives would contribute to their flourishing.

It is still common to say, often referring to Nagel's 'What it's like to be a bat', that we don't know what goes on inside animal minds. Contrasted with explanations of animal behaviour reduced to instinctive, automatic responses to situations, it may seem a very benign thing to grant animals an inner life. But not so when everything in the lives of animals is then still to be solely understood in terms of strictly individual experience, considered to be ultimately an inner, mental affair of which we can gain little knowledge and definitely no 'access'.

For humans it seems easy to claim how the importance of caring for offspring is revealed in our everyday practices, besides the formally verbalized discourses we generate around those: By granting this behaviour the status of meaningful, as partaking in purposeful interaction and exchange, in what we for humans are happy to consider an expression of interest and belief. For humans we would not say 'well

---

[3]In a sad twist of irony, it seems especially the species which are endangered due to their body parts being considered an aphrodisiac who tend to be administered Viagra to foster their breeding activities (see for example: http://www.dailymail.co.uk/news/article-2921925/How-rhino-got-horn-Zoo-couple-given-animal-viagra-struggling-breed-naturally-ll-hear-patter-not-tiny-footsteps.html).

they may spend a lot of time and effort and clearly display bodily affection, but we can only be sure about what they really mean/want/think when they state unequivocally ‘I love my child’? When considering caring for animals we could as well risk in erring on the safe side, and believe in our critical but benevolent interpretations of animal acts. Otherwise our fear of anthropomorphizing easily leads to mindless behaviourist interpretative schemes in which animals do not live meaningful lives in for them meaningful worlds, and in meaningful relations to others.

Another ground for granting certain rights to animals, again with Donaldson and Kymlicka, for Kasperbauer comes “by virtue of their membership in human communities”. This membership is conceived as defined by the dependence of (domesticated) animals on humans, and a possible inability to survive outside of spaces of human care. The central relation is then between an individual animal and the human collective. This however is only one and a rather limited way of understanding shared lives between humans and domesticated others. What further could it mean to think of humans, dogs, horses, sheep and cows to take part in a communally shared lifeform? When animals are not just made dependent, but also actively participate in a (partially) common life? This to some extent may be the case only in an ideal world, or at least a much improved one. At the same time, human animal relations already can have this character (or arguably until recently as in the case of animal use in agriculture), since living with animals involves a much more complicated set of relations and experiences than reductionist representations of these would have it (Wolch and Emel 1998; Jones 2003; Driessen 2012).

Again it may seem just a matter of philosophical nit-picking: what does it matter ‘why’ we should let apes be merry, have sex and procreate without us preventing them from doing so, as long as we agree they do? Because the argument as presented by Kasperbauer is limiting the kind of claims we can make, the kind of language we can use, and the kind of aspects of animal lives we are attuned not just to value but also to discern. My point is that a reduced liberal individualism—whether grounded in utilitarian or rights based conceptions of the good—tends to offer a bleak, narrow and unconvincing idea of what it is to be an individual. Humans and a large set of animals (though of course very much depending on the species; including gender roles and temporary mixings and groupings of these) it is safe to say only can self-realize through meaningful relations to others—whether conspecifics or other species.

A multispecies community can also be considered as an intricate network of relations between individuals, that helps generate and at the same time is the product of an always emergent multispecies ‘culture’; a mode of relating that involves mutual learning, adaptation and the development of meaning when learning to respond to signals of others. This does not have to be limited to domesticated animals, bred over millennia to be attuned to our lifeform and semiotics. The distinction between domesticated and wild is no longer, or has never been, that strict anyway (e.g. Haraway 2015). And also for those that still can be considered wild animals living among us, especially in captivity, their wildness does not necessarily mean the absence of (possibly) meaningful communication. Something to turn to in the remainder of this chapter.

4 Individualist Capabilities?

Would an animal ethics based on Nussbaum's capabilities approach more easily help us think beyond reductionist individualism? Keulartz's starting point is the question of animal ethics in the Anthropocene, the era of mass extinction, climate change, invasive species, habitat fragmentation and the disappearance of wild spaces untouched by humans. He identifies the need to develop an alternative to utilitarian and deontological animal ethics, in view of current day problems of wildlife conservation and extinction, with nature so thoroughly messed up by human activities. Conserving wild animals nowadays is no longer a matter of fencing an area and calling it off limits to humans.[4] The practicalities of conservation in the Anthropocene include capturing, transporting and releasing animals in novel habitats, captive breeding and ex situ conservation such as in zoos, and other modes of active (meta) population management.

This, according to Keulartz, blurs the very distinction between in situ and ex situ conservation, and thus between captivity and freedom. Every animal now lives in an enclosure. Some spaces are just a bit bigger than others, and moments of relative freedom for some may be alternated with living in sites under more intensive forms of care and control. Captivity is a condition for a number of wild animals that may be considered necessary—at least temporarily—e.g. in relocating animals to other habitats or awaiting future restoration or creation of habitats. Animal 'liberation', is in this view a complex and partly unfeasible aim. It is better to think of how to organize spaces and regimes of care to optimally cater to animals.[5] Arguments for this, and how to ground guidelines for doing so, Keulartz derives from Nussbaum's capabilities approach, which starts from (a duty for) facilitating the development of species specific capabilities. With Nussbaum he argues against the 'moral individualism' of common approaches to animal ethics. But the only ethically relevant membership for Nussbaum is being a 'member' of a species. This then results in granting an individual a set of generic species characteristics, in order to define species specific norms of individual flourishing.

Does Keulartz's take on Nussbaum's capabilities offer a view of animal life that includes experiences of social relations? The capabilities approach could allow for a

[4]Which was a dubious practice anyway of course, assuming the wild is beyond the human, and displacing local and indigenous human communities to make it so.

[5]Others have discussed this more critically, in terms of biopolitics (e.g. Biermann and Mansfield 2014; Srinivasan 2014; Chrulew 2011). Emphasizing how, in the name of populations and the preservation of life, all kinds of harm are deemed legitimate to produce viable populations and save genetically or otherwise defined (sub)species. Conserving sets of genes as the key to species survival and biodiversity to be released again in some future restored Eden.

move beyond a traffic model of relations to others as more than mere obstacles to pursuing what one wants. Nussbaum's generic list of human functionings and capabilities includes besides bodily health, practical reason, and emotions, also "(7) affiliation with others and the social bases of self-respect, (8) a meaningful relationship with other species and with nature" (Nussbaum 2004, p. 314; 2006, p. 76).

Also for animals, sociality can of course be considered a capability that a just society should cater for. But, in Keulartz's chapter, the proposed ideal of developing capabilities goes only so far as to give a fuller account of meaningful animal lives. It is still very much a particular, individualist ideal of the good life, as an expression of individual affordances. The capabilities approach as proposed by Keulartz thus involves a way of viewing the world in which animals live and experience meaning; but mainly as individual projects of self-realization. Maintaining close relations to others only appear in service of that. More intimate bonds seem not in themselves central. Keulartz's view of flourishing tends to exclusively focus on individual athletic feats—whereas flourishing could just as well be considered to involve feedbacks offered by relations. These can indeed, as Keulartz argues, be provided not only by members of the same species but also by humans, depending on the species and the situation.

When Keulartz promotes the role of training and interaction however, he does so not in the name of meaningful communication and relations, but as exercise of abilities, as personal animal experiences of achieving preset feats to which the animal is genetically predisposed. Thereby not including much of what Hearne, Despret, and Haraway (on whose work he builds his argument) would emphasize: the experience of engaging with others. Keulartz in his chapter thus only goes so far in helping to see relations and animal others as not just traffic obstructions on our ideally private road, but meaningful presences we could engage with in mutual interactions, in relations of learning, of developing a shared material culture in an animated lifeworld. Besides promoting a bureaucracy of listing and facilitating capabilities—important it may be to regulate and promote improvements—what also seems necessary is a more open stance of seeking to explore others and ourselves in new combinations, in new modes of communal living.

Indeed, circuses and modes of entertainment, or seriously improved versions of these institutions that would provide for the basic needs and more elaborate interests of these individuals, could play a role in generating these opportunities for intra- or interspecies encounters conducive to mutual learning. A less predefined and pre-given mode of living with other species has been proposed by Haraway. In her words: "they made an ethical semiotic material claim on us rooted in their capacity for response—response-ability—in what I sometimes call "the open," where what is to come is not yet—is not fixed by teleology or function, whether malignant or benign—and might still be otherwise" (Haraway 2010). Training animals is then not just a matter of providing them with appropriate exercise, but also holds the potential of self-transformation of the trainer and the emerging of new interspecies relationships. And thereby new experiences and new forms of both human and animal subjectivity.

Keulartz in his chapter draws on both Nussbaum and Haraway in his defence of animal training. But Haraway's mutually transformative take on this seems in tension with Nussbaum's much more static approach to animals and their relations to us. With Haraway, the point is to create spaces for interaction that are not predefined and predesigned. Not to assume a pre-existing set of capabilities to flourish in a single way, but to see these as emergent properties of interactions. This could be more conducive to a notion of community that is more than providing for individual members in a zero sum game, but is productive of new modes of sociability and subjectivity.

5 Relational Welfare and Rights

Brando in her chapter further discusses the potential of enrichment and performing to improve the welfare of wild captive animals. She sets out with formulating a proviso to animal training practices: of (human) entertainment[6] to be a matter of 'asking' "animals to voluntarily participate and showcase and highlight their physical and cognitive capabilities." Thus at the start of her chapter, animal well-being seems again a matter of a one-way 'showing off' of their talents. She emphasizes how training and performance should be respectful to the individual and the species.

In a (conspecific) relational view, the question is how this training may impact their everyday behavior, their biosemiotic abilities in relating to conspecifics, their group structure and shared living experience. Then it is not just a matter of perhaps rendering sea lions 'happy' to learn new tricks, but their possibly becoming too 'unnatural' in their behaviours to still relate to each other. I am no expert on sea lion communal life, but would emphasize the impacts of (certain types of) training and performing also for their implications for 'animal cultures', especially as perhaps 1 day, they could bring their new tricks to the wild.

It's not just a bear's dignity (in the eyes of humans) that is disrespected or harmed when he is made to ride a bicycle in a tutu—but also his potential to be taken seriously by other bears—and perhaps his sense of self? But this does not rule out animals developing meaningful novel behaviours after being involved in human (material and behavioural) cultures. With Gruen, Brando agrees that if we only stick to 'natural behaviour' as meaningful to animals, that "assumes a static, essentialistic or naturalistic view of species-specific characteristics." For Brando it is better when the animal is not just expected to display innate or species generic behaviour. Animal life then becomes a matter of learning, as a dynamic and flexible form of responding to their own experience of their environment.

[6]One could wonder why we do not define the purpose of zoos as entertainment of their animal inhabitants rather than human visitors.

In Brando's view, taking the perspectives of animals involves more than health, capabilities (and perhaps dignity), but also social relations with conspecifics and sometimes, potentially, some humans. Wild animal dignity is proposed as a relational concept: 'wild dignity' is seen "within their own social networks as evaluated through ours". In Brando's account of animal welfare, their 'social life' is an integral part—thus exposing how the differences between a welfare, flourishing or relational perspective are not so absolute; More a matter of focus, attention and sensibility, than mutually exclusive principled moral commitments, or competing views of the good life for animals.

Brando mentions various situations in which sociability and relationality of animals is already a theme. In the farm animal welfare oriented 'welfare quality' project, 'good social behaviour' as well as 'good human-animal interactions' are listed besides more bodily aspects of welfare. In her discussion of best practices, e.g. the elephant enclosure in Chester Zoo (UK), their living in a family group, resting and sleeping in a social setting and being outfitted with water cannons to play with, is described as part and parcel of other aspects of their quality of life. Brando's example of the social life of sea lions gives an impression of the complexities involved in facilitating animal sociability under conditions of captivity (and training). With wild behaviours occurring during different times of the year, depending on gender and involving group dynamics in different environmental contexts. Finding appropriate housing and forms of training would then best be approached as an ongoing process of interaction and mutual learning. Which at some point is then to be critically evaluated for meeting basic demands for welfare, a flourishing life in terms of development and expression of capabilities, and a rich social life which potentially could involve human caretakers.

What Brando did not go into in her chapter is some of the harsher realities of living in a zoo that may lead to a questioning of the quality of communal animal life (cf Chrulew 2011; Acampora 2010; Hribal 2011). If wild animals can only cope with life in modern urban institutions of confinement by being treated with antidepressants or sedatives, perhaps these places meet rather minimal utilitarian demands for non-suffering. But then beyond not flourishing, also lacking meaningful relations can be a good indicator of compromised welfare; or a ground for discussing the legitimacy, desirability, or justice of captivity in these situations.

6 Conclusion

In addition to suffering from a lack of freedom to move in unhampered ways, the plight of many animals also involves the disruption of their communities, and/or forced interactions with others. Without looking at most animals that now live in zoos as communal, relational beings, also the main ethical legitimations of the institution of the zoo—conservation and education—become pointless: animals not able to live in groups are just breeding stock, stored genetic material, and will never have the prospect (however remote) of being released again. And what's to learn about animals by looking at an assortment of individuals without ample opportunities to express their social capabilities? Then visitors are just looking at barely living natural history exhibits, not yet stuffed specimens stereotypically swinging their heads or plucking their feathers.

For all their commendable emancipatory intentions, individualizing modes of ethically theorizing the animal subject—as found in parts of the preceding chapters—actually reinforce reductionist understandings of animal life and render invisible the potential for human-animal sociability. Even though concerned with animal suffering, this is an individualism that confines the animal in an ideal of uninhibited solipsistic loneliness.

Animals living in the age of humans thereby also live in (neo-)liberal times, where decisions over what constitutes a good or acceptable life are taken in terms of individual preference satisfaction—as if they are alienated consumers like us. The isolated individualism of various strands of animal ethics thereby seems less a solution to than a symptom of our modern disconnect to living with others, or living in more than human communities. Imagined—or imaginary—communities for sure, like all social and cultural phenomena, but no less real in their impacts on who we are or can become.

Read in this way, these chapters induce the appreciation and validation of a set of experiences that could in more intricate and attuned ways inform processes of rethinking and redesigning the kinds of animal lives we are willing to cater for, or to allow to emerge; and to literally and symbolically relay our roads or slow down our traffic for. Otherwise animal confinement practices and animal ethics debates both contribute to a sadly exclusive human culture in which animals only figure as generic decoration, never as active participants who co-shape a world of meaning, imagination and community.

References

Acampora, R. (ed.). 2010. *Metamorphoses of the zoo: Animal encounter after Noah*. Lanham, MD: Lexington Books.

Berger, J. 2009. *Why look at animals?*. London: Penguin.

Biermann, C., and B. Mansfield. 2014. Biodiversity, purity, and death: Conservation biology as biopolitics. *Environment and Planning D: Society and Space* 32(2): 257–273.

Braverman, I. 2013. *Zooland: The institution of captivity*. Stanford, CA: Stanford University Press.

Burt, J. 2005. John Berger's "Why look at animals?": A close reading. *Worldviews* 9(2): 203–218.

Callicott, J.B. 1980. Animal Liberation: A triangular affair. *Environmental Ethics* 4: 311–338.

Chrulew, M. 2011. Managing love and death at the zoo: The biopolitics of endangered species preservation. *Australian Humanities Review* 50: 137–157.

Donaldson, S., and W. Kymlicka. 2011. *Zoopolis*. Oxford: Oxford University Press. doi:10.1007/978-3-319-44206-8_17

Driessen, C. 2012. Farmers engaged in deliberative practices; An ethnographic exploration of the mosaic of concerns in livestock agriculture. *Journal of Agricultural and Environmental Ethics* 25(2): 163–179.

Driessen, C. 2014. Animal deliberation. In *Animal politics and political animals*, ed. M. Wissenburg, and D. Schlosberg, 90–104. London: Palgrave MacMillan.
Haraway, D. 2010. Staying with the trouble: Xenoecologies of home for companions in the contact zones. In *David Schneider Memorial Lecture at Meetings of the Society for Cultural Anthropology, Santa Fe, New Mexico, May 7.*
Haraway, D. 2015. Anthropocene, Capitalocene, Plantationocene, Chthulucene: Making Kin. *Environmental Humanities* 6: 159–165.
Hribal, J. 2011. *Fear of the animal planet: The hidden history of animal resistance*. Oakland, CA: AK Press.
Jones, O. 2003. The restraint of beasts: Rurality, animality, actor network theory and dwelling. In *Country Visions*, ed. P. Cloke, 283–307. London: Pearson Education.
Nimmo, R. 2012. Animal cultures, subjectivity, and knowledge: Symmetrical reflections beyond the great divide. *Society & Animals* 20(2): 173–192.
Nussbaum, M. 2004. Beyond compassion and humanity: Justice for nonhuman animals. In *Animal Rights: Current Debates and New Directions*, ed. Cass R. Sunstein and Martha Nussbaum, 299–320. Oxford: Oxford University Press.
Nussbaum, M. 2006. *Frontiers of Justice*. Harvard: Belknap Press.
Russell, B. 1959. *My philosophical development*. London: Allen and Unwin.
Srinivasan, K. 2014. Caring for the collective: Biopower and agential subjectification in wildlife conservation. *Environment and Planning D: Society and Space* 32(3): 501–517.
Wolch, J., and J. Emel (eds.). 1998. *Animal geographies: Place, politics and identity in the nature —culture borderlands*. London: Verso.

Part IV
Between Animal Ethics and Conservation Ethics

Captivity for Conservation? Zoos at a Crossroads

Jozef Keulartz

Abstract This chapter illuminates a variety of issues that speak to the question of whether 'captivity for conservation' can be an ethically acceptable goal of the modern zoo. Reflecting on both theoretical disagreements (animal protectionists versus wildlife conservationists) and practical challenges (the small percentage of endangered species actually exhibited in zoos, disappointing success of reintroduction programs), the paper explains why the 'Noah's Ark' paradigm is being replaced by an alternative 'integrated approach.' It explores the changes in the zoo's core tasks that the new paradigm implies. And it pays special attention to the changes that would have to be made in zoos' collection policies: connection with in situ projects, emphasizing local species and the local biogeographical region, exchange of animals among zoos and between zoos and wildlife, and a shift towards smaller species. Finally the question will be addressed whether the new paradigm will achieve a morally acceptable balance between animal welfare costs and species conservation benefits.

1 Introduction

Today, the animal world is under severe attack as a result of two strongly interconnected global processes. On the one hand, global environmental changes such as climate change, land use and land cover change, deforestation and desertification have a disruptive impact on plant and animal life. Entire populations are being confronted with the alternative to abandon their original habitat or to go extinct. On the other hand, globalization causes massive dislocations of entire populations. As trade, travel, transport and tourism boom, the world is becoming more and more

---

This is a reprint from the *Journal of Agricultural and Environmental Ethics* 28 (2015), 335–351.

J. Keulartz (✉)
Radboud University Nijmegen, Nijmegen, The Netherlands
e-mail: J.Keulartz@science.ru.nl

B. Bovenkerk and J. Keulartz (eds.), *Animal Ethics in the Age of Humans*,
The International Library of Environmental, Agricultural and Food Ethics 23,
DOI 10.1007/978-3-319-44206-8_20

borderless and, by the same token, it is becoming increasingly vulnerable to invasive species. Since globalization took off, more plants and animals have become globetrotters than ever before (Keulartz and Swart 2012).

Because animals are constantly on the move worldwide as a result of these global processes, traditional in situ (place-based) conservation methods seem no longer sufficient to save threatened species (Sandler 2012). The magnitude of anthropogenic environmental stress from bioinvasion, habitat fragmentation, nitrogen deposition, biodiversity loss, and, above all, climate change, makes it unavoidable to replace the hands-off approach that has guided mainstream species conservation until recently by a more proactive and interventionist strategy.

However, this new strategy has led to manifold conflicts between wildlife conservationists and animal protectionists (Minteer and Collins 2013). As Michael Soulé has remarked in his presidential address at the third annual meeting of the Society for Conservation Biology in 1989: "Conflicts between animal rights groups and management agencies are increasing in frequency and cost—the cost being borne by endangered species and ecosystems as well as by the public that pays for expensive rescue operations and time-consuming court battles" (Soulé 1990, p. 235).

In his address, Soulé claimed that among the many environmental challenges of the coming decades, "the onslaught of alien species" would be the most revolutionary. And he foresaw that attempts by conservationists to control destructive exotics would meet resistance from "well-meaning animal welfare enthusiasts", who oppose eradication programs that involve techniques such as hunting and trapping or make use of pesticides such as piscicides, chemical substances which are poisonous to fish (Keulartz and Van der Weele 2008).[1]

Another potential area of conflict between wildlife conservationists and animal activists concerns managed relocation (also known as assisted colonization or assisted migration). The human-aided relocation of threatened species may be required when their historical ranges have become inhospitable due to climate change or habitat fragmentation and destruction, and when moving on their own to other regions where environmental conditions are more suitable is impossible. Relocated animals will inevitable experience chronic stress at all stages of the process, from capture and captivity to transport and release to novel areas. Such relocation-induced chronic stress increases the overall vulnerability of the individuals and, as a result, decreases the probability that the population will become self-sustaining (Dickens et al. 2010).

Here, I will focus on yet another major battlefield between wildlife conservationists and animal activists: ex situ conservation through zoos and aquaria. As a response to the ongoing decline in effectiveness of in situ conservation and the

---

[1]A famous example of the clash over programs to eradicate invasive species is the controversy about the feral pigs in Hawaii, between the Nature Conservancy and People for the Ethical Treatment of Animals (PETA). While conservation biologists have argued that the pigs should be killed and removed because they threaten Hawaii's biodiversity, animal activists argued that it is wrong to harm and kill the pigs because they are sentient animals (Woods and Moriarty 2001).

accompanying loss of biodiversity, zoos began to turn their attention to the conservation of endangered species and wildlife in the 1970s and 1980s. *Captivity for Conservation* became a crucial slogan for the modern zoo. A major milestone in this development was the Convention on Biodiversity which was signed at the Earth Summit in Rio de Janeiro in 1992. In the wake of the Earth Summit the first *World Zoo Conservation Strategy* was launched in 1993. Its conclusion explicitly stated that, at a time when species, habitats and ecosystems worldwide are threatened with extinction, modern zoos must commit to the conservation of species and wildlife.

> Caring for our planet's biological systems is one of the greatest challenges to humankind. Consequently, conservation is being seen as the central theme of zoos, and zoos should thus further evolve into conservation centers (WAZA 1993, p. 3).

In this scheme of things the zoo was envisaged as a kind of Noah's Ark which owed its *raison d'être* primarily to its contribution to the conservation of species through breeding and reintroduction programs. As the main institution for ex situ conservation of wild animal species, the zoo was now confronted head-on with the potential conflicts between animal protectionists and wildlife conservationists.

In this essay, I will analyze the moral issues at stake in these conflicts over the zoo with an eye to possibilities to bridge the differences between the conflicting parties. I will argue that both sides of the controversy may find common ground in the view that zoos will be morally justifiable only if the costs in terms of animal welfare and freedom are clearly outweighed by the benefits to species preservation. Next, I will argue that the Noah's Ark paradigm does not meet this standard, that it therefore has lost credibility and has gradually given way to a new paradigm; the 'integrated approach' in which the zoo is primarily seen as a conservation park or center. I will then set out the implications of the new approach for the zoo's core tasks and explore the collection policy options that are open to zoos in order to carry out these tasks in the best possible way. Finally, I will address the question whether the new paradigm may achieve a morally acceptable balance between animal welfare costs and species conservation benefits. But in order to understand what is at stake in the battle over the zoo's legitimacy and right to existence I will first take a closer look at the heated philosophical debate between animal ethicists and environmental ethicists.

2 The Animal Ethics/Environmental Ethics Debate

The philosophical debate between animal ethicists and environmental ethicist erupted around 1980 (Hargrove 1992). Before that time it was expected that an adequate environmental ethics would develop as a natural extension of animal ethics. Both Peter Singer's animal liberation theory and Tom Regan's animal rights theory denounced traditional morality for its 'human chauvinism' and its 'speciesism'. The time seemed ripe for a moral rehabilitation of the rest of animate nature, and animal ethicists and environmental ethicists were supposed to join forces in fighting for wildlife preservation. But, as Sagoff (1984) has remarked somewhat sarcastically, this was a “bad marriage”, followed by a “quick divorce”.

In 1980, Baird Callicott published a highly polemical article to counter the widespread view that the existing animal ethics of Singer and Regan were fully capable of answering all environmental ethical questions. In particular, Callicott opposed their view that only individual animals have intrinsic value and direct moral standing, not collective entities such as species or ecosystems. While Singer believes that we have no duties to species because as such they “are not conscious entities and so do not have interests above and beyond the interests of the individual animals that are members of the species” (Singer 1979, p. 203), Regan holds that “the rights view does not recognize the moral rights of species to anything, including survival” (Regan 1983, p. 359).

According to Callicott, animal ethicists demonstrate their ‘ecological illiteracy’ by pleading for a responsibility to maintain individual animals irrespective of whether these are tame or wild, rare or common, indigenous or exotic. In line with Aldo Leopold’s ‘land ethic’, he advocated a holistic approach in which individual organisms are perceived as parts of the biotic community. Such an approach does not accord equal moral worth to each and every member of the biotic community; “the moral worth of individuals is relative, to be assessed in accordance with the particular relation of each to the collective entity which Leopold called ‘land’” (Callicott 1980, p. 327).

Holmes Rolston, one of the founding fathers of environmental ethics, based his holistic view on a neo-Darwinian account of the evolutionary history of life on earth. He writes:

> In an evolutionary ecosystem it is not mere individuality that counts; species is also significant because it is a dynamic life form maintained over time by an informed genetic flow. The individual represents (re-presents) a species in each new generation. It is a token of a type, and the type is more important than the token (Rolston 1988, p. 143).

It follows that “the individual is subordinate to the species, not the other way around” (idem. 149).

Tom Regan responded to these attacks against the animal ethicists’ basic principle with the accusation that environmental ethicists were committing the crime of ‘environmental fascism’ by subordinating the rights of individuals to the interests of the greater whole. “Environmental fascism and the rights view are like oil and water, they don’t mix” (Regan 1983, p. 362).

Animal ethicists and environmental ethicists usually not only differ with regard to the locus of moral concern—individual organisms or greater wholes—, they also tend to use different species concepts (Sandler 2012, p. 4). Animal ethicists have generally adopted a *conventionalist* species concept; they see a species merely as a category or class with arbitrarily drawn boundary lines that may serve as a convenient mapping device for theoretical purposes only.[2] Environmental ethicists, on the other hand, generally hold a *realistic* species concept. Rolston, for instance, argues that a species is a real historical entity, a "dynamic historical lineage" that can persist as a discrete, vital pattern over time (Rolston 1988, p. 151).[3]

While it is evident that conflicts between individualistic animal-centered and holistic species-centered ethicists and activists will increase as conservation strategies will inevitable become more and more interventionist and hands-on, the gap between these divergent views simply seems too wide to bridge, even in cases where advocates of both sides have common cause (Minteer and Collins 2013, p. 43). Such seemingly intractable controversies are often not resolvable by recourse to facts and unlikely to be settled by compromise. They require what Donald Schön and Martin Rein, inspired by John Dewey's idea of 'reconstructive thinking', have called "frame restructuring", i.e. the attempt to integrate conflicting frames.

As a necessary first step towards such an integration, both sides of the controversy regarding the moral acceptability of zoos have to develop a 'double vision', namely "the ability to act from a frame while cultivating awareness of alternative frames" (Schön and Rein 1994, p. 207). They should learn to 'squint' so to speak in order to see things from both angels simultaneously. Only then will it be possible to find a morally defensible balance between animal welfare concerns on the one hand and species conservation commitments on the other.

3 Balancing Animal Welfare/Rights Against Species Conservation

Most animal rights proponents will resist any attempts at such value balancing. They consider infringing an individual's right to freedom for the sake of the preservation of the species as morally wrong. For Regan any type of captivity or manipulation of a sentient animal is morally unacceptable, irrespective of the possibly beneficial consequences for the protection of rare or endangered species. The rights view's answer to the question whether zoos are morally defensible, "not surprisingly, is No, they are not" (Regan 1995, p. 46).

Utilitarians like Singer, on the other hand, do allow for value balancing and accept reductions in animal welfare when the survival of entire populations or species is at stake. A case in point is Singer's approach to the problem of the South American cane toad that was introduced to Australia as a method of agricultural pest control, but became itself a serious pest for native wildlife. Whereas animal

[2]According to Dale Jamieson "the notion of a species is an abstraction; the idea of its welfare is a human construction. While there is something that it is like to be an animal there is nothing that it is like to be a species" (Jamieson 1995, p. 61).

[3]For a recent contribution to the animal ethics/environmental ethics debate, see McShane (2014).

rights groups fiercely opposed the eradication of cane toads because they believe that killing an animal, unless to end suffering, is always bad, Singer argued that "where killing can be done humanely, and is necessary to preserve endangered species, it may be defensible."[4]

When Singer was recently asked whether zoos should exist for the sake of species preservation, he answered as follows, "I think if a species is likely to become extinct in the wild and you can capture the animals humanely and recreate the physical and behavioral conditions, then could release them or their progeny in the wild, then that function of zoos is defensible."[5] Singer feels, however, that most zoos today fail to live up to their conservation mission. They tend to confine animals for our amusement in ways that are contrary to their interests. Even if these zoos do occasionally preserve an endangered species, "what is the point of preserving animals if they are having miserable lives?"[6]

Dale Jamieson, another animal ethicist working in the utilitarian tradition,[7] is even more skeptical about zoos than Peter Singer. As starting point for his assessment of the moral defensibility of the zoo, Jamieson claims that keeping animals in captivity is wrong, unless a case can be made that the benefits outweigh the moral presumption against depriving animals of liberty. In his classic essay 'Against Zoos' from 1985, Jamieson concludes that the supposed benefits of zoos—amusement, education, scientific research and species preservation—are outweighed by the moral presumption against keeping animals. Ten years latter, he repeated his analysis with a strong focus on the modern zoo's most compelling reason for its existence, its contribution to species conservation. And again Jamieson's final judgment proved unfavorable for the zoo; he considered the benefit of preservation "not significant enough to overcome the presumption against depriving an animal of its liberty" (Jamieson 1995, p. 60).

4 Noah's Ark in Rough Water

Jamieson's verdict was, to a considerable extent, shared by an increasing number of critiques within the zoo-community. The vision of the zoo as a Noah's Ark started to shipwreck as the breeding programs ran into some substantial problems. Many of the animals exhibited in zoos do not belong to groups designated for conservation. Because the space for all the zoo animals in the world could easily fit within New York's 212.7 km^2 borough of Brooklyn (Conway 2011, p. 4), zoos can only maintain a limited number of endangered species. This number will even decrease

[4]http://www.abc.net.au/environment/articles/2011/01/17/3113142.htm.

[5]https://www.thedodo.com/peter-singer-on-the-animal-rig-726248280.html.

[6]http://www.mkhumanists.org.uk/node/73.

[7]Jamieson calls his brand of utilitarianism 'progressive consequentialism' (Jamieson and Elliot 2009).

as exhibits increase in size to meet animal welfare, while zoos are usually also very reluctant to give up popular animals that are not threatened with extinction. But even if zoos were to dedicate half their space to breeding programs for the conservation of endangered species, they would still—according to the most optimistic estimates—be unable to accommodate more than around 800 of the 7.368 species of vertebrates that are threatened with extinction according to the 2013 IUCN Red List.[8] Due to lack of space zoos are increasingly being pressed to separate grain from chaff and devote more resources to a chosen few. As Leslie Kaufman has strikingly remarked, breeding endangered animals "feels less like Noah building an ark and more like Schindler making a list" (Kaufman 2012a).

Research has shown that zoos currently hold roughly one in seven (15 %) threatened species of terrestrial vertebrates (Conde et al. 2011).[9] Moreover, zoos even struggle to breed these few species because the populations are usually too small. As Sarah Long, director of the Population Management Center in Chicago, has remarked, "Noah got it all wrong. One or two or even a dozen of each species is not enough" (Kaufman 2012b). Initially, the target of zoo breeding programs was to maintain 90 % of genetic variability of a species for a period of 200 years (Soulé et al. 1986). Because this time frame requires very large numbers of animals per species, it has been reduced from 200 to 100 years in the mid-1990s. But the majority of breeding programs do not have sufficient space to meet even this objective.

However, not only are the success rates of breeding programs disappointing, the prospects of reintroduction programs are also low, largely because ecological, social, economic and political aspects were not taken into consideration. Reintroduction is a costly business which often diverts attention from other, more cost-effective options. In captivity, animals risk losing the skills they need to survive in the wild. Lastly, the ecosystems into which they are eventually released are dynamic systems which have often undergone dramatic changes in the time span between the breeding program and the reintroduction, sometimes as a result of anthropogenic disturbances such as CO_2 emissions and deforestation. A review by Beck (1995) estimated that only 16 out of 145 reintroduction projects using captive-born animals were successful. It also showed that most animals for reintroduction do not come from zoos but from other specialized facilities. Although the situation improved after the development of the *Guidelines for the Placement of Confiscated Animals* (IUCN 2000), the performance of zoos regarding the reintroduction of captive-bred animals still fell far short of expectations. Instead of a guiding light, reintroduction proved to be a shooting star, "providing an

[8]The list cites 1,140 species of mammals, 1,313 species of birds, 847 species of reptiles, 1,948 species of amphibians, and 2,110 species of fish.

[9]Invertebrates are almost absent in zoos (see the section on Collection policy options).

eye-catching attraction but not long-term illumination for conservation (Stanley Price and Fa 2007, p. 173).[10]

By the turn of the century, Noah's Ark seemed to have become irretrievably shipwrecked as the zoo community started to realize the limitations of ex situ (zoo-based) conservation as a prelude to in situ (field-based) conservation (Lees and Wilcken 2009). The vision of the zoo as a Noah's Ark has gradually given way to a new paradigm, the 'integrated approach'. This transition becomes apparent when the first *World Zoo Conservation Strategy* of 1993 is compared to the new *World Zoo and Aquarium Conservation Strategy* of 2005. The first document explicitly describes reintroduction as the ultimate goal of ex situ conservation. The second document, on the other hand, recognizes the reintroduction of captive-bred animals as a useful instrument for the conservation of wildlife, but cautions against high expectations because of the complexities of returning these animals to the wild (Stanley Price and Fa 2007, p. 156). It outlines a much broader conservationist role for zoos, including research, training, education, awareness campaigns and direct support for in situ projects. In the latest strategy, the primary mission of zoos is to integrate all these elements with their efforts to protect endangered species and conserve healthy ecosystems (Mace et al. 2007; Bowkett 2008; Lees and Wilcken 2009). Insofar as captive-breeding for reintroduction is considered necessary and appropriate, it should be accomplished as part of such a larger, integrated, holistic program (Hutchins 2003, p. 18).

This paradigm shift in the zoo's mission raises the question whether Jamieson's verdict will remain valid that the supposed benefits of the zoo, in particular its contribution to species conservation, are outweighed by the moral presumption against keeping animals. Will the integrated approach make it possible to strike a balance between animal welfare and rights on the one hand and species conservation on the other that leads to a morally more favorable evaluation of the zoo? To answer this question, I will proceed in two steps: first I will spell out the implications of the new paradigm for the zoo's core tasks; and next I will explore the collection policy options that are open to zoos in order to perform these tasks optimally.

---

[10]Furthermore, many of the successful reintroductions have their own history of struggle and strife. Take, for example, the Black-footed ferret (McCarthy 2004, p. 196/7). In 1986, the Black-footed ferret population had diminished to a mere 18 individuals, but thanks to a captive breeding program, more than 220 now roam the prairie of the US state of Wyoming. The program was not, however, entirely plain sailing. When the kits were released they were far too blasé to make themselves scarce when predators such as eagles, coyotes and badgers arrived on the scene. The researchers tried to resolve this problem by building a mock predator. They attached wheels to a stuffed badger, which would win fame as RoboBadger. The only way the ferrets could escape RoboBadger was to find a burrow. The researchers then tried to increase the ferrets' aversion to RoboBadger by firing rubber bands at them.

5 Implications of the Integrated Approach for the Zoo's Core Tasks

In this section, I will successively discuss the impact of the integrated approach on education, awareness and advocacy; financial support and fund-raising; research and training; and population management.

5.1 *Education, Awareness and Advocacy*

Education in the context of the integrated approach must be fully geared to the conservation of species and wildlife. With over 700 million visitors a year, zoos have a unique opportunity to provide for such an education to large numbers of people (Gusset and Dick 2011). Research on the impact of education on zoo visitors is still in its infancy (Davey 2006; Falk et al. 2007; Marino et al. 2010), but it seems that two courses of action are absolutely essential to achieve the desired effect.

First of all, the animals should be shown in an environment which resembles their natural environment as closely as possible—where they get the best chance to develop behavior typical of their own species. This will help to prevent stereotypical behavior, which is a poor advertisement for a zoo and merely undermines the educational message. The much-desired 'naturalization' of zoos is well underway; nowadays, when zoos are designed, habitat takes precedence over taxonomic group. But the process runs up against limits. True habitats cover huge expanses of territory (look at the area covered by tigers). Moreover, it is virtually impossible to create realistic simulations of some forms of predatory behavior, such as chasing and killing prey, in captivity. Likewise, in the absence of predators, prey behavior, such as vigilance, may not be adequately exhibited in captivity (Kreger et al. 1998). When Seattle's Woodland Park Zoo started feeding whole sheep and goats to the big felids and whole rabbits and chickens to the small felids in the 1970s, many members of the public were aghast at the sight of flesh being torn from recognizable carcasses (Hancocks 2001, p. 249). They are used to it now, but feeding live animals to the cats is probably a step too far.

Secondly, the visitor must not be overcome by a numbing sense of helplessness. Awareness alone will not change behavior and it might even prove counter-productive if visitors are not afforded an opportunity to act (Sterling et al. 2007; Gwynne 2007). In his 1999 book *Beyond Ecophobia*, David Sobel contends that we must allow people to connect with nature and love the Earth before we ask them to save it. Most people know by now that the natural world is in trouble. When they hear that yet another ten thousand acres of rainforest were being lost while they were spending time at the zoo, they may distance themselves from, rather than connect to, the natural world which is so painful, unless they get a chance to make a difference—no matter how small.

Congo Gorilla Forest, a project of 2.7 ha in Bronx Zoo dedicated to wild animals and habitats from central Africa, offers such a chance. It provides visitors with an opportunity to make a direct contribution to the conservation of the African rainforest by allowing them to name the project that will benefit from the three-dollar supplement on their admission fee. The initiative raises one million dollars a year for fieldwork (Gwynne 2007). Another good example is the *They're Calling on You* campaign at Melbourne Zoo. Visitors to the gorilla enclosure are asked to donate their old mobile phones, which are then sent off for recycling. The idea is to save coltan, an ore that is mined at the expense of the gorillas' habitat, and to generate funding for their conservation.[11]

5.2 Financial Support and Fundraising

This is the simplest way for zoos to assist in situ conservation. Recent research has shown that zoos do indeed make a financial contribution to this cause. But investment in conservation by zoos is generally still low. It has been suggested that zoos devote at least 10 % of their income to in situ conservation (Tribe and Booth 2003). Available data point to less than 5 % of income being spent on conservation (Fa et al. 2011). This finding could reinforce the view that the mission of zoos is primarily 'window-dressing'.

Various fundraising ideas have recently been circulating. For instance, visitors could be asked to contribute 'on-the-spot' to a project of their choice (as in the case of the *Congo Gorilla Forest* project); 'conservation-contribution' machines could be installed so that visitors can donate cash to conserve certain species; groups of schoolchildren could be asked to adopt projects. In Australia, valued donors are taken on guided tours behind the scenes. Some feel that the time has come to set aside a percentage of the admission fee for in situ conservation projects (Conway 2003, p. 12).

5.3 Research and Training

The basic principle of research and training in the context of conservation efforts is: "Export expertise rather than repatriate animals" (Stanley Price and Fa 2007, p. 169). The tools and technologies developed by zoos are becoming increasingly relevant for in situ conservation as natural habitats continue to being damaged and destroyed at the current pace. Habitat loss and habitat fragmentation lead to an

[11]http://www.zoo.org.au/get-involved/act-for-wildlife/theyre-calling-on-you.

ongoing conversion of what were originally continuous populations to so-called 'metapopulations'. Metapopulations are collections of subpopulations, that are spread geographically over patches of habitat. Because these patches are usually small and because the movement of the animals between these patches is restricted for lack of connectivity, local extinction of subpopulations is a common event. This situation asks for metapopulation management, by which one may try to artificially perform the function of dispersal and recolonization of patches of locally endangered or extinct species.

With metapopulation management the distinction between classic in situ and ex situ conservation is gradually breaking down. As the size and genetic diversity of remaining wildlife populations is progressively declining, these populations are becoming more and more similar to ex situ populations. Accordingly, zoo-based expertise in genetic management of small populations of captive animals may be useful to the conservation of small and declining populations in the wild, whereas zoo-based skills in animal handling may be helpful in wild-to-wild translocations of animals from one site to another, to repopulate habitats where species have extirpated (Hutchins 2003; Hatchwell et al. 2007).

5.4 *Population Management*

Apart from research on management of small populations and wild-to-wild translocation, zoos can also contribute to metapopulation management through what has been called "integrated species conservation planning" (IUCN 2014). The new approach to conservation is increasingly replacing captive-breeding for reintroduction that has fallen out of favor due to the growing recognition of its limits (Stanley Price and Fa 2007, p. 173). It refers to the exchange of animals between in situ populations (in the wild) and ex situ populations (in human care) and has also been termed a 'hybrid' (Redford et al. 2012) or 'pan situ' approach (Minteer and Collins 2013). On the one hand, captive populations can be used for restocking in areas with declining populations or for reintroduction in areas where populations have gone extinct; on the other hand, the demographic and genetic viability of ex situ populations can be boosted by supplying genetic founders from wildlife populations (Byers et al. 2013).

6 Collection Policy Options

In the previous section, I have shown how the paradigm shift in the zoo's mission will affect its core tasks. However, in order to fulfill these tasks in an optimal way some substantial changes have to be made with respect to the zoo's collection policy.

In this section, I will discuss some of the most important options that have been proposed and that have often already been put into practice in one combination or another: creating a link between the collection and in situ conservation projects; putting more emphasis on local species and the local biogeographical region; exchanging animals among ex situ populations and between ex situ and in situ populations; and replacing big charismatic mammals with smaller species.

6.1 Link Between Collection and In Situ Conservation Projects

The species in the collection should match the zoo's conservation goals. This can be achieved by the creation of explicit connections between the animals on display and the in situ projects that are being supported, so that visitors can learn about the living conditions of the exhibited animals in the wild. This argument returns in WAZA's global strategy *Building a Future for Wildlife*, which also presents a good example of a link between ex situ and in situ conservation:

> *Pongoland* in Leipzig Zoo has created a link between the ex situ conservation and breeding of chimpanzees and the attempts at in situ conservation of the Wild Chimpanzee Foundation (WCF) in Ivory Coast. The zoo is acting as guarantor for the long-term funding of projects in Tai National Park. Projects have been set up specifically to raise local awareness of the plight of chimpanzees, currently an endangered species. Visitors to Leipzig Zoo are told about the collaborations with the WCF; meantime, the local population is told about the activities at Leipzig Zoo and the research in Pongoland for the conservation of the chimpanzee (WAZA 2005, p. 10).

Another example is the spectacular 11,000 m^2 rainforest at Zurich Zoo. This exhibition was developed in cooperation with Masoala National Park in Madagascar. Over the years, a whole string of direct and indirect links were forged between the Swiss zoo and the Madagascan park. Zurich Zoo provides the funding for small-scale development projects in communities around the park. These projects have proven highly successful in winning the support of the villagers and the local government for the park. Two nurseries have been established nearby to help the local community and to supply the zoo with seeds. Support was provided for field research in Masoala. The park was promoted in Madagascar and in Europe as a place of international importance for the conservation of biodiversity (Hatchwell and Rübel 2007).

These kinds of alliances between zoos in developed countries and protected areas in developing countries are in everyone's interests: on the one hand, they help zoos to strengthen the impact of their activities on in situ wildlife conservation; on the other, they secure long-term funding for protected areas.

6.2 Emphasis on Local Species and the Local Biogeographical Region

The link between the collection and in situ conservation projects is easier to make when more emphasis is put on local species and the local biogeographical region. A shift in breeding programs for reintroduction from exotic to indigenous species is entirely in keeping with Article 9 of the Convention on Biological Diversity. This article states that ex situ conservation should take place preferably in the country of origin of the biological component. Likewise, the *International Union for the Conservation of Nature* (IUCN) has recommended that regional zoo associations work with threatened species in their own biogeographical area (IUCN 2002; Dickie et al. 2007).

A stronger emphasis on local species and regional problems closer to home is also important from an educational perspective, given that education should preferably address problems with direct relevance for the target group. Education can encourage local involvement and action. "If the Giant panda is going to be saved, the most important audience for educational initiatives is undoubtedly in China" (Hutchins 2003, p. 23).

6.3 Exchange of Animals Among Zoos and Between Zoos and Wildlife

Because of limits of space, zoos can only maintain a small fraction of currently threatened species. To address this problem, zoos have a number of options. They can reduce the number of species they maintain that are not threatened and specialize in species that are. "Specialization is key to every successful threatened species propagation program" (Conway 2011, p. 5). In addition to specialization zoos can increase and improve regional and global cooperation. Only very few captive populations managed in isolation are self-sustaining since population sizes generally need to be very large to retain 90 % of the genetic diversity over 100 years. Populations with less than 50 individuals "might have a high likelihood to be managed inadvertently or implicitly for extinction" (Fa et al. 2011, p. 271). The problem of low numbers can be addressed by collaborative management and the exchange of animals among zoos.

To combat the problem of numbers the interactive exchange of animals between captive and wild populations in the context metapopulation management is also a very effective strategy. The integration of in situ and ex situ programs opens the possibility to simultaneously improve the demographic stability and genetic diversity of the wild and captive populations of endangered species.

6.4 A Shift Towards Smaller Species

But the most effective strategy to combat the problem of limited space is without any doubt a shift away from the large charismatic mammals towards smaller species, particularly amphibians, invertebrates and some species of fish, which occupy less space, are relatively inexpensive to keep, have a high birth rate and are easy to reintroduce.[12] Several initiatives have already been launched on this front, not least the Amphibian Conservation Action Plan, a partnership involving the World Association of Zoos and Aquariums (Gewin 2008).

The ever-present collections of charismatic megafauna—lions, tigers, giraffes, elephants, zebras, bears, hippos, and rhinos—are a poor reflection of the rich diversity of the animal kingdom. There are around 30 million animal species on this planet, 1,640 of which are mammals. The average American zoo collection features 53 well-known mammals, a ratio of 1:31. The ratio for birds, at 1:98, is less than a third of this. The ratio for reptiles, at 1:104, is less still. The disproportion becomes even more alarming when it comes to very small creatures. Amphibians in the average US zoo are represented in a ratio of only 1:2,000. And the ratio for invertebrates is an incredible one to several million. Over 95 % of all fauna are small enough to hold in the palm of your hand, but in zoos, they are conspicuous by their absence (Hancocks 2001, p. 165).[13]

Some fear that turning the spotlight on small species will weaken the attraction of zoos. Zoos need to balance conservation credibility with commercial viability; to reach the aim of species conservation they need to attract visitors. The focus on charismatic mammals is considered to be appropriate because these animals are supposed to act as flagship species or ambassadors that raise public awareness and support for in situ conservation (Baker 2007, p. 147; Leader-Williams et al. 2007, p. 237). However, the assumption that zoos will not attract enough visitors without large mega-vertebrates is far from uncontroversial. Recent findings even suggest that small mammal displays yield a higher cost to benefit ratio, in terms of exhibit popularity per unit cost, than large mammal displays. They also suggest that imaginative displays of small-bodied species can substantially increase zoo attendance (Fa et al. 2011, p. 79).

A case in point is *Micropia*, the first museum of micro-organisms such as moulds, yeasts, (micro)algae, bacteria, archaea, and viruses. *Micropia*, located in

---

[12]A recent and also very promising strategy to tackle the problem of limited space concerns the creation of walkways between enclosures that allow animals greater freedom of movement. Building a network of trails, in particular top tree trails, gives animals the opportunity to rotate between various interconnected display and off-display areas. Animals may spend mornings in one area and afternoons in another. This design strategy was first applied in Philadelphia Zoo, with only 42 acres a relatively small zoo. http://theconversation.com/zoos-of-the-future-break-down-the-enclosure-walls-26605

[13]Edward O. Wilson once said that it cannot be stressed enough "that, as a whole, invertebrates are more important than vertebrates for the conservation of ecosystems. If invertebrates were to die out, I fear that the human race would survive for just a few months" (Wilson 1987, p. 345).

Artis Royal Zoo in the center of Amsterdam, the Netherlands, opened its gates in October 2014. The museum uses 3D viewers, allowing visitors to see how living microbes move around, eat and reproduce. It has become a popular venue that has plenty of interactive displays on offer, including a body scanner which can show you what types of microbes live on your body, and a *Kiss-o-meter* which can count the number of microbes transferred during a kiss.

7 Towards a New Balance Between Animal Welfare/Rights and Wildlife Conservation?

Will the integrated approach, if rigorously applied, tip the balance between animal welfare and species conservation concerns in favor of the zoo? Most animal rights proponents will deny this possibility because they are opposed to such value balancing. However, this abolitionist position will lose normative force as the borderline between in situ and ex situ conservation will more and more be blurred, i.e., as zoos will increasingly become more like national parks and wildlife reserves and, vice versa, parks and reserves will take on some of the character of zoos, and hence be subject to zoo dilemmas. Like zoo populations, wild populations are increasingly becoming too small to be demographically and genetically viable and will inevitably go extinct without continuous monitoring and management. In such a situation abolitionism is tantamount to capitulation to species extinction (Minteer and Collins 2013).

On the other hand, a focus on smaller species such as reptiles, amphibians and fish might to some extent address the concerns raised by adherents of the rights view. After all, Tom Regan's rights theory does not extend to all animals, but only to those animals, notably mammals, that can be regarded as subjects-of-a-life because they have capacities for emotion, memory, belief, desire, intentional action, self-awareness, as well as conceptual abilities and a sense of the future, including their own future.

Unlike animal rights advocates, animal welfare proponents generally do allow for trade-offs between animal welfare and species conservation concerns. But most of them agree with Dale Jamieson's verdict that the zoo's contribution to species conservation is not significant enough to overcome the presumption against keeping animals in captivity. They usually also endorse Jamieson's view that we cannot save wild nature by bringing it indoors but only by setting aside large tracts of land and change our present environmentally unfriendly behavior.[14] "Should zoos breed animal populations that have no home to return to?" (Hanson 2002, p. 171), these animal ethicists and activists ask.

---

[14]Jamieson even blames zoos for being deeply implicated in causing the problem that they purport to be addressing; they undermine the case for preserving wild nature by taking more and more animals out of the wild (Jamieson 1995, p. 62).

But, on the other hand, one might ask if it does make any sense to preserve or create wild lands when there are only few populations left to inhabit these places. As David Hancocks has remarked, simply setting aside wild lands will not always be sufficient. He illustrates this with the example of the management plan to save the Javan tiger that was published in 1980 by the Indonesian government, assisted by the WWF and the IUCN. "It overruled any efforts at captive propagation, relying solely upon habitat protection. Today the Javan tiger is extinct" (Hancocks 2001, p. 175). Moreover, radically altering our present lifestyle might take too long for many endangered species to survive. All in all, an ethical position that focuses solely on the preservation of habitat will carry little normative force in a situation where in situ conservation is no longer sufficient to slow down or stop the current species extinction rate. Preserving wild lands and saving endangered species need not be exclusionary, but should be pursued together to effectively meet widespread threats such as climate change, habitat loss, poaching, invasive species and disease.

Moreover, Jamieson's unfavorable judgment of the zoo might need to be revised in light of the paradigm shift towards the integrated approach. Under this approach, the prospects for the zoo to achieve a morally acceptable balance between animal welfare costs and species conservation benefits look rather good, provided that the zoo's core tasks are all geared to wildlife conservation and the species collection clearly reflects the zoo's conservation goals. A shift towards small species, which generally experience less welfare problems in captivity and fewer behavioral problems that make return to the wild difficult than large animals, would certainly tip the scales in favor of the zoo. This also applies to the adoption of integrated metapopulation management. Interactive exchange of animals between captive and wild populations will greatly enhance our capacity to sustain the genetic and demographic viability of both populations. Reductions of animal welfare due to capture, research, captive breeding and reintroduction will be all the more ethically justified as the risk of extinction of small and fragmented populations in the wild will be significantly minimized.

Pie in the sky, critics of the zoo will say—and not without reason. Today, the zoo is standing at a crossroads—and has to decide if it will fully commit to the new paradigm and develop into a conservation center or if it will degenerate (further) into a venue for entertainment that will provoke increasing criticism, not only from animal protectionist but also from wildlife conservationists.

References

Baker, A. 2007. Animal ambassadors: An analysis of the effectiveness and conservation impact of ex situ breeding efforts. In *Zoos in the 21st century. Catalysts for conservation*, ed. A. Zimmermann et al., 139–154. Cambridge, UK: Cambridge University Press.

Beck, B. 1995. Reintroduction, zoos, conservation, and animal welfare. In *Ethics on the Ark*, ed. B. Norton, et al., 155–163. Washington: Smithsonian Institution Press.

Bowkett, A.E. 2008. Recent captive-breeding proposals and the return of the Ark Concept to global species conservation. *Conservation Biology* 23(3): 773–776.

Byers, O., C. Lees, J. Wilcken, and C. Schwitzer. 2013. The one plan approach: The philosophy and implementation of CBSG's approach to integrated species conservation planning. *WAZA Magazine* 14: 2–5.

Callicott, J.B. 1980. Animal liberation: A triangular affair. *Environmental Ethics* 2: 311–338.

Conde, D.A., N. Flesness, F. Colchero, O.R. Jones, and A. Scheuerlein. 2011. An emerging role of zoos to conserve biodiversity. *Science* 331(6023): 1390–1391.

Conway, W. 2003. The role of zoos in the 21st century. *International Zoo Yearbook* 38: 7–13.

Conway, W. 2011. Buying time for wild animals with zoos. *Zoo biology* 30: 1–8.

Davey, G. 2006. Visitor behavior in zoo: A review. *Anthrozoos* 19(2): 143–157.

Dickens, M.J., D.J. Delehanty, and L.M. Romero. 2010. Stress: An inevitable component of animal translocation. *Biological Conservation* 143: 1329–1341.

Dickie, L.A., J.P. Bonner, and C. West. 2007. *In situ* and *ex situ* conservation: blurring the boundaries between zoos and the wild. In *Zoos in the 21st century. Catalysts for conservation*, ed. A. Zimmermann et al., 220–235. Cambridge, UK: Cambridge University Press.

Fa, J.E., S.M. Funk, and D. O'Connell. 2011. *Zoo conservation biology*. Cambridge: Cambridge University Press.

Falk, J.H., et al. 2007. *Why zoos & aquariums matter: Assessing the impact of a visit to a zoo or aquarium*. Silver Spring, MD: Association of Zoos & Aquariums.

Gewin, V. 2008. Riders of a modern-day Ark. *PLoS Biology* 6(1): 18–21.

Gusset, M., and G. Dick. 2011. The global reach of zoos and aquariums in visitor numbers and conservation expenditures. *Zoo Biology* 30: 566–569.

Gwynne, J.A. 2007. Inspiration for conversation: moving audiences to care. In *Zoos in the 21st century. Catalysts for conservation*, ed. A. Zimmermann et al., 52–62. Cambridge, UK: Cambridge University Press.

Hancocks, D. 2001. *A different nature. The paradoxical world of zoos and their uncertain future*. Berkeley: University of California Press.

Hanson, E. 2002. *Animal attractions. Nature on display in American zoos*. Princeton: Princeton University Press.

Hargrove, E. (ed.). 1992. *The animal rights/environmental ethics debate. The environmental perspective*. Albany: State University of New York Press.

Hatchwell, M., and A. Rübel. 2007. The Masola rainforest: a model partnership in support of *in situ* conservation in Madagascar. In *Zoos in the 21st century. Catalysts for conservation*, ed. A. Zimmermann et al., 205–219. Cambridge, UK: Cambridge University Press.

Hatchwell, M., A. Rübel, L.A. Dickie, C. West, and A. Zimmermann. 2007. Conclusion: The future of zoos. In *Zoos in the 21st century. Catalysts for conservation*, ed. A. Zimmermann et al., 343–360. Cambridge, UK: Cambridge University Press.

Hutchins, M. 2003. Zoo and aquarium animal management and conservation: Current trends and future challenges. *International Zoo Yearbook* 38: 14–28.

IUCN. 2000. *Guidelines for the placement of confiscated animals*. Switzerland: Gland.

IUNC. 2002. *Technical guidelines on the management of ex situ populations for conservation*. Switzerland: Gland.

IUCN. 2014. *Guidelines on the use of ex situ management for species conservation*. Version 2.0. Gland, Switzerland.

Jamieson, D. 1985. Against zoos. In *In defense of animals*, ed. P. Singer, 108–117. New York: Basil Blackwell.

Jamieson, D. 1995. Zoos revisited. In *Ethics on the Ark*, ed. B. Norton, et al., 52–66. Washington: Smithsonian Institution Press.

Jamieson, D., and R. Elliot. 2009. Progressive consequentialism. *Philosophical Perspectives* 23: 241–255.

Kaufman, L. 2012a. To save some species, zoos must let others die. *New York Times*, May 27.

Kaufman, L. 2012b. Zoos struggle to breed endangered animals. *New York Times*, July 5.

Keulartz, J., and C. Van der Weele. 2008. Framing and reframing in invasion biology. *Configurations* 16(1): 93–115.

Keulartz, J., and J. Swart. 2012. Animal flourishing and capabilities in an era of global change. In *Ethical adaptation to climate change*, ed. A. Thompson, and J. Bendik-Keymer, 123–144. Cambridge, MA: The MIT Press.

Kreger, M.D., M. Hutchins, and N. Fascione. 1998. Context, ethics, and environmental enrichment in zoos and aquariums. In *Second nature. Environmental enrichment for captive animals*, ed. D. Shepherdson et al., 59–82. Washington: Smithsonian Institution Press.

Leader-Williams. N., A. Balmford, M. Linkie, G.M. Mace, R.J. Smith, M. Stevenson, et al. 2007. 'Beyond the ark: conservation biologists' views of the achievements of zoos in conservation. In *Zoos in the 21st century. Catalysts for conservation,* 236–254. Cambridge, UK: Cambridge University Press.

Lees, C.M., and J. Wilcken. 2009. Sustaining the Ark: The challenges faced by zoos in maintaining viable population. *Zoo Yearbook* 43: 6–18.

Leopold, A. 1949. *A Sand County Almanac*. London: Oxford University Press.

Mace, G.M., A. Balmford, N. Leader-Williams, A. Manica, O. Walter, C. West, and Zimmermann, A. 2007. Measuring conservation success: assessing zoos' contribution. In *Zoos in the 21st century. Catalysts for conservation*, ed. A. Zimmermann, 322–342. Cambridge, UK: Cambridge University Press.

Marino, L., S.O. Lilienfeld, R. Malamud, N. Nobis, and R. Brogliod. 2010. Do zoo and aquariums promote attitude change in visitors? A critical evaluation of the American zoo and aquarium study. *Society and Animals* 18: 126–138.

McCarthy, S. 2004. *Becoming a tiger. How baby animals learn to live in the wild.* New York: Harper Perennial.

McShane, K. 2014. Individualist Biocentrism vs. Holism Revisited. *Les Ateliers de l'Ethique/The Ethics Forum* 9(2): 130–148.

Minteer, B.A., and J.P. Collins. 2013. Ecological ethics in captivity: Balancing values and responsibilities in zoo and aquarium research under rapid global change. *ILAR* 54(1): 41–51.

Redford, K., D. Jensen, and J. Breheny. 2012. Integrating the captive and the wild. *Science* 338: 1157–1158.

Regan, T. 1995. Are zoos morally defensible. In *Ethics on the Ark*, ed. B. Norton, et al., 38–51. Washington: Smithsonian Institute Press.

Regan, T. 1983. *The case for animal rights*. Berkeley: University of California Press.

Rolston, H. 1988. *Environmental ethics. Duties to and values in the natural world.* Philadelphia: Temple University Press.

Sagoff, M. 1984. Animal liberation and environmental ethics: Bad marriage quick divorce. *Osgoode Hall Law Journal* 22: 297–307.

Sandler, R.L. 2012. *The ethics of species. An introduction*. Cambridge, UK: Cambridge University Press.

Schön, D., and M. Rein. 1994. *Frame reflection*. New York: Basic Books.

Singer, P. 1979. Not for humans only: The place of nonhumans in environmental issues. In *Ethics and problems of the 21st century*, ed. K.E. Goodpaster, and K.M. Sayre, 191–206. Notre dame: University of Notre Dame Press.

Soulé, M. 1990. The Onslaught of alien species, and other challenges in the coming decades. *Conservation Biology* 4(3): 233–239.

Soulé, M., M. Gilpin, W. Conway, and T. Foose. 1986. The Millenium Ark: how long a voyage, how many staterooms, how many passengers. *Zoo Biology* 5: 101–113.

Stanley Price, M.R., and J.E. Fa. 2007. Reintroductions from zoos; a conservation guiding light or a shooting star? In *Zoos in the 21st century. Catalysts for conservation*, ed. A. Zimmermann et al. 155–177. Cambridge, UK: Cambridge University Press.

Sterling, E., J. Lee, and T. Wood. 2007. Conservation education in zoos: An emphasis on behavioral change. In *Zoos in the 21st century. Catalysts for conservation*, ed. A. Zimmermann et al., 37–50. Cambridge, UK: Cambridge University Press.

Tribe, A., and R. Booth. 2003. Assessing the role of zoos in wildlife conservation. *Human Dimensions of Wildlife* 8: 65–74.

Wilson, E.O. 1987. The little things that run the world (the importance and conservation of invertebrates). *Conservation Biology* 1: 344–346.

Woods, M., and V. Moriarty. 2001. Strangers in a strange land: The problem of exotic species. *Environmental Values* 10: 163–191.

World Association of Zoos and Aquariums (WAZA). 1993. *Executive summary of the world zoo conservation strategy—the role of the zoos and aquaria of the world in global conservation.* Gland: WAZA.

World Association of Zoos and Aquariums (WAZA). 2005. *Building a future for wildlife—the world zoo and aquarium conservation strategy*. Gland: WAZA.

The Flights of the Monarch Butterfly: Between In Situ and Ex Situ Conservation

Brendon M.H. Larson and Stephanie Barr

Abstract For several decades, conservation biologists seeking to prevent the extinction of rare species have invoked a distinction between conservation 'in the wild' (in situ) and conservation in captivity (ex situ). They prefer the former because it maintains the species in a 'natural' state rather than one 'contaminated' by human culture. Drawing on long-standing critiques of the nature-culture dichotomy, however, several critics have recently challenged the in situ/ex situ dichotomy on the grounds that it is both theoretically and practically untenable. Here, we extend these critiques through a case study of the Monarch butterfly (*Danaus plexippus*), an iconic and beautiful species that summers in Canada and the U.S. and winters in southern California and Mexico. This case not only demonstrates several ways that in situ and ex situ conservation interweave, but necessitates further reconsideration in the face of climate change (specifically in terms of the question of assisted migration). Through this case study, we unveil several ecological complexities that further emphasize the need for thinking across the in situ/ex situ divide.

1 Introduction

Traditional conservation is founded on a separation between nature and culture.[1] This belief has given rise to diverse practices that maintain such a distinction, primarily in the form of conservation reserves and parks that seek to maintain natural processes and species in a state relatively free from human activities and impacts (known as in situ conservation). As human impacts have grown globally, however, it has often been necessary to manage species and processes within these

[1]This statement applies more in North America than in Europe, where there is a longer history of human entanglement with nature (e.g., Drenthen and Keulartz 2014). We address the case of the Monarch butterfly in its North American context.

B.M.H. Larson · S. Barr (✉)
University of Waterloo, Waterloo, Canada
e-mail: blarson@uwaterloo.ca

B. Bovenkerk and J. Keulartz (eds.), *Animal Ethics in the Age of Humans*,
The International Library of Environmental, Agricultural and Food Ethics 23,
DOI 10.1007/978-3-319-44206-8_21

parks to maintain their presence and function. Sometimes this management has necessitated removing a rare or declining species from its natural habitat (in situ) so that it can be sustained in captivity (so-called ex situ conservation, e.g., botanical gardens, seed banks, and zoos), potentially for reintroduction of its descendants into the wild at a later date. Although the distinction between in situ and ex situ conservation may appear to be almost self-evident, it is theoretically problematic.

The distinction is problematic for several reasons (reviewed in Braverman 2014). Most significantly, this terminology is incompatible with emerging views that question a dichotomy between nature and culture and instead highlight their entanglement (e.g., Cronon 1996 and many others). On the one hand, many wild areas are strongly influenced or managed by humans. For example, the intensive management of populations of animals in reserves somewhat resembles the treatment of animals in captivity. On the other hand, captive populations often exist in 'naturalized' conditions (e.g., in zoos or botanical gardens) that are in some cases more natural than the degraded conditions found in the wild[2] and they are often continually exchanged with natural populations. Given these exchanges, it makes little sense to assume that 'pure' nature, on one side of a fence, is good and that culture, on the other side, is a negative contaminating influence.

A particular challenge to the in situ/ex situ distinction is that we are now living in an epoch, which geologists propose to christen the Anthropocene (e.g., Monastersky 2015), 'the age of humans,' where human impacts are omnipresent. At the largest scale, the greatest of these impacts result from human-caused climate change (IPCC 2014). Anthropogenic climate change means that clear dichotomies between nature/culture and in/out are no longer so simple (if they ever were!): Boundaries instead become grey and murky and contested, and the idea of nature (as the logical binary opposing 'culture') may be a handicap rather than a boon.

In this chapter, we explore a rich case study, concerning the Monarch butterfly, to demonstrate the entanglement of in situ/ex situ conservation in the Anthropocene. We examine how simple dichotomies may erode in the face of the specificities of an organism's life history and ecological interactions. We perhaps think too much in terms of a simple model of in situ/ex situ conservation that applies best to relatively large charismatic mammals (a common bias in conservation), where individuals of a species are removed from the wild, bred in captivity, and then either they or their offspring are returned. This does not necessarily apply well to species with much shorter generation times, more complex life stories (or "flight ways" as Van Dooren 2014 calls them), migratory movement among distant locations, and critical interspecific interactions—all of which must be understood in the context of climate change.

Before introducing the Monarch, we turn to a schematic model that we will use to facilitate our thinking through the complexities of in situ/ex situ conservation in the Anthropocene (Fig. 1). The two left panels represent traditional thinking about

[2]On some occasions, animal liberationists have cut the fence in zoo enclosures in an attempt to free animals from captivity, but the animals remain in the enclosure (Braverman 2014).

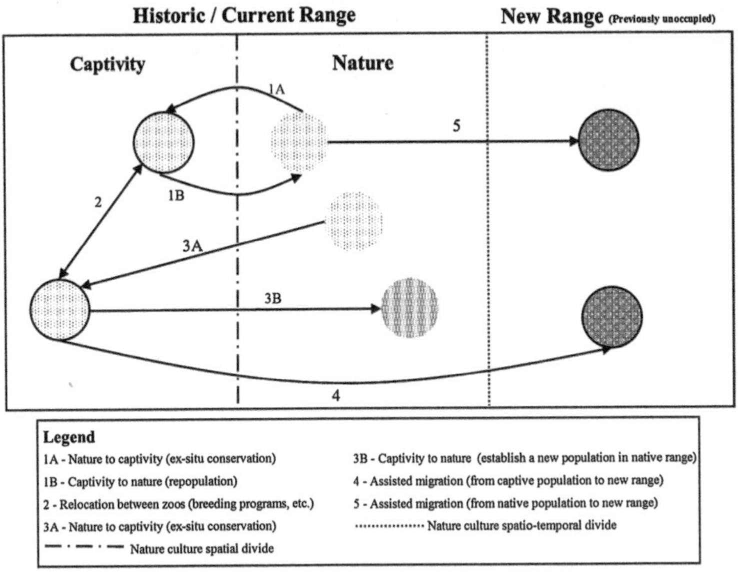

Fig. 1 A model for trans situ conservation in the Anthropocene. See text for elaboration

in situ/ex situ conservation, in terms of the reciprocal movements of species between nature and captivity (arrows 1A and 1B, 3A and 3B), including movements among captive populations (arrow 2). Together, these movements form a meta-population, though one that crosses the nature/culture boundary given that it includes both natural and captive populations. The occurrence of such metapopulations is widely recognized (e.g., Braverman 2014 and Byers et al. 2013 on the IUCN's One Plan approach), and often specifically attends to issues of genetic diversity. Some have proposed that this deconstruction of the in situ/ex situ distinction should be called pan situ conservation (Minteer and Collins 2013).

To our knowledge, however, the literature stops at this point, effectively treating nature as static. But in an era of climate change, we know this is misleading. As the climate warms, a population of a species in a given location may either die, adapt, or migrate. The problem, however, is that the latter two (and more optimistic) options take time, time that species may not have. Although there is of course great uncertainty in climate models (in large part owing to uncertainty about the extent of human mitigation), it is commonly predicted that species will not be able to adapt or migrate quickly enough given the expected rate of climatic warming (e.g., Hoegh-Guldberg et al. 2008).

Accordingly, conservationists propose to help species migrate to keep pace with climate change, an adaptation measure variably known as assisted migration, assisted colonization, or managed relocation (McLachlan et al. 2007; Richardson et al. 2009; Schwartz et al. 2012). The general idea is to move them northward or upward, as the case may be, into a location where the climate of the future will suit them. Although this measure is highly contested, with critics arguing that it is risky, hubristic, or impractical, others counter that we face a 'do-or-die' situation (for ethical analyses, see Minteer and Collins 2010; Aubin et al. 2011; Palmer and Larson 2014).

We are struck by the parallel concerns about moving species into captivity or moving them 'into the future' with assisted migration in the sense that we move them into a location where they will survive under expected climate change. Assisted migration also adds an important temporal 'adaptation' dimension to prior conceptions of in situ and ex situ conservation. Returning to Fig. 1, assisted migration involves moving a species from either natural (arrow 5) or captive (arrow 4) populations to new locations to increase the odds of persistence into the future. Although assisted migration may apply at various scales (within, just beyond, or much beyond the current range limit), it is most contentious at the largest scale, the movement of a species beyond its existing range limit (the very definition of a non-native species, as normally understood). We therefore set aside inter situ conservation (arrow 3B), another proposed bridge between in situ and ex situ conservation, which uses paleoecological information for "The establishment of species by reintroduction to locations *outside the current range but within the recent past range* of the species" (Burney and Burney 2007, p. 485, italics added). We propose that this broader set of conservation considerations, including not only natural and captive populations, as traditionally conceived, but 'populations adapted to the future' as well, could be referred to as trans situ conservation.

We now turn to a case study of the Monarch to consider in more depth this form of conservation. We begin with an overview of the biology of the Monarch and in situ/ex situ approaches to its conservation before considering in more detail how they interweave. We note from the outset that there is currently little ex situ conservation of the Monarch per se, though for reasons that beg the very category of ex situ conservation.

2 The Flights of the Monarch

The Monarch (*Danaus plexippus*) is a large (10 cm wingspan) and beautiful butterfly with a striking pattern of black stripes and white spots on a bright orange background, an aposematic pattern that serves to warn potential predators of its toxicity (passed on from the milkweed host of its larvae, see below). It occurs in

North America as two main populations, one east and one west of the Rocky Mountains, which have together numbered historically in the 100s of millions.[3] Butterflies from the eastern population travel to a small region of overwintering sites in Mexico while the western population overwinters in more numerous aggregation sites along the coast of central and southern California (Solensky 2004; Howard and Davis 2009). These migrations are considered to be "one of the unique and spectacular biological phenomena in North America" (Luna and Dumroese 2013, p. 6).

We largely focus on the eastern population of the Monarch given particular conservation concerns about it. These Monarchs perform the longest known insect migration in the world, and Monarchs are the only butterfly species that is known to make a round-trip migration (Fig. 2, which we describe based on Howard and Davis 2009; Luna and Dumroese 2013; Species at Risk Public Registry 2015). It takes 4–5 generations of Monarchs to complete one round-trip migration, but one generation completes the entire 3,000 km trip back to Mexico in the fall (arriving in November and December). Beginning in March or early April, Monarchs that overwintered in Mexico—which are now 7–9 months old—begin their migration north and the females lay eggs on milkweeds along the Gulf Coast of the southern United States. Butterflies that emerge from these eggs continue to journey north, producing new offspring along the way. After several generations, by late May or early June, the species reaches its apogee in the northern United States and southern Canada, where 2–3 additional generations may be produced over the summer.[4] In the fall, individuals from the final generation of Monarchs, wherever they originate, make the 3,000 km trip back to Mexico, to begin the cycle anew.

Once Monarchs arrive in Mexico they cluster together almost exclusively in a specific type of tree, the Oyamel Fir (*Abies religiosa*, also sometimes called the Sacred Fir) (Species at Risk Public Registry 2014, 2015). This coniferous tree is found in montane cloud forest at elevations from 2400–3600 m in the Transverse Neovolcanic Belt in central Mexico. Monarchs from the eastern population are restricted to about 12 massifs west of Mexico City within 800 km of one another. These sites were unknown until 1975, when they were found—after 38 years of searching—by two Canadian entomologists (Species at Risk Public Registry 2014). Although about 30 overwintering sites are known, spread over an area of about 6400 km^2, appropriate habitat occurs over only 562 km^2 (Slayback et al. 2007). This may seem to be extensive, but for several synergistic reasons the Monarch is considered to be a species at risk.

---

[3]Additional, smaller and more sendentary, populations occur in central America, Cuba and southern Florida (Species at Risk Public Registry 2014).

[4]Though there are occasional reports of individual butterflies that overwintered in Mexico returning all the way to southern Canada (Species at Risk Public Registry 2014).

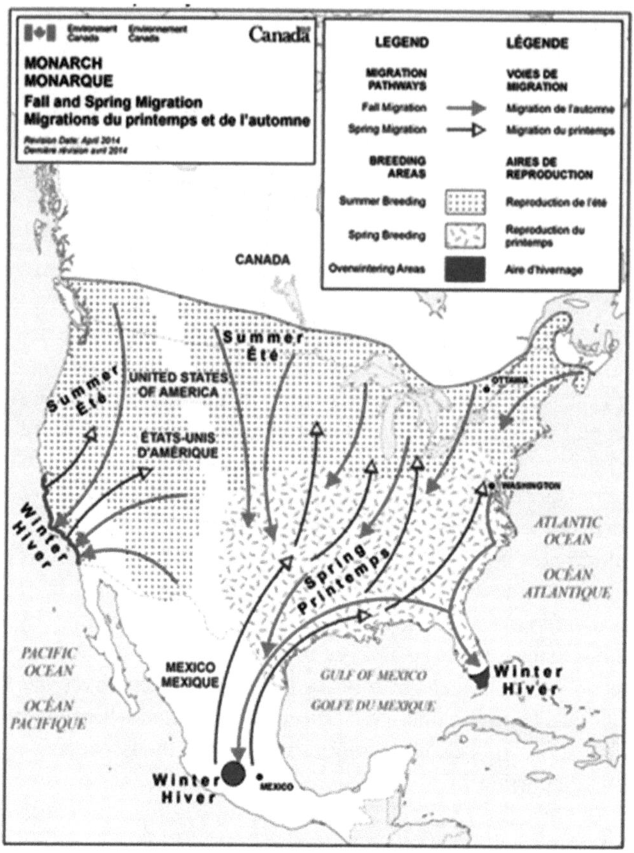

Fig. 2 The flights of the Monarch (from Species at Risk Public Registry 2014)

3 In Situ and Ex Situ Conservation of the Monarch

There are tremendous concerns about the Monarch, which is for example classified as a species of 'special concern' by the Committee on the Status of Endangered Wildlife in Canada (COSEWIC) because current trends suggest it could become threatened by extinction (Species at Risk Public Registry 2014). Yearly population sizes are estimated from the area occupied by their roost in central Mexico, which has declined significantly in recent years. As recently as 1996–1997, the area occupied, about 45 ha, multiplied by an average density of about 5,000 butterflies/m^2 led to a population estimate of $\sim$1 billion butterflies. In contrast, only 0.67 ha were occupied in 2013–2014, leading to an estimate of about 15 million butterflies, representing "the smallest population observed since the Monarch colonies began to be monitored in 1993" (Species at Risk Public Registry 2014; and see Rendón-Salinas et al. 2014; Vidal and Rendón-Salinas 2014).

Until recently, then, Monarch conservation has been focused on in situ conservation of its overwintering habitat. Not only is this area small, but it is also under pressure from illegal logging, whether for commercial uses, agriculture, or the production of charcoal. Accordingly, conservation organizations have purchased land in the region, focused on the Monarch Butterfly Biosphere Reserve, a 56,259 ha reserve in central Mexico that protects about 70 % of overwintering sites (Gustaffson et al. 2015; UNESCO 2015). The reserve was established by a 1986 decree by the Mexican president and in 2008 it became a UNESO world heritage site (Vidal et al. 2014). After the creation of the reserve, the World Wildlife Fund and the Mexican Fund for Nature Conservation established a Monarch conservation trust fund to support communities who had lost the right to log the forests (Honey-Roses 2009; Vidal et al. 2014). There have been additional efforts to involve local peoples in Monarch conservation and to demonstrate the value of Monarch-based ecotourism.

Nonetheless, there are still significant threats to this forest (Navarrete et al. 2011; Species at Risk Public Registry 2014). In addition to continued pressure from logging, there are concerns about catastrophic loss of overwintering Monarchs to freezing. Overwintering Monarchs lower their metabolic rate, so they are dependent on the protection provided by Oyamel Fir trees. Yet heavy rains, followed by cold temperatures, which are precisely the type of erratic weather predicted to increase under climate change, have already contributed to recent Monarch population declines (Species at Risk Public Registry 2014). Recent modeling efforts, based on climate models, in fact suggest that habitat suitable for the Oyamel Fir will move further up mountainsides in the coming decades, to the extent that the Monarch Butterfly Biosphere Reserve will no longer contain suitable habitat by the turn of this century. Overall, the amount of habitat suitable for the Oyamel Fir is projected to decline by 69.2 % by 2030, 87.6 % by 2060, and 96.5 % by 2090 (Sáenz-Romero et al. 2012).

But there has been growing recognition that the breeding habitat of the Monarch, further north, is also crucial to its survival. This habitat supports not only wildflowers that provide nectar sources along its migratory route, which fuel its migration (and allow it to overwinter by relying on built-up fat stores), but perhaps more importantly the sole food for its larvae, several species of milkweed (genus *Asclepias*).[5] Milkweed species vary in their habitat requirements, but as a whole—especially given the widespread occurrence of common milkweed (*Asclepias syriaca*), they are found across a wide range of habitats, from 'high-quality' prairies, wetlands and woodlands to agricultural fields and roadsides. Milkweed populations have been declining, however, in part due to habitat loss and urbanization, but most

[5]In southern parts of their range, Monarchs may feed on additional species in the milkweed family (now, technically, the subfamily Asclepiadoideae of the family Apocynaceae). In some regions of eastern North America, at least, Monarchs sometimes lay their eggs on members of a genus of non-native invasive plants, known as dog-strangling vine (genus *Vincetoxicum*), which is a relative of milkweed. However, they cannot survive on this host, so it is an additional 'sink' for Monarchs (see Casagrande and Dacey 2007).

significantly from the planting of genetically-modified herbicide-resistant crops. Flockhart et al. (2015) conclude that the consequent reduction in milkweed plants to herbicides—especially in the so-called Corn Belt in the U.S.A.—is the most significant threat to long-term survival of the Monarch (though they emphasize that this does not discount the importance of conservation in Mexico as well).

Both the in situ and ex situ conservation of the Monarch are driven by its iconic status (see Gustaffson et al. 2015). To North American conservationists, the potential disappearance of Monarchs raises the spectre of the early-20th-century extinction of the Passenger Pigeon, once considered the most abundant bird in North America if not the world. Speaking personally, the senior author's (BL) interest in the natural world in part originated through raising white-black-and-yellow striped Monarch caterpillars as a child, an experience shared by many North American naturalists. Elsewhere, people are inspired by the large numbers of Monarchs that congregate to rest in staging areas along the north shore of the Great Lakes (especially Lakes Erie and Ontario) prior to crossing them.

The iconic status of the Monarch has therefore united scientists, citizen scientists, school children, non-governmental organizations, and politicians in a transboundary effort to preserve the species using both in situ and ex situ approaches (Gustafsson et al. 2015). For example, Monarch Live—A Distance Learning Adventure, has used the internet to connect 500,000 students in Canada, the U.S.A. and Mexico and efforts to create and improve habitat for Monarchs (and other pollinators) were conducted on 405,000 ha in more than 20 National Forests across the U.S.A. (Luna and Dumroese 2013). Plant nurseries and botanical gardens engage in extensive efforts to promote plantings of milkweeds, whether along roadsides or in backyard 'butterfly gardens' (Luna and Dumroese 2013; Gustaffson et al. 2015).

Citizen science is a vital part of the effort to conserve the Monarch, and it is led by organizations such as Monarch Watch (http://www.monarchwatch.org). Monarch Watch facilitates the tagging and release of Monarchs each fall during their southward migration and tracks where they are found. Additionally, many zoos and butterfly conservatories run Monarch education and conservation programs. For example, the Cambridge Butterfly Conservatory in southern Ontario has several programs to inspire people to participate in Monarch conservation and to educate them about pollinators in general (A. Tofflemire, personal communication).[6] During the summer months, Monarch eggs are collected from nearby milkweed plants and the caterpillars are raised for use in interactive children's programs about the life history of Monarchs. For the past 13 years the Conservatory has also provided kits to children who are interested in raising their own Monarchs to release during the fall migration. Each fall, the Conservatory organizes a tagging weekend when the public is invited to join in tagging and releasing both captive-bred and wild Monarchs: in 2015, 180 were tagged and released. There are many such programs, with the Toledo Zoo aiming for 1200 releases in fall 2015

[6]As pointed out by Gustafsson et al. (2015, p. 8), however, Monarchs themselves are not thought to be important pollinators.

(Dungjen 2015). In short, many people are motivated by concern for the Monarchs' future.

4 Interweaving In Situ and Ex Situ Conservation

The Monarch has a complex life history ranging across the North American continent, and includes dependencies on actors ranging from a fir tree found in a highly specialized and restricted habitat, a widespread plant genus found in agricultural landscapes, and conservationists seeking to maintain its numbers (in both locations). These dependencies create challenges common to many migratory species, yet which again contrast greatly with the type of species that is the stereotypical focus of ex situ conservation. It is truly an *ecological* story in that multiple species, in different places, are interwoven. It could be seen as largely a question of in situ conservation, insofar as it primarily concerns Oyamel Fir in Mexico to support overwintering Monarchs and milkweed (as well as nectar plants) further north to support summering Monarchs. However, to the extent that concerns about captivity derive from concerns about human agency, it is worth pointing to the ways in which human agency is entangled in the conservation of the Monarch (as with many other rare species). We next consider in more detail a few of the ways in which the plight of the Monarch interweaves in situ and ex situ conservation.

We begin by again emphasizing that traditional ex situ conservation is limited in the case of the Monarch. Although Monarch eggs are collected and their larvae are raised and released, as described above, these actions mainly serve as a means of outreach and perhaps inspire citizen science activities. In fact, the release of Monarchs could even be a threat to them, through "the transmission of disease and parasites, impacts on migratory patterns, [and] harmful genetic mixing …" (Species at Risk Public Registry 2014). Consequently, the Monarch is not so much a model for rethinking ex situ conservation per se as one for demonstrating the extent to which in situ and ex situ conservation always already intermingle.

To begin, consider how in situ conservation of the Oyamel Fir forest in Mexico is threatened by climate change. Climate change is perhaps the strongest omen of the inadequacy of traditional thinking about in situ and ex situ conservation. That distinction is founded on the idea that in situ conservation provides a static place for populations to return to even after time in captivity. But climate change means that in situ conservation is always-already subject to the influence of human agency (even if indirectly, through human causation of climate change). In short, in situ conservation begins to require continuous management—in a semi-cultural landscape, so it looks more and more like ex situ conservation.

Consider the Oyamel Fir a little further. These trees are critical for the survival of overwintering Monarchs (e.g., Anderson and Brower 1996), so they are a means to the end of Monarch conservation. Yet they are already dying in response to climatic change, as their suitable habitat shifts higher and higher on the mountains. The trees cannot migrate upward quickly enough on their own. Accordingly, based on their models, Sáenz-Romero et al. (2012, p. 98) recommend “assisted migration of (Oyamel Fir) upwards in altitude by 275 m so that populations of 2030 would occupy the same climates as today.” As noted for other montane tree species facing a similar situation (e.g., the whitebark pine in the Rocky Mountains of North America, Palmer and Larson 2014), this altitudinal shift would pose significant challenges for conservation management, for example if it needs to exceed the tree line and thus the presence of adequate soil for tree growth (Sáenz-Romero et al. 2012). Although Sáenz-Romero et al. (2012, p. 103) note that assisted migration “has the added advantage of establishing a founder population that eventually could serve as a seed source for natural migration,” whether this would allow the recovery of wildness with time depends on one’s interpretation of the concept (see Palmer and Larson 2014).

Several alternative management options have been proposed, each focusing on issues related to the Monarch’s philopatry (that is, the tendency for its offspring to return to the same overwintering location(s) each winter). For example, Oyamel Fir might be transplanted to mountains with higher peaks, and thus potentially suitable habitat under climate change, yet it is unknown whether Monarchs would adopt these new sites. Alternatively, Oyamel Fir trees could be replaced by another, similar species, such as an endangered spruce tree, *Picea martinezii*, whose suitable habitat will shift into the Monarch Butterfly Biosphere Reserve. This could be a win-win outcome for this rare tree and the Monarch, if the tree could in fact grow in the Monarch Butterfly Biosphere Reserve, if it would be accepted by the Monarch, and if it would equally protect overwintering Monarchs from the elements (Sáenz-Romero et al. 2012). It is worth noting, in this context, that the Monarch has been introduced to locations around the world, so it may be more adaptable than we think.

The in situ conservation of the Monarch involves not just wild landscapes, especially the Oyamel Fir forests in Mexico, but more challengingly, widespread habitats suitable for milkweed across North America. The Monarch is unusual given the extent to which it is threatened, even in its summer range, despite the fact that its milkweed food plant is so widespread. Its conservation, then, requires a nuanced consideration of the role of cultivated landscapes to its survival, whether in agricultural fields or along roadsides.

There’s another interesting twist: the range of milkweeds (especially the common milkweed, mentioned above) has expanded and contracted over the past few centuries as a function of human activity (see Brower 1995 and summary in Species at Risk Public Registry 2014). Their distribution, then, has long reflected the influence of people, and thereby influenced the distribution of Monarchs. Until the late 1800s, it appears that the eastern population of the Monarch mainly bred in the central prairie region of North America. Its range shifted eastward around the turn of the century in response to the eastward spread of common milkweed into agricultural lands created as forests were cleared. In fact, “the historically cleared deciduous forest corresponds with the main current breeding range of the Eastern population” (Species at Risk Public Registry 2014 citing Urquhart 1960). Later in the 20th century, however, many farms were abandoned, which created early-successional fields that could also support milkweed plants (along with roadsides and the populations remaining in agricultural fields). Yet, overall, milkweed population sizes are likely past their peak—despite recent efforts to propagate them, not only as a result of habitat loss and herbicide-resistant crops, but because "abandoned farmlands … are at risk of being lost, as they either regenerate into forest or are developed and converted for residential or industrial development" (Species at Risk Public Registry 2014). While this decline may be counter-balanced by northward expansion of milkweed in response to climate change, it remains difficult to predict how climate change will affect the Monarch overall (e.g., Zipkin et al. 2012).

Wintering populations of the western population of Monarch, in southern California, pose another challenge to an ideal of natural in situ conservation. Most of these populations occur in stands of eucalyptus trees, non-native species introduced from Australia in the middle of the 19th century (see Coates 2006 for elaboration), rather than in stands of trees native to California (Species at Risk Public Registry 2014). These trees appear to provide excellent overwintering sites for the Monarchs, yet they are often eradicated as an invasive species (see Davis 2013). The result is that Monarchs in southern California find themselves in a fraught position: they are now the focus of conservation efforts even while their adopted overwintering trees are potentially at risk because conservationists devalue them because their presence reflects human agency ('non-nativeness', see Larson 2007 among others for critical discussion). By their own agency, they again remind us of the entanglements of nature and culture.

5 Concluding Thoughts

Returning to our model for species movements in the Anthropocene, Fig. 1, the Monarch exhibits many boundary crossings between the domains of nature and culture (e.g., the left two panels) even if it is not the focus of ex situ conservation efforts. The more significant contribution of this chapter is to introduce the ecological boundary-crossings involved in the conservation of this species, which is noteworthy as a charismatic invertebrate species and exemplary of many species that are migratory or which have complex life histories. Humans and butterflies interweave in many ways, whether through the story of the Oyamel Fir, the history of human-milkweed interactions, or the efforts by so many people to track the fate of the Monarch.

Yet climate change adds a further dimension. Just as ex situ conservation was once denigrated (e.g., Braverman 2014), not least because it subjects a species to human nurture and culture, part of the concern about assisted migration as a conservation strategy is that it contaminates a species with human agency. In both cases, species are moved outside of their current range and into the care of humans, so in that respect they both break down the nature/culture divide. Whether a species is brought into captivity or moved by assisted migration, it is essentially being brought from nature into the realm of culture, at least insofar as a species moved by assisted migration bears the 'fingerprints' of human agency and the landscape into which it is moved similarly bears the effects of human-caused climate change.

Under climate change, there is now no longer a 'pure nature' to return to, if there ever was, but instead only a nature that is more-or-less affected by humans. Assisted migration is almost in a sense a form of captivity in the wild—once we move species, we will have to play a role in maintaining them into the future (see Palmer and Larson 2014). As part of modernist risk society, humans face, more and more, the repercussions of our actions. If we raise and release Monarchs and conduct assisted migration of the Oyamel Fir, we must potentially manage this species in perpetuity.

This paper raises further questions. Where does ex situ conservation end and in situ conservation begin? Where does culture end and nature begin? Are the forests in Mexico still 'natural' when they have been 'degraded' by human activity and when they are under the thrall of climate change? Are the agricultural landscapes that support Monarchs in North America 'cultural' when they are so important to this and many other species. As Braverman (2014) and others point out, perhaps we must remove the problematic term 'nature' from conservation discussions, thereby not only removing the dualistic conception of in situ and ex situ conservation but facilitating more challenging, nuanced and contextual decisions about what species matter and why (e.g., Sandler 2013).

References

Anderson, J.B., and L.P. Brower. 1996. Freeze-protection of overwintering monarch butterflies in Mexico: Critical role of the forest as a blanket and an umbrella. *Ecological Entomology* 21: 107–116.

Aubin, I., C.M. Garbe, S. Colombo, C.R. Drever, D.W. McKenney, C. Messier, J. Pedlar, et al. 2011. Why we disagree about assisted migration: Ethical implications of a key debate regarding the future of Canada's forest. *The Forestry Chronicle* 87: 755–765.

Braverman, I. 2014. Conservation without nature: The trouble with *in situ* versus *ex situ* conservation. *Geoforum* 51: 47–57.

Brower, L.P. 1995. Understanding and misunderstanding the migration of the monarch butterfly (*Nymphalidae*) in North America: 1857–1995. *Journal of the Lepidopterists Society* 49: 304–385.

Burney, D.A., and L.P. Burney. 2007. Paleoecology and "inter-situ" restoration on Kaua`i, Hawai`i. *Frontiers in Ecology and the Environment* 5: 483–490.

Byers, O., C. Lees, J. Wilcken, and C. Schwitzer. 2013. 'The "One Plan" approach: The philosophy and implementation of CBSG's approach to integrated species conservation planning. *WAZA Magazine* 14: 2–5.

Casagrande, R.A., and J.E. Dacey. 2007. Monarch butterfly oviposition on swallow-worts (*Vincetoxicum* spp.). *Environmental Entomology* 36: 631–636.

Coates, P. 2006. *American perceptions of immigrant and invasive species: Strangers on the land.* Berkeley, CA: University of California Press.

Commission for Environmental Cooperation. 2008. *North American Monarch. Conservation Plan.* Commission for Environmental Cooperation, Montreal. http://www3.cec.org/islandora/en/item/2350-north-american-monarch-conservation-plan-en.pdf. Accessed 27 Oct 2015.

Cronon, W. 1996. The trouble with wilderness; or, getting back to the wrong nature. In *Uncommon ground: Rethinking the human place in nature*, ed. W. Cronon, 69–90. New York: W.W. Norton & Company.

Davis, G. 2013. The Eucalyptus is part of California. In *The East Bay Express*. http://www.eastbayexpress.com/oakland/the-eucalyptus-is-part-of-california/Content?oid=3660640. Accessed 3 Nov 2015.

Drenthen, M., and J. Keulartz (eds.). 2014. *Old world and new world perspectives in environmental philosophy: Transatlantic conversations*. New York: Springer.

Dungjen, T. 2015. Toledo Zoo releases 600 monarchs: Total of 1,200 to be released before Oct. migration, up from 250 in '14. *The Blade*. http://www.toledoblade.com/local/2015/09/10/Toledo-Zoo-releases-600-monarchs.html. Accessed 29 Oct 2015.

Flockhart, D.T.T., J.-B. Pichancourt, D.R. Norris, and T.G. Martin. 2015. Unravelling the annual cycle in a migratory animal: Breeding-season habitat loss drives population declines of monarch butterflies. *Journal of Animal Ecology* 84: 155–165.

Gustafsson, K.M., A.A. Agrawal, B.V. Lewenstein, and S.A. Wolf. 2015. The monarch butterfly through time and space: The social construction of an icon. *BioScience* 65(6): 612–622.

Hoegh-Guldberg, O., L. Hughes, S. McIntyre, D. Lindenmayer, C. Parmesan, H. Possingham, and C. Thomas. 2008. Assisted colonization and rapid climate change. *Science* 321: 345–356.

Honey-Roses, J. 2009. Disentangling the proximate factors of deforestation: The case of the monarch butterfly biosphere reserve in Mexico. *Land Degradation and Development* 20: 22–32.

Howard, E., and A.K. Davis. 2009. The fall migration flyways of monarch butterflies in eastern North America revealed by citizen scientists. *Journal of Insect Conservation* 13: 279–286.

IPCC. 2014. *Climate change 2014: Synthesis report*. Contribution of Working Groups I, II and III to the Fifth Assessment Report of the Intergovernmental Panel on Climate Change. IPCC, Geneva, Switzerland.

Larson, B.M.H. 2007. Who's invading what? Systems thinking about invasive species. *Canadian Journal of Plant Sciences* 87: 993–999.

Luna, T., and R.K. Dumroese. 2013. Monarchs (*Danaus plexippus*) and milkweeds (*Asclepias* species): The current situation and methods for propagating milkweeds. *Native Plants* 14: 5–15.

McLachlan, J.S., J.J. Hellmann, and M.W. Schwartz. 2007. A framework for debate of assisted migration in an era of climate change. *Conservation Biology* 21: 297–302.

Minteer, B.A., and J.P. Collins. 2010. Move it or lose it? The ecological ethics of relocating species under climate change. *Ecological Applications* 20: 1801–1804.

Minteer, B.A., and J.P. Collins. 2013. Ecological ethics in captivity: Balancing values and responsibilities in zoo and aquarium research under rapid global change. *ILAR Journal* 54: 41–51.

Monastersky, R. 2015. The human age. *Nature* 519: 144–147.

Navarrete, J.L., M.I. Ramírez, and D.R. Pérez-Salicrup. 2011. Logging within protected areas: Spatial evaluation of the monarch butterfly biosphere reserve, Mexico. *Forest Ecology and Management* 262: 646–654.

Palmer, C., and B.M.H. Larson. 2014. Should we move the whitebark pine? Assisted migration, ethics, and global environmental change. *Environmental Values* 23: 641–662.

Rendón-Salinas, E., A. Fajardo-Arroyo, and G. Tavera-Alonso. 2014. Forest surface occupied by monarch butterfly hibernation colonies in December 2014. http://www.worldwildlife.org/publications/forest-surface-area-occupied-by-monarch-butterfly-hibernation-colonies-in-december-2014. Accessed 3 Nov 2015.

Richardson, D.M., J.J. Hellmann, J.S. McLachlan, D.F. Sax, M.W. Schwartz, P. Gonzalez, E. J. Brennan, et al. 2009. Multidimensional evaluation of managed relocation. *Proceedings of the National Academy of Sciences USA* 106: 9721–9724.

Sáenz-Romero, C., G.E. Rehfeldt, P. Duval, and R.A. Lindig-Cisneros. 2012. *Abies religiosa* habitat prediction in climatic change scenarios and implications for monarch butterfly conservation in Mexico. *Forest Ecology and Management* 275: 98–106.

Sandler, R. 2013. Climate change and ecosystem management. *Ethics, Policy and Environment* 16: 1–15.

Schwartz, M.W., J.J. Hellmann, J.M. McLachlan, D.F. Sax, J.O. Borevitz, J. Brennan, A.E. Camacho, et al. 2012. Managed relocation: Integrating the scientific, regulatory, and ethical challenges. *BioScience* 62: 732–743.

Slayback, D.A., L.P. Brower, M.I. Ramírez, and L.S. Fink. 2007. Establishing the presence and absence of overwintering colonies of the Monarch butterfly in Mexico by the use of small aircraft. *American Entomologist* 53: 28–40.

Solensky, M.J. 2004. Overview of monarch migration. In *The Monarch Butterfly: Biology and conservation*, ed. K.S. Oberhauser, and M.J. Solensky, 79–83. Ithaca, NY: Cornell University Press.

Species at Risk Public Registry. 2014. Management Plan for the Monarch (*Danaus plexippus*) in Canada—2014 [Proposed]. http://www.sararegistry.gc.ca/default.asp?lang=En&n=D5AAB693-1. Accessed 28 Oct 2015.

Species at Risk Public Registry. 2015. Species Profile: Monarch. http://www.sararegistry.gc.ca/species/speciesDetails_e.cfm?sid=294. Accessed 28 Oct 2015.

UNESCO. 2015. Monarch Butterfly Biosphere Reserve. http://whc.unesco.org/en/list/1290. Accessed 30 Oct 2015.

Urquhart, F.A. 1960. *The Monarch Butterfly*. Toronto: University of Toronto Press.

Van Dooren, T. 2014. *Flight ways: Life and loss at the edge of extinction*. New York: Columbia University Press.

Vidal, O., and E. Rendón-Salinas. 2014. Dynamics and trends of overwintering colonies of the monarch butterfly in Mexico. *Biological Conservation* 180: 165–175.

Vidal, O., J. López-García, and E. Rendón-Salinas. 2014. Trends in deforestation and forest degradation after a decade of monitoring in the monarch butterfly biosphere reserve in Mexico. *Conservation Biology* 28: 177–186.

Zipkin, E.F., L. Ries, R. Reeves, J. Regetz, and K. Oberhauser. 2012. Tracking climate impacts on the migratory monarch butterfly. *Global Change Biology* 18: 3039–3049.

Between Individualistic Animal Ethics and Holistic Environmental Ethics Blurring the Boundaries

Bernice Bovenkerk and Marcel Verweij

Abstract Due to its emphasis on experiential interests, animal ethics tends to focus on individuals as the sole unit of moral concern. Many issues in animal ethics can be fruitfully analysed in terms of obligations towards individual animals, but some problems require reflection about collective dimensions of animal life in ways that individualist approaches can't offer. Criticism of the individualist focus in animal ethics is not new; it has been put forward in particular by environmental ethics approaches. However, the latter tend to be so far removed from the concerns of animal ethicists that both groups talk at cross purposes. We think the gap between environmental and animal ethics could be bridged by on the one hand focusing more on the collective dimensions of our concerns with animals—after all, individuals are constituted by the collective of which they are a part—and on the other hand, by showing that moral status can also be attributed to groups in an indirect way, related to the moral status of their individual members. In our paper we explore various (novel) ways of conceptualising the moral relevance of collectiveness in animal life. We draw on insights from public health ethics, as this field of inquiry has also developed—at least partly—in response to individualist approaches in human bioethics, creating more room for recognizing the value of population health, interpersonal relations, solidarity, and ways in which a collective is constituted.

1 Introduction

Most animal ethicists take individual animals as the sole locus of moral concern. While they have forcefully argued that we need to take the interests of sentient animals into account in our moral decisions, some such decisions have an impact beyond the interests of individual animals. Think for example of the decision to rewild previously domesticated animals or decisions about the in situ or ex situ

B. Bovenkerk (✉) · M. Verweij
University of Wageningen, Wageningen, The Netherlands
e-mail: bernice.bovenkerk@wur.nl

B. Bovenkerk and J. Keulartz (eds.), *Animal Ethics in the Age of Humans*,
The International Library of Environmental, Agricultural and Food Ethics 23,
DOI 10.1007/978-3-319-44206-8_22

conservation of animal species. Such situations seem to call for reflection on collective dimensions regarding our dealings with animals. Environmental ethicists argue that collectives such as populations or ecosystems should be the locus of moral concern. However, with their primary focus on collectives, environmental ethicists seem to have a blind spot for the interests of individual animals. Animal and environmental ethicists appear to talk at cross purposes. We will argue that the gap between the two approaches may be—at least partially—bridged by focusing more on the collective dimensions in our concerns with animals. We will first examine the discussion between individualistic animal ethicists and holistic environmental ethicists about the question of the appropriate level of moral concern, including recent additions to this discussion. We certainly do not aim to settle this dispute once and for all, but we conclude that no convincing argument for the attribution of moral status to wholes has been given yet. However, neither do we think that the attribution of direct moral status to wholes is necessary in order to include collectives in our moral reasoning. We will argue that in many cases it is difficult to speak about the moral status of the individual without reference to the group and vice versa. As we will explain, even when they hold that only individual animals could reasonably be attributed moral status, animal ethicists should take into account collective dimensions.

In order to show the mutual dependence of individuals and groups, we think it is fruitful to seek analogies to the various ways in which human bioethics—notably ethics of public health—is increasingly acknowledging the collective dimensions of ethical problems in relation to human health and welfare. In the second part of this chapter, we therefore present several ways to conceptualize the moral relevance of collectivities in public health ethics and seek possible analogies in animal ethics. We will discuss four conceptualizations of the moral relevance of collectives: (1) the group as aggregation of individual members, (2) group welfare as something pertaining to the group as a whole, (3) considerations about the just distribution of goods within a group, and (4) an emphasis on collectivist or communitarian values. As we will show, some ways in which collectives could be conceptualized correspond better with the individualistic perspective than others, but as we will also show, in many cases the status of individuals is constituted by their relation to the group. Finally, we briefly discuss collective dimensions in animal ethics that go beyond the simple focus on moral status of the individual, but that still have repercussions for our treatment of individual animals: those dimensions that have to do with the species concept. While our primary aim in this chapter is to make a conceptual analysis, at the end we will briefly reflect on the practical consequences of this focus on collective dimensions and address the question to what extent this focus is indeed able to bridge the gap between animal and environmental ethics. We conclude that while we will not convince die hard ecocentrists and thereby do not wholly succeed in building a bridge between atomists and holistists, if we start from the moral status of individuals, we still have several options to attribute value to collectives. Hereby we hope to lay a basis for incorporating some of the intuitive implications of ecocentric ethics into animal ethics.

2 Individualism in Animal Ethics

In animal ethics we encounter a strong focus on the question of which animals possess moral status. This is often framed as a question about what entities possess those capacities necessary to have interests. Interests, in turn, tend to be understood as experiential interests. For example, Singer (1999) takes as a basic premise that equals should be treated equally and unequals unequally. Equality is here to be understood in terms of interests. If a specific action will cause an equal amount of suffering to two different animals (or to an animal and a human being), the animals have an equal interest in avoiding that action and therefore we should treat these animals equally. An entity that is not conscious cannot experience interests and in Singer's view therefore does not have interests. Regan (1983) argues that all beings who can subjectively experience their lives (in other words, all subjects-of-a-life) have inherent value and should be treated with equal respect. Palmer (2010) argues that our specific relationship with an animal influences the question of how we should treat this animal. For example, when we have domesticated or captured an animal we have specific duties of care towards this animal that we do not have towards wild animals. Yet, when it comes to the question of what class of animals we have duties towards in the first place, she takes the capacity to have experiential welfare as the determining criterion. In other words, according to all these authors, we only need to think about the question of how to treat an animal if that animal can subjectively experience its own interests. Due to this emphasis on experiential interests, animal ethics tends to focus on individuals as the sole unit of moral concern (Palmer 2010). After all, groups or collectives do not have conscious experiences.

Many issues in animal ethics can be fruitfully analyzed in terms of obligations towards individual animals, but some problems require reflection about collective dimensions of animal life in ways that individualist approaches can't offer. For example, the question of whether it is morally permissible to genetically modify or breed animals for specific characteristics addresses the level of animal species or populations, rather than individuals. When we selectively breed animals to exhibit particular characteristics we often unintentionally breed along unwanted characteristics, which may lead to health problems. Think of English Bulldogs that due to our preference for big heads and flat faces have breathing difficulties and cannot easily give birth naturally. As Palmer (2011) has pointed out, the intervention in the genome of these animals is carried out before the animals are born and therefore (as long as the animals still have a life worth living) we cannot say we have harmed these particular animals. After all, had we not bred them in this way they would not even exist. If we want to argue that we have harmed these animals, for example because we have violated their integrity (see Bovenkerk et al. 2002), we need to make a claim about the species or population that we have changed. Similarly, the concern that many have with conserving animal biodiversity is not directed at the preservation of the lives of individual animals, but of their species. Think of the initiative to establish gene banks—biodepositories of genetic material of animals and plants that face extinction; the idea behind this seems to be that even if all individual specimens of a particular species go extinct, the species itself can still survive by bringing it back into existence. Disregarding for the moment the question of whether this is a desirable way to go about biodiversity conservation, the underlying assumption seems to be that we have reasons to preserve a species that surpass the reasons that we have to protect all the currently living individuals of that species. However, only these individuals have experiential welfare, whereas the species that they belong to do not.

3 Animal Versus Environmental Ethics

Criticism of the individualist focus in animal ethics is not new; it has been put forward in particular by environmental ethics approaches. However, the latter tend to be so far removed from the concerns of animal ethicists that we are witnessing a dialogue of the deaf. The two groups appear to be operating in different worlds. In fact, as Varner has pointed out, "the one point on which environmental ethicists reached a general consensus during the field's first decade was that both the theoretical foundations and practical implications of animal rights views were inconsistent with those of environmentalism" (Varner 2003, p. 95). In his diagnosis, this is because environmental ethicists believe a number of implications of animal ethics views run counter to environmental concerns: those regarding wildlife population control (in particular the prohibition on hunting and the prevention of predation) and those regarding biodiversity preservation (in particular killing invasive or exotic species and breeding endangered species in captivity). These are all situations where the interests of individual animals appear to be opposed to the interest of a collective. Environmental ethicists, in particular ecocentrists, aim to formulate duties that we may have towards collectives such as populations, species, and ecosystems. They attribute interests to natural entities, but these are interests of a non-experiential kind, such as an interest in living or flourishing. They argue that collectives can have interests that surpass, or even run counter to, the composite interests of their individual members. Johnson (1992) gives the example of a population of prey animals. While it is in the interest of the population that old or weak animals are killed by predators, it is obviously not in the interest of the individual victims in question. However, this example does not prove that populations are morally relevant, simply that a group can have independent interests; an extra argument is needed to make the step from simply interests to morally relevant interests. Ecocentrists argue that if we were not to take into account the interests of collectives, a whole host of problems could not be addressed satisfactorily. For example, if we had to choose between killing two members of an endangered species or a whole group of individual animals of an abundant species, according to animal rights views we should (supposedly) favor the latter—larger—group, whereas surely, we should favor the members of the endangered species. In general, they argue, animal ethicists cannot show why it would be important to preserve biodiversity. And with their focus on individuals with experiential interests, can they even argue that we need to take the interests of future generations into account? Finally, the holistic assumptions of ecocentrists appear to have a strong intuitive appeal (Nelson 2010).

In the disagreement between individualist animal ethicists and holist environmental ethicists two, related, questions are central: (1) are only conscious interests or goods morally relevant or also non-conscious ones? (2) can only individual entities or also wholes be morally considerable? This is an ongoing and seemingly inconclusive debate, which we do not aim to settle here. On the basis of more recent literature we do here briefly want to show firstly, that this is indeed still an ongoing debate, secondly, how attempts have been made to refine the ecocentric view, and thirdly, why in our view this is still unconvincing. In his intriguing dissertation '*Taking Genes Seriously*', Dierks (2014) aims to show that not only sentient animals, but also individual living beings, species, and future generations have direct moral considerability. He does so by arguing that genes are morally considerable as they are goal-directed, that therefore genes have interests and can be harmed in the pursuit of these interests. Dierks uses a variety of the so-called 'marginal cases argument' (see also Tanner 2006, 2009) to argue that entities that do not have conscious or subjective interests nonetheless have interests and that we therefore have moral obligations towards them. He focuses on the case of the comatose person: "Although a comatose human does not take an (active, conscious) interest in nourishment, he or she, nonetheless, has a (biological, well-being) interest in it" (Dierks 2014, p. 64). He goes on to argue that even though they do not have conscious desires, plants possess the same morally considerable interests as humans, understood as biological and well-being interests. Similar to former ecocentrists or biocentrists he therefore argues that interests do not have to be experiential in order to count as morally relevant interests. His reasoning assumes that it is biological and well-being interests, rather than subjective or experiential interests, which make humans morally considerable. However, this is not self-evident. Is it their biological or well-being interests that make us think of comatose patients as worthy of moral concern, or is it rather the concern for the significant others of the patient or is it respect for the person the patient used to be or could again be in the future (see Nolt 2006)? In other words, we can wonder whether we should really rely on our intuition that coma patients deserve respect here to support the claim that they are direct subjects of moral consideration. Perhaps they deserve protection or respect because they are indirect subjects of moral consideration only.

It has been argued that animal ethicists put too much emphasis on the ability to suffer or enjoy as an entrance criterion for moral status. Isn't there something more fundamental underlying pain and pleasure, namely the will to live, the drive to sustain oneself and to procreate? Aren't the ability to feel pain and pleasure not simply mechanisms through which these more fundamental drives are achieved in sentient animals? And don't we share these more fundamental drives with non-sentient creatures with the ability to self-renew (Eckersley 1992)? As McShane reformulates this point:

> Once we understand the function of subjective states in their biological context, we will see that subjectivity is just one of the many tools that organisms use to process information about their environment and to adjust their behavior accordingly.... subjective or conscious states are just one of the ways that organisms have evolved to pursue their basic biological goals: self-defense, metabolism, reproduction, etc. (2015, p. 134).

We should, then, not take the ability to experience pain and pleasure as the morally relevant criterion, but rather the possession of biological goals[1] (idem, 135). This raises the question, however, what exactly it is about biological goals that makes them morally relevant. If these goals are simply out there, but are not experienced by anyone, and in other words, there is no one to whom they directly matter, then an extra argument is still needed to take us from fact to value.

To counter such subjectivist arguments, Dierks puts forward that something can be in a person's interest without the person actually being interested in it. For example, the ingestion of vitamin C was in the interest of 18th century sailors, despite the fact that they were unaware of this, as they had never heard of vitamin C (an example Dierks borrows from Varner 1998). Dierks also relies on Taylor's (1986) distinction between someone's apparent good and someone's actual good. Someone can have a preference for copious amounts of alcohol, but this is clearly not in this person's interest regarded objectively. In other words, something can be in someone's interest even if this person is not aware of this, or does not actually regard it as being in her interest. However, what Dierks fails to show in our view, is how the distinction between someone's apparent and someone's actual good pertains to the question of whether an entity is morally considerable. After all, we are already talking about morally considerable beings here; sentience could be seen as the threshold condition for membership in the moral community and once one belongs to the community all interests of that being should be taken into account, including the ones the being is unaware of. In other words, the fact that objective interests exist that are not subjectively experienced as such does not prove anything in this case, because this person at least has *some* goods that she experiences subjectively and these can form the basis for her moral status quite independently of any goods or interests that she might have and that she is unaware of. Dierks quotes Attfield (1981, p. 41) as saying that, "Truistically [plants] are unaware of their interests: but even creatures with cognition are often unaware of theirs, whether they are flourishing or not." Yet, creatures with cognition are at least some of the time conscious of their interests; sufficiently often to warrant the claim that their interests matter to them in general. This is not something that, according to today's available knowledge, can be said of plants.

[1]Biological goals are seen as "the fulfilment of a thing's biological functions" (McShane 2015, p. 139).

4 From Is to Good to Ought

In environmental ethics a move is often made, first from 'is' to 'good' and then from 'good' to 'ought' (Nolt 2006, 2009). It is argued that there are goods in nature and that therefore we ought to protect these goods. For example, Rolston (1988, pp. 98–100) argues that "organisms are evaluative systems that seek a valued state" and that because "moral agents should consider the consequences of their actions for evaluative systems", we need to take into account the consequences of our actions for organisms (Nolt 2006). Similarly, Taylor (1986) argues that by virtue of being alive all living entities have a good of their own and that we ought to respect whatever has a good of its own, hence we owe moral respect to all living entities. However, the step from something having a good or seeking a valued state to being worthy of our moral concern needs to be supported with arguments. Nolt (2006) examines a number of possible argumentation lines and finds those arguing from goods to direct obligations to natural entities wanting. This type of reasoning starts with the premise that we find a specific 'good-making characteristic' of ourselves worthy of direct moral consideration and then argues that since we have to treat like cases alike, natural entities that have these characteristics are also worthy of direct moral consideration. It is then argued that natural entities share with us a will to live or a good of their own, for example. Yet, as Nolt argues, there are other characteristics, most notably sentience, that plants or genes or species do not share with us and it is not clear that if we were to take away sentience, we ourselves would still be morally considerable. In other words, it is not self-evident that characteristics like being alive or having a good of one's own, without a characteristic like sentience, makes moral agents obligated towards us. As Nolt summarizes this argument:

> Such extrapolating justifications are often plausible for human or animal ethics, but they fare poorly, I think, in a broad biocentric ethic… and in environmental ethics generally… Because the goods of different natural entities vary widely in character, it is difficult to extrapolate obligations convincingly from one sort to others… To employ the principle of parity- that like cases should be treated alike – we need like cases (2006, p. 362/3).

Summing up, an extra argument is needed to make the step from an entity having a particular good or characteristic to moral agents having obligations towards this entity. In the case of sentience the argument could be that sentient creatures take a subjective interest in their own welfare in the sense that it matters to them what happens to them. If we want to make the step from having a will to live or having a good of one's own to direct moral obligations, we need to argue what it is about these characteristics that make them morally relevant. We think that until such an argument has been convincingly made, this debate is inconclusive. But then, how could animal ethicists respond to the charge that if they cannot attribute direct moral status to species they cannot account for the widely held intuition that we should preserve biodiversity or favor the protection of endangered animals? Of course, they can point to indirect moral status: it is in our and other animals' interest to preserve species, because diverse ecosystems tend to be more robust or because we simply tend to value rare things more than common things. Nevertheless, this type of moral status, which is contingent on our own interests, would probably not suffice for ecocentrists.

The second point of dispute between animal and environmental ethicists is about whether individuals or collectives should be the locus of concern. More particularly, ecocentrists argue that as wholes are not reducible to just the sum of their parts, wholes should be granted an independent status. Some animal ethicists have feared that such a move to holism will result in the subordination of the members of a community to the good of the community, causing Regan (1983) to term ecocentrism 'environmental fascism'. Ecocentrists argue that this idea of subordination relies on a mistaken picture of humans (or animals) as being completely separated from nature, while they are in fact intimately interconnected with nature (Lema 2014). Individualists, on their part, argue that only individual organisms and not wholes have biological goals (whether or not these are conscious goals) and that therefore wholes are not morally considerable (McShane 2015, p. 135). A problem for this view, according to McShane (idem, 139) is that organisms are in fact "a lot more like wholes than like individuals… [Even humans are in a functional sense] more like ecosystems than like discrete individuals". It has even been argued that an individual's behavior is partly determined or constituted by interaction with microbes in her body (see for example Dinan et al. 2015). While this may be true in a biological sense, the question is how this translates to moral relevance. At least the sentientist could still claim that the unit of concern is the unit where goods are experienced.

Despite the fact that we so far remain unconvinced by the ecocentrists' defence of direct moral status of ecosystems, species, and populations, we do acknowledge that holism has a strong intuitive appeal and at least at first sight could answer certain practical questions more adequately than individualist animal ethics. When we think of specific clashes between animal liberationists and ecocentrists, we cannot help but think that both have a valid point of view, but both emphasize different aspects of the same problem. For example, in the Netherlands we have witnessed a heated debate about the introduction of large herbivores into a newly established nature area, the Oostvaardersplassen (see Klaver et al. 2002). The horses and cattle in question are part of a program of dedomestication. Because the aim is to create a wild ecosystem and the animals are perceived to be part of such an ecosystem, nature managers have decided that the animals should become wild by leaving them alone as much as possible. This means that when their numbers have exceeded the carrying capacity of their habitat, particularly in winter when insufficient food is available, they may starve. Rather than using a strategy of supplementary feeding, nature managers use reactive culling, or mercy killing, to release them from their suffering. Moreover, they are very reticent to provide veterinary assistance. Animal protection groups and many members of the public strongly disagree with these strategies and argue that when we put a fence around a nature area, prohibiting the animals from moving elsewhere in search of food, we have a duty to give supplementary food and veterinary care when the animals are starving or ill. The value of wildness that ecocentrists champion in this case clashes with the value of animal welfare that animal liberationists want to protect. How could we reconcile these different viewpoints? We think that even though there is as of yet no knockdown argument for the moral relevance of wholes, there are still collective dimensions that individualistic animal ethics needs to take into account and that could reconcile seemingly opposing viewpoints in dilemmas such as these.

5 Ways to Conceptualize Collective Dimensions

We think the gap between environmental and animal ethics could, at least partially, be bridged by focusing more on the collective dimensions in our concerns with animals. A variety of considerations exist between the idea of animal ethics that every concern should be reducible to respect and care for individual animals on the one hand, and the ecocentric focus on care for only collectives such as ecosystems on the other hand. In an attempt to blur the boundaries between individualism and holism, we ask what types of moral concerns may be linked to the collective dimensions of moral problems, even if we were to only attribute direct moral status to individuals. We think that there are different ways to conceptualize collectivity that could possibly be morally relevant. Drawing on insights from public health ethics, we explore various (novel) ways of conceptualizing the moral relevance of collectiveness in animal life. We do this because public health ethics has also developed partly in response to mostly individualist approaches in human bioethics, creating more room for recognizing the value of population health, interpersonal relations, solidarity, and ways in which a collective is constituted.

What do we mean when we are speaking of collective dimensions in our concerns with animals? In public health ethics, collective dimensions have been conceptualized in at least four ways, and most of these may have relevant analogies in animal ethics. We do not wish to argue that all insights from public health ethics can be translated directly to the animal ethics case; rather, we use these analogies as a heuristic device to try to track down collective dimensions in animal ethics. What we aim to show with these analogies is that we cannot think adequately about individuals without also thinking about group structures. The four collective dimensions we will distinguish are not meant to be mutually exclusive; overlaps between them are possible.

Firstly, in public health ethics it is natural to be concerned with the health and welfare of the public, understood as just the aggregation of the health of all individuals in the group (Verweij and Dawson 2007). In ethical concerns about human health, numbers make a difference. If we need to choose between an intervention that would make either a large number of people better off, or just a smaller number of people, then, other things being equal, we should choose the former. This will be obvious for utilitarians, but numbers are also relevant in many non-consequentialist approaches, such as contractualism (Hirose 2001). A similar line of reasoning holds for our treatment of animals; an animal experiment, for example, is regarded as more problematic the more animals are used. This is the case both when we argue from a utilitarian and from a rights perspective. According to Regan's (1983) 'miniride principle', when we have several groups with comparable harms at stake, we should harm the group that has fewer individuals. So one way to be morally concerned with collectives is through the aggregation of welfare or claims of individuals.

Secondly, certain characteristics of the group can have repercussions on the level of the individual. We can think of group welfare as something that goes beyond the mere aggregate of individual members' welfare. In public health as well as for animal health we can see the value of realizing and maintaining herd immunity (Fine 1993). Collectives with herd immunity are, in some senses, better off, than groups without herd immunity. Each individual benefits from being protected within the herd, even those who are not immune, or those who are ill and weak, and therefore more vulnerable to certain infections. The value of herd immunity is not necessarily reducible to aggregate welfare, because, at least in theory, the aggregate health of the group might be better if the weaker individuals were not protected and only the healthier part of the population would survive. This is most clear if aggregate health is understood as average health. Precisely because we are not only concerned with numbers or averages, but also with how 'strong' the group is *as a group* this concern cannot be merely reduced to concern for the group as consisting of the aggregate of many individuals. In the animal domain, the idea of herd immunity is equally applicable. Similarly, sanitation, clean water and adequate housing conditions are measures that benefit a whole group, both of humans and animals. But there are more reasons for being concerned with group health in distinction to the (aggregation of) health of individuals. For example, it is possible that certain risk factors only exist because of the genetic composition of the group. A population of animals can be healthy on average, but because its gene pool is not very diverse, the group as a whole could be considered unhealthy, or at least vulnerable. Also, there are situations in the human context where the health or welfare of individual members of a group are sacrificed for the health or welfare of the whole group. For example, when a soldier gets sent to war to fight for the good of the whole community (Tannenbaum 1991). In the animal context, a farmer may sometimes have to choose to improve the welfare of the whole herd by diminishing the welfare of a number of individuals in the herd, for example when the cost of medically treating individual animals compromises the farmer's ability to look after the welfare of the whole herd (Tannenbaum 1991). Apparently, here the welfare of the group exceeds the welfare of the aggregate. Another example where the collective dimension goes beyond the mere aggregate is provided by the social structure of prey animals. This structure can only exist due to the fact that they are predated on. This may not be good for the individuals that are preyed upon, but it is good for the collective, indeed it constitutes the characteristics of the group. For example, prairie dogs' quite sophisticated communication has evolved in a specific way due to the variety of predators they encounter; they use different alert calls for eagles than they do for snakes, for example. This way their calls can elicit the most effective response to danger (Slobodchikoff 2012). This group characteristic of being a prey animal has had important consequences for the development of communication styles and thereby has influenced the constitution of the individual members.

Thirdly, the structure of groups can greatly influence individuals, indeed it can even play a constitutive role for the characteristics of individuals. One example in the animal context is that of the social animal. For social animals interaction with conspecifics is so important that they would suffer greatly when they were held in isolation. Moreover, groups often have internal characteristics that surpass the aggregate of the characteristics of the individual members. In particular, inequality and hierarchy can exist within groups and one's role in the group can determine one's fate. Hierarchies and inequalities can be important for the internal organization of the group and in the end be in all the members best interest, for example when strong leaders maintain the peace in the group. While inequalities can be meaningful, they can of course also be problematic. In the realm of public health it is often considered ethically problematic if a population shows structural inequalities in health, notably where such inequalities are linked to societal and economic factors (Marmott and Bell 2012). The question of which inequalities are indeed to be considered as *inequities* is a complex matter, but the focus on distribution of health and quality of life within a population is clearly another ethically relevant concern about that population—at least so in relation to human health and quality of life.

Could a parallel be drawn with animal ethics? In other words, does it make sense to see unequal distributions of health, welfare or freedom within a group of animals as unjust? Arguably, if we have an obligation to care for the well-being of those animals, we have good reason to give priority to the condition of animals that are worst off: those that are ill or in pain, that are suffering, or otherwise lack in quality of life. Whether this also means that we should take away inequalities within a group of animals is a matter of debate. Are such inequalities within the group to be considered morally problematic in relation to the group as such, or in relation to the question of how the group (e.g. a group of cattle) is treated by relevant care-givers (e.g. farmers)? In other words, is distributive justice among animals a meaningful concept? At first sight one might think it is not, as animals have no concept of unfairness. However, we could ask whether one needs to have a concept of unfairness in order to be able to be treated unfairly. What about children? Young children have a very rudimentary conception of fairness at best, one, moreover, that they share with certain animals (most notably primates; see Brosnan and de Waal 2003). In fact, it has been argued that playing fairly among animals teaches animals (and humans) to manage conflicts and to promote peaceful relations within the group; something that benefits the group as a whole (Palagi et al. 2016). On the other hand, a relation seems to exists between justice and self-respect. Being treated unequally is a greater affront to those creatures that can experience (diminished) self-respect, and this seems to be a capacity that is closely tied to human social relations (although we do not wish to exclude the possibility that future research demonstrates that the basic conditions for self-respect can be found among animals). Yet, even if being treated unequal to one's neighbors is worse for humans as it is an offense to their self-respect, this does not prevent us from saying that there is also an element of justice involved in distribution among a group of animals.

Fourthly, a number of authors in bioethics reject a mere individualistic approach to public health and emphasize more collectivist or communitarian values, such as solidarity. Solidarity can be considered in various ways, but in its strongest form it refers to a value that is intrinsically linked to the identity of the group as a whole, where individuals within the group also, in an important sense, see themselves as part of a group rather than as solitary individuals (Dawson and Verweij 2012; Prainsack and Buyx 2011; Verweij 2015). Personal identity is then at least partly experienced as being constituted by one's role in, or membership of, the group. In public health and many other areas of societal life we have good moral reasons to strengthen and honor such social bonds, as they are a necessary basis for flourishing of the group as well as individuals within the group. Similar forms of community can be found in the animal kingdom. For example, there are many accounts of groups of animals that protect their weaker individuals. A threat to one individual can evoke a counter-attack of the entire group. In some cases, individuals can even be identified by their role in the group in a much more extreme sense. In a beehive, individual worker bees sacrifice themselves for the other bees, in particular the queen bee, in order to safeguard the continued existence of the whole group.

A second way in which we can speak about group identity is when we are dealing with capacities that can only be understood in collective terms. For example, ants or bees appear not to possess a high degree of intelligence on an individual level, but collectively they have a type of intelligence with which they can pursue highly advanced goals. The collective could in this case be regarded as an organism in its own right (See also Huebner 2014 on what he terms 'macrocognition', arguing that some collectives have cognitive capacities similar to those of individuals). Another way in which in animal ethics the individual's identity is strongly linked to that of the group is when we consider the idea of species-specific flourishing. With her capabilities-approach, Nussbaum (2006) argues that we have a duty to contribute to the flourishing of all sentient beings and therefore we need to take into account what capabilities are central to the specific species that individual sentient beings are a member of. After all, there is a lot of diversity in the animal kingdom and different kinds of animals need to develop different capabilities to be able to flourish. Without giving direct moral standing to a species, therefore, we might still have reason to give value to species because if we want species-specific flourishing we need a concept of species. As Nussbaum phrases it: "Creatures cannot flourish in isolation, and thus for animals, as for humans, the existence of suitable groups and communities is an important part of the flourishing of individuals" (2006, p. 357). While in the second dimension that we distinguished—that of herd immunity—characteristics of the group influenced those the individual, in this fourth dimension the individual derives their very identity from that of the group.

6 Integrity and Wildness

We have explored possible analogies between moral concerns in public health- and animal ethics, seeking various ways of explicating how collectives or collective dimensions can be morally relevant. The upshot of this exercise is that in many cases it is difficult to think of the status of individuals without thinking about the collective and vice versa. Of course, in environmental ethics it is not uncommon to look at collective dimensions, but in this exploration we are particularly interested in approaches that can be related to the moral status of *individual* animals, as we concluded in the first section of this paper that the arguments for attributing moral status directly to collectives are unconvincing. By emphasizing the collective dimensions of our thinking about individual animals, we also hope to have opened an avenue for realizing the moral status of collectives, albeit an indirect moral status. We should point out that while the first three dimensions can be reduced to individual concerns, the fourth dimension cannot be completely individualized.

In the realm of animal ethics we can also find other considerations that we can understand as collective dimensions, that just like the species-specific flourishing idea, have to do with the species concept. For example, in the debate about the genetic modification of animals the 'integrity argument' is sometimes referred to (Bovenkerk et al. 2002). This argument states that interfering with an animal's 'essence' violates that animal's integrity. It is not obvious that violating animal integrity is violating that particular individual's integrity. Integrity appears to be a notion that refers to 'how we think a type of animal should be'; it refers to the 'cowness of a cow'. This is a view about what animals are like as a group. By referring to some kind of shared essence of a species or population or breed, we are making claims on a collective level. Nevertheless, seeing that argument as fully 'collective' or essentialist and ignoring the link with particular individuals could be an oversimplification. After all, in practice the genetic modification is done to specific individuals and gets expressed through the phenotype of individual animals. Similarly, the value of wildness can pertain to a group of animals rather than to individuals. For example, when animals are bred in captivity they tend to become less wild (for different interpretations of wildness see the contribution by Clare Palmer in this volume) and this is perceived as a loss of value. On the level of each individual the changes might be small, but cumulatively, after generations of breeding, the wildness of the group may have decreased to such an extent that they become domesticated. This in the end will have consequences for future individuals. While arguments based on wildness or integrity seem to be difficult to sustain within an individualist-focused animal ethics, analyzing them in terms of collective values might offer possibilities to reinterpret them and give them a stronger basis.

7 Practical Implications

While the focus of this chapter was primarily to make a conceptual contribution to the debate and to show that animal- and environmental ethics are not as diametrically opposed as is often assumed, we will give a number of examples here where we think the gap between the two ways of thinking can be bridged by focusing more on collective dimensions. In the aforementioned Dutch nature area Oostvaardersplassen Heck cattle and konik horses are largely left to their own devices because nature managers want to dedomesticate them as part of a project of rewilding. They have met with fierce opposition from the public and animal rights advocates for withholding necessary care from the animals. However, when we consider the collective as an aggregate and also when we think about the collective health of the population, we come to see that for most individuals in the population non-interference is beneficial. The population is strengthened when more robust individuals survive and when the group becomes better able to sustain itself and this is ultimately in the interest of many current and future individuals of the group. On the other hand, a focus on the interests of the individual animals shows that we need to take care to harm the individuals as little as possible. This encourages us to look for possible alternatives; are there other ways in which the health and self-sustaining power of the group can be strengthened whilst avoiding unnecessary suffering? One could think of euthanizing weak or sick individuals, birth control measures in case of overpopulation, or the creation of corridors, so that the animals have a larger area to roam and stand a better chance to survive when resources are scarce. In this example, we can see that the value that we attach to wildness can also be in the interest of the animals, both regarded as individuals and as a group.

Another example is provided by the practice of 'therapeutic hunting' of species that tend to regularly exceed the carrying capacity of their habitat (so-called obligatory management species; see Varner 2003). From an ecocentrist point of view this is allowed in order to protect ecosystems from being destroyed. Varner (2003) argues that it should also be allowed by animal ethicists in order to protect the aggregate welfare of the hunted species, both of the current and the future individuals of that species, when no viable alternative forms of population control are present.[2] This is the case for utilitarians, as they have the obligation to minimize suffering and more suffering would result from not hunting obligatory management species than from letting nature run its course. After all, when these animals overshoot the carrying capacity of their natural range, many individuals will die a slow hunger death. Varner argues that even for animal rights defenders who follow Regan (1983) therapeutic hunting should be preferable, when no alternatives exist:

> If fewer animals will die if therapeutic hunting is used to regulate a wildlife population than if natural attrition is allowed to take its course then… therapeutic hunting is not only permissible but a morally mandatory expression of respect for animals' rights (2003, p. 106).

---

[2]We acknowledge that the creation of wildlife corridors is an option, but unfortunately one that has too little effect in practice.

He bases this on Regan's use of the 'miniride principle' which states that if comparable cases of harm (in this case animal death) will take place we should harm the smaller group rather than the larger.[3] While animal and environmental ethicists might have different justifications for these practices, they can nevertheless reach similar conclusions. This means that animal ethicists would need to have more eye for the collective dimensions of animal life and ecocentrists would need to acknowledge that concern for individual animals does not necessarily run counter to collectivist aims.

8 Conclusion

We have argued that the discussion between atomistic animal ethicists and holistic environmental ethicists is too polarized. While we remain unconvinced by ecocentric arguments that aim to attribute direct moral status to collectives, we have argued that in many cases such direct moral status is unnecessary if we want to take into consideration collective dimensions. While we realize that this does not create a solution for all involved—in particular not for die hard ecocentrists—we hope at least that animal ethicists will in future have more eye for collective values and that environmental ethicists will become more open to individual values. We have presented a diversity of ways of thinking about collectives. Between the two extremes of on the one hand a sole focus on individuals and on the other hand a sole focus on collectives a variety of positions exist. We hope to have shown that thinking about collectives can co-exist, at least up to a certain point, with an individualistic perspective on ethics. The individual and collective dimensions are not necessarily mutually exclusive. In fact, in some of our examples the boundaries between the individual and the collective become blurry, as it is impossible to think of the status of individuals without also thinking about the collective and vice versa. The position and role of an individual in the group and the group characteristics are constitutive of the individual and its situation. We conclude, then, that a partial bridge has been built between the two camps. The next step will be to investigate the ways in which an exploration such as this will help to clarify practical dilemmas where animal and environmental ethics clash.

Acknowledgments We would like to thank Jozef Keulartz and the members of the Philosophy Group of Wageningen University for their helpful comments.

[3]It could be countered that a moral difference could be made between killing and allowing to die, but Varner disagrees. This is a complicated debate that we cannot hope to settle here.

References

Attfield, R. 1981. The good of trees. *The Journal of Value Inquiry* 15: 35–54.

Bovenkerk, B., F.W.A. Brom, and B.J. Van den Bergh. 2002. Brave new birds: The use of "Animal integrity" in animal ethics. *Hastings Center Report* 32(1): 16–22.

Brosnan, S., and F. De Waal. 2003. Monkeys reject unequal pay. *Nature* 425(6955): 297–299.

Dawson, A.J., and M.F. Verweij. 2012. Solidarity: A moral concept in need of clarification. *Public Health Ethics* 5(1): 1–5.

Dierks, J. 2014. *Taking genes seriously. An interest-based approach to environmental ethics and biodiversity conservation*. Dissertation, Greifswald University.

Dinan, T.M., R.M. Stilling, C. Stanton, and J.F. Cryan. 2015. Collective unconscious: How gut microbes shape human behavior. *Journal of Psychiatric Research* 63: 1–9.

Eckersley, R. (1992). *Environmentalism and political theory: Toward an ecocentric approach*. London: UCL Press.

Johnson, L.E. 1992. Toward the moral considerability of species and ecosystems. *Environmental Ethics* 14(2): 145–157.

Fine, P.E.M. 1993. Herd immunity: History, theory, practice. *Epidemiologic Reviews* 15(2): 265–298.

Hirose, I. 2001. Saving the greater number without combining claims. *Analysis* 61(4): 341–342.

Huebner, B. 2014. *Macrocognition. A theory of distributed minds and collective intentionality*. Oxford: Oxford University Press.

Klaver, I., J. Keulartz, H. Van den Belt, and B. Gremmen. 2002. Born to be wild: A pluralistic ethics concerning introduced large herbivores in the Netherlands. *Environmental Ethics* 24(1): 3–21.

Lema, S. 2014. The ethical implications of organism-environment interdependency. *Environmental Ethics* 36: 151–169.

Marmott, M., and R. Bell. 2012. Fair society, helathy lives. *Public Health* 126(1): S4–S10.

McShane, K. 2015. 'Individualist biocentrism vs. holism revisited. *The Ethics Forum* 9(2): 130–148.

Nelson, M.P. 2010. Teaching holism in environmental ethics. *Environmental Ethics* 32(1): 33–49.

Nolt, J. 2006. The move from "good" to "ought" in environmental ethics. *Environmental Ethics* 28 (4): 355–375.

Nolt, J. 2009. The move from "is" to "good" in environmental ethics. *Environmental Ethics* 31(2): 135–154.

Nussbaum, M. 2006. *Frontiers of justice. Disability, nationality, species membership*. Cambridge, MA: The Belknap Press of Harvard University Press.

Palagi, E., G. Cordoni, E. Demuru, and M. Bekoff. 2016. Fair play and its connection with social tolerance, reciprocity and the ethology of peace. *Behaviour*. doi:10.1163/1568539X-00003336.

Palmer, C. 2010. *Animal ethics in context*. New York: Columbia University Press.

Palmer, C. 2011. Is something wrong with my bulldog? Breeding, ethics and harm to animals. *Old World and New World Perspectives on Environmental Philosophy*. Conference proceedings, ISEE conference Nijmegen, the Netherlands.

Prainsack, B., and A. Buyx. 2011. *Solidarity: Reflections on an emerging concept in bioethics*. Nuffield Council on Bioethics. http://nuffieldbioethics.org/wp-content/uploads/2014/07/Solidarity_report_FINAL.pdf.

Regan, T. 1983. *The case for animal rights*. Berkeley, CA: University of California Press.

Rolston, H. 1988. *Environmental ethics: Duties to and values in the natural world*. Philadelphia, PA: Temple University Press.

Singer, P. 1999. *Practical ethics* (2th ed., reprinted). Cambridge: Cambridge University Press.

Slobodchikoff, C. 2012. *Chasing Doctor Doolittle: Learning the language of animals*. New York: St. Martin's Press.

Tannenbaum, J. 1991. Ethics and animal welfare: The inextricable connection. *Journal of the American Veterinary Medical Association* 198(8): 1360–1376.

Tanner, J. 2009. The argument from marginal cases and the slippery slope objection. *Environmental Values* 18: 51–66.

Tanner, J. 2006. Marginal humans, the argument from kinds and the similarity argument. *Facta Universitatis Series: Philosophy, Sociology and Psychology* 5(1): 47–63.

Taylor, P. 1986. *Respect for nature. A theory of environmental ethics*. Princeton, NJ: Princeton University Press.

Varner, G. 2003. Can animal rights activists be environmentalists? In *Environmental Ethics. An Anthology*, ed. A. Light, and H. Rolston III, 95–113. Malden, MA: Blackwell Publishing.

Verweij, M.F., and A.J. Dawson. 2007. The meaning of "public" in "public health". In *Ethics, prevention, and public health*, ed. A.J. Dawson, and M.F. Verweij, 13–29. Oxford, UK: Oxford University Press.

Verweij, M.F. 2015. How (not) to argue for the rule of rescue. Claims of Individuals vs Group Solidarity. In *Identified versus statistical victims. An interdisciplinary perspective*, ed. I.G. Cohen, N. Daniels and N. Eyal, 137–149. New York: Oxford University Press.

Compassion as a Practical and Evolved Ethic for Conservation

Daniel Ramp and Marc Bekoff

1 Introduction

Every day, environmental policymakers, conservation practitioners, and land managers, make decisions that affect the lives of wild animals. The societal norms that shape the context for those decisions are complex, entrenched, and often opaque to the wider community. Despite the complexity, there is growing awareness that conservation decisions—particularly those aimed at resolving conflicts between humans and other animals but also conflicts between other animals—often result in the harm and death of other animals. That these animals carry the burden of achieving conservation objectives through their harm and death is often ignored or else justified on utilitarian grounds. It is common for some members of a single species to be killed for the "good" of their species or for members of one species to be killed for the "good" of another species (e.g., golden hamsters for black-footed ferrets; Bekoff 2010). Animals are also routinely killed to prevent them from moving from protected areas to private land (e.g., wolves and dingoes; Treves and Karanth 2003) and because they are considered to be invasive or pests (Littin 2010). These tradeoffs often result in harm and death for captive and wild animals, justified in the name of conservation and human benefit (Callicott 1990; Bekoff 2003; Ben-Ami et al. 2014). The question is where the trade-offs stop and the protection of indi-

This is a reprint from Jeffrey Moussaieff Masson and Susan McCarthy (1995), *When Elephants Weep. The Emotional Lives of Animals* (pp. 24–44). New York: Dell Publishing.

D. Ramp (✉)
University of Technology Sydney, Sydney, NSW, Australia
e-mail: Daniel.Ramp@uts.edu.au

M. Bekoff
University of Colorado, Boulder, CO, USA
e-mail: Marc.bekoff@gmail.com

B. Bovenkerk and J. Keulartz (eds.), *Animal Ethics in the Age of Humans*,
The International Library of Environmental, Agricultural and Food Ethics 23,
DOI 10.1007/978-3-319-44206-8_23

viduals begins. How many individuals are acceptable to kill and harm in the name of conservation (Vucetich and Nelson 2007) and at what point does their well-being matter (Bekoff 2013)? Conservation practice and policy have historically addressed this question uncertainly and often without virtue (Vucetich and Nelson 2013). This has not been an easy question to resolve, primarily because the problem has not been with conservation itself, but rather with the manner in which it has been performed.

In turn, this situation can, in part, be attributed to the perceived difficulty in comparing complex and competing sets of values (e.g., individual rights and values versus ecosystem health), notions of human exceptionalism and dominance, and the overly strong focus on species welfare (as opposed to individual welfare) in conservation metrics. The result is that without consensus on how to accommodate what appear to be contradictory values, moral confusion and ethical dilemmas are common. Consequently, there has traditionally been a lack of animal welfare concern in conservation decision making when addressing the stark but real question of which animal lives and which animal dies, in part because of past disagreements between conservation and animal welfare scientists on the role of individual welfare in conservation (e.g., Soulé 1985). Environmental legislation and policy is almost wholly grounded in measures of species and ecosystem welfare (Gaston and Fuller 2007), with little emphasis on individual welfare. In contrast, animal welfare legislation often has tight control over the individual welfare of domestic animals and livestock. Wild animals are frequently exempted from animal welfare legislation by categorizing them as *pests*, a term used to define a species that is a nuisance, out of balance, invasive, or exotic (Nagy and Johnson 2013). Often, the issues faced by conservation managers are in relation to species that are placed in this category, and, perhaps not coincidentally, transgressions of humane treatment are frequently justified by the application of this label (Littin 2010).

Language and culture are strongly linked to our treatment of other animals (Webb and Raffaelli 2008). Societal norms are often rendered with euphemisms to surmount the impression of transgressions of equity and justice, therefore obscuring the obvious ethical conflict our actions cause. Bias in the selection of flagship species to protect, such that flagship species are used to leverage support for broader conservation objectives, is a complex mix of charisma (Lorimer 2007; Tisdell and Nantha 2007) and politics (Yeo and Neo 2010). Animal protection is strongly influenced by prevailing views on sentience and cognition (Bekoff and Jamieson 1990), position on the phylogenetic scale (Harrop 1999), and utility to humans (O'Sullivan 2011). Of course, not all species can be saved from extinction, nor can all individuals be spared from dying or suffering needlessly at the hands of humans. But there is currently little deliberate ethical enquiry in how tradeoffs in lives are made (Larson 2007), and there has been little room for empathy (Bekoff 2013).

2 Blockages in Conservation Decision-Making

The focus on species welfare and anthropocentric views of nature has been an impediment to scientifically validating the inclusion of individual welfare in conservation. Although species welfare, with its focus on the prevention of extinction, is a vital and admirable conservation objective, the welfare of individuals and their social groups should also be considered as important. Harm to animals is more than just the extinction of species and subsequent declines in biodiversity. This awareness stems from the dramatic rise in human–wildlife conflicts (Redpath et al. 2013) and a growing recognition of the intrinsic value of conscious and sentient animals (Bekoff 2007, 2014). Harm also encompasses the suffering experienced by individuals and associated costs to social units and the populations to which they contribute (Fraser 1993; Paquet and Darimont 2010). Although many of these harms result from global conservation problems, such as habitat loss, climate change, and pollution, many harms paradoxically result from humans engaging in proactive conservation measures (Fraser and MacRae 2011).

Although clear and inclusive ethical foundations for conservation have been articulated numerous times, these foundations have been gradually eroded by human needs and benefits and prevailing views of human exceptionalism. Motivations for humans to engage in conservation encompass a wide variety of societal values, incorporating ethical, aesthetic, and economic concerns (Doak et al. 2013). These often competing values are increasingly resulting in conflict and present considerable philosophical, moral, and practical problems for conservation planners and policymakers. Conservation interventions require decision-makers to make tradeoffs among philosophical and ethical value sets and practical limitations. Problematically, conflict resolution between conservation and other human endeavors is, more often than not, dictated by decisionmaking aimed at economic and utilitarian values (Artelle et al. 2014), to the detriment of aesthetic and ethical values (Jepson and Canney 2003). The denial of the intrinsic value and sentience of nonhuman animals is also strongly influential. Despite having the protection of animals and conflict resolution at its heart, conservation science as a discipline has struggled to establish a strong environmental ethic and moral philosophy, to the detriment of individual animals.

3 Conservation Is Ethically Challenged

Discourse surrounding the ethical position of conservation is not new (Ehrlich 2002, 2009; Minteer and Collins 2005; Dunlop 2006). Although the dominant paradigm governing conservation policy is based on a limited anthropocentric version of utilitarianism, which is a form of consequentialism, there are several other ethical positions with implications for conservation and animal protection, such as deontology (Regan 1983) and deep ecology (Naess 1973; Routley 1973) that reject the utilitarian view of the environment (i.e., the shallow view). The importance of ethics in conservation was noted as part of the founding principles of the Wildlife Society in 1937 in the United States, in which the "development of all types of wildlife management along sound biological lines" was seen as a crucial move away from indiscriminate exploitation for human benefits (Bennitt et al. 1937). Aldo Leopold further championed this idea in 1949 with his land ethic

(Leopold 1949). At that time, Leopold firmly embedded the sharing of land between humans and other animals in his land ethic, allowing for both the preservation of wilderness where possible and for economic exploitation (Callicott 1990). Importantly, the land ethic approach encompassed the view that individuals and their wellbeing mattered (Davradou and Namkoong 2001). The implementation of this ethic, however, has remained stymied by the more dominant anthropocentric utilitarian ethic that espouses decisions that promote overall well-being (actions for the greater good). Not only has exclusive concern for human benefits taken a strong hold of motivational mechanisms (e.g., as proposed in the new conservation; Kareiva and Marvier 2012), but the grounding of conservation in species welfare has, perhaps unintentionally, subjugated individuals for collective benefits.

The challenge for conservation science is to decide whether current decision-making in practice and policy evolves to accommodate contemporary moral systems and to incorporate the rapidly developing scientific understanding of the sentience and consciousness of nonhuman animals or whether, at a practical level, the trumping of welfare outcomes for individual animals by species (or population or ecosystem) welfare or by human benefits continues to be permissible (Fraser 2012). Evidence from conservation policy suggests that this challenge is often sidestepped and that the *status quo* remains by default (Artelle et al. 2014). Why has this happened? Negativity toward individual welfare has become entrenched because it has been viewed as an impediment to holistic decisionmaking. Dominant and entrenched values are maintained through individual and collective aspirations (Jepson and Canney 2003) and through cultural norms and perceptions (Lejano and Fernandez de Castro 2013), whereas disengagement of moral selfsanctions allow for harmful practices by removing self-restraint (Bandura 2007). Entrenched views can have a significant and wide-ranging influence on conservation policy and decisionmaking, such that vocal stakeholders often assert valuebased rhetoric to the detriment of alternative value sets (e.g., Ramp 2013). This is evident in a wide variety of conservation issues that result in direct conflict between advocates of alternative positions (Redpath et al. 2013) but, conversely, can also drive efforts to alleviate conflicts via education programs and comanagement strategies (e.g., Gelcich et al. 2008).

4 Compassion in Conservation

The good news is that the asserted dichotomy between conservation and animal welfare has waned, and the unifying aim of reducing harm to nonhuman animals in both these disciplines has gathered momentum (Fox and Bekoff 2011; Fraser 2010; Littin 2010; Paquet and Darimont 2010; Bekoff 2013, 2014). The challenge has been for decision-makers to establish a practical framework (i.e., one that is ideologically sound and able to be implemented through on ground activities) that accommodates the ethical and moral position of individual animals within conservation practice (Bekoff 2002). In a case of widespread convergent evolution, examples of conservation initiatives aimed at achieving both conservation goals and reduced harm to individuals have emerged over the last few decades. Despite the dominant lethal paradigm underlying many conservation initiatives (Bergstrom et al. 2014), compassionate approaches to solving conflicts between humans and wildlife have evolved. These highlight the rise of what has been termed *compassionate conservation* (Bekoff 2010, 2013), a rapidly growing international and cross-disciplinary movement that stipulates that we need a conservation ethic that incorporates the protection of other animals as individuals, not just as members of populations of species but valued in their own right.

The evolutionary underpinnings for the ethical behavior of compassion lie in the evolutionary processes that reward empathic adaptations (Goetz et al. 2010), but compassion can be practically defined as reflecting empathy in humans for nonhuman animals and a drive to alleviate harm and suffering. Unlike the dominant utilitarian approach to conservation, which puts the cost of reaching conservation targets squarely on the shoulders of other animals, a compassionate ethic for conservation brings empathy into decisionmaking alongside other values. It is not a rights position but, rather, puts forward a scientific and evidence-based conceptual approach that stipulates that conservation initiatives should first do no harm (Bekoff 2010). This is important not only because of what we now know about the cognitive and emotional lives (consciousness and sentience) of other animals (Bekoff 2007; Bekoff and Pierce 2009) but also as a moral imperative for providing modern solutions for sharing space with nature and for fostering the possibility for diverse species to live in peaceful coexistence (Hinchliffe et al. 2005). Compassionate conservation allows for—but does not prescriptively dictate—outcomes in which the interests of others supersede those of humans.

Examples in which compassionate conservation principles are being applied have been steadily growing. Predation by carnivores is being successfully managed using fencing, fladry, and guard animals (Fox and Bekoff 2011) rather than through shooting, trapping, and poisoning programs. Pack structures in unexploited predator populations show evidence of declining rogue behavior and pack stability (Purcell 2010), thereby reducing negative interactions with humans (Bekoff and Jamieson 1996). Issues in which species interfere with human endeavors are being solved by focusing on holistic ecological solutions and educational programs (e.g., restricting access to rubbish tips and putting roofs on city rubbish bins for ibis in Australia, as in Martin et al. (2012), and the use of fencing and local community engagement in the management of brown bears in Turkey; Ambarli and Bilgin 2008). The costs of ignoring individual welfare in scientific programs are also gaining attention (Jewell 2013), because mainstream invasive techniques for studying animals are now known to be stressful and change species-typical behavior (Bekoff 2000). Human–wildlife conflicts can be exacerbated by a lack of science, transparency, and concern for harm to animals in decisionmaking (Artelle et al. 2014).

Compassionate conservation is challenging decision-makers to have clear objectives where the lives of animals are affected. If interventions are necessary, the range of values of different stakeholders (human, nonhuman) should be articulated so that trade-offs may be transparently evaluated. Adaptive management principles are a must, in which scientifically credible monitoring programs measure key performance indicators. For situations in which harm to animals, either from an individual or species welfare perspective, is part of an intervention option, clear data are needed on the minimum number of animals that will be affected to achieve the desired outcome. Rating systems for harm exist for domestic animals and livestock and should be integrated into conservation interventions targeting wildlife.

5 Cause for Optimism and a Call to Action

Although millions of animals are killed in the name of conservation around the world each year, there is good cause to be optimistic. Acceptance of the intrinsic value of wild animals in nature and an associated motivation to prevent harm to those animals, is often reported as an almost universal ethic among a wide variety of stakeholders (Butler and Acott 2007; Dubois and Fraser 2013). Compassionate conservation can help by providing solutions to conservation conflicts that have either proved intractable to resolve or incurred a heavy welfare cost to wild or captive individuals. That is not to say that individual welfare supersedes species or ecosystem welfare, only that we cannot continue to ignore individuals in conservation practice. By considering animal welfare alongside animal conservation, it becomes possible to establish wildlife conservation frameworks that are explicitly oriented toward the management of the lives of individuals and their social groups and not just the species or population as a whole (Fraser 2010; Paquet and Darimont 2010). The word management reflects the fact that, in today's world, humans must interfere in the lives of other animals to resolve conflicts that inevitably arise from sharing space.

Despite clear progress, decision-makers have only just begun to recognize the importance and utility of a compassionate and practical framework for conservation decision-making and triage. Difficult questions need to be asked about how best to engage with nature to resolve conservation conflicts. The aspirations underlying compassionate conservation have been long held by those interested in protecting nature, but as a movement, with a clear framework of operation, compassionate conservation is in its infancy. Factoring compassion into conservation presents a simple and morally acceptable approach to resolving issues of land sharing by using the universal ethic of a concern for the suffering of others and attempts to alleviate it. Peaceful coexistence with other animals and their homes, grounded in compassion, is needed in an increasingly human-dominated world if society is to preserve and conserve nature in holistic and humane ways.

Acknowledgments We are indebted to the Born Free Foundation, particularly Will Travers and Chris Draper, for supporting and promoting the conceptualization of compassionate conservation. Liv Baker, Dror Ben-Ami, Louise Boronyak, and Kate Litten provided helpful discussion. The manuscript was greatly improved by comments from Paul Paquet and anonymous reviewers.

References

Ambarli, H., and C.C. Bilgin. 2008. Human–brown bear conflicts in Artvin, north-eastern Turkey: Encounters, damage, and attitudes. *Ursus* 19: 146–153.

Artelle, K.A., J.D. Reynolds, P.C. Paquet, and C.T. Darimont. 2014. When science-based management isn't. *Science* 343: 1311.

Bandura, A. 2007. Impeding ecological sustainability through selective moral disengagement. *International Journal of Innovation and Sustainable Development* 2: 8–35.

Bekoff, M. 2000. Field studies and animal models: The possibility of misleading inferences. In *Progress in the reduction, refinement, and replacement of animal experimentation*, ed. M. Balls, A.M. van Zeller, and M.E. Halder, 1553–1559. Amsterdam: Elsevier.

Bekoff, M. 2002. The importance of ethics in conservation biology: Let's be ethicists not ostriches. *Endangered Species Update* 19: 23–26.

Bekoff, M. 2003. Minding animals, minding Earth: Old brains, new bottlenecks. *Zygon* 38: 911–941.

Bekoff, M., D. Ben-Ami, K. Boom, L. Boronyak, C. Townend, D. Ramp, and D.B. Croft. 2007. *The emotional lives of animals*. Novato, CA: New World Library.

Bekoff, M. 2010. Conservation lacks compassion. *New Scientist* 207: 24–25.

Bekoff, M. (ed.). 2013. *Ignoring nature no more: The case for compassionate conservation*. Chicago: University of Chicago Press.

Bekoff, M. 2014. *Rewilding our hearts: Building pathways of compassion and coexistence*. Novato, CA: New World Library.

Bekoff, M., and D. Jamieson. 1990. Cognitive ethology and applied philosophy the significance of an evolutionary biology of mind. *Trends in Ecology & Evolution* 5: 156–159.

Bekoff, M., and D. Jamieson. 1996. Ethics and the study of carnivores: Doing science while respect-ing animals. In *Carnivore behaviour, ecology, and evolution*, vol. 2, ed. J.L. Gittleman, 15–45. Ithaca, NY: Cornell University Press.

Bekoff, M., and J. Pierce. 2009. *Wild justice: The moral lives of animals*. Chicago: University of Chicago Press.

Ben-Ami, D., et al. 2014. The welfare ethics of the commercial killing of free-ranging kangaroos: An evaluation of the benefits and costs of the industry. *Animal Welfare* 23: 1–10.

Bennitt, H.R., J.S. Dixon, V.H. Cahalane, W.W. Chase, and W.L. McAtee. 1937. Statement of policy. *Journal of Wildlife Management* 1: 1–2.

Bergstrom, B.J., L.C. Arias, A.D. Davidson, A.W. Ferguson, L.A. Randa, and S.R. Sheffield. 2014. License to kill: Reforming federal wildlife control to restore biodiversity and ecosystem function. *Conservation Letters* 7: 131–142.

Butler, W.F., and T.G. Acott. 2007. An inquiry concerning the acceptance of intrinsic value theories of nature. *Environmental Values* 16: 149–168.

Callicott, J.B. 1990. Whither conservation ethics? *Conservation Biology* 4: 15–20.

Davradou, M., and G. Namkoong. 2001. Science, ethical arguments, and management in the preservation of land for grizzly bear conservation. *Conservation Biology* 15: 570–577.

Doak, D.F., V.J. Bakker, B.E. Goldstein, and B. Hale. 2013. What is the future of conservation? *Trends in Ecology & Evolution* 29: 77–81.

Dubois, S., and D. Fraser. 2013. Rating harms to wildlife: A survey showing convergence between conservation and animal welfare views. *Animal Welfare* 22: 49–55.

Dunlop, B.N. 2006. Conservation ethics. *Society* 43: 13–18.

Ehrlich, P.R. 2002. Human natures, nature conservation, and environmental ethics. *BioScience* 52: 31–43.

Ehrlich, P.R. 2009. Ecoethics: Now central to all ethics. *Journal of Bioethical Inquiry* 6: 417–436.

Fox, C.H., and M. Bekoff. 2011. Integrating values and ethics into wildlife policy and management: Lessons from North America. *Animals* 1: 126–143.

Fraser, D. (1993). 'Assessing animal well-being: Common sense, uncommon science,' in Food Animal Well-Being Conference Proceedings and Deliberations, 13–15 April, Indianapolis (pp. 37–54). Purdue University Office of Agricultural Research Programs.
Fraser, D. 2010. Toward a synthesis of conservation and animal welfare science. *Animal Welfare* 19: 121–124.
Fraser, D. 2012. A "practical" ethic for animals. *Journal of Agricultural and Environmental Ethics* 25: 721–746.
Fraser, D., and A.M. MacRae. 2011. Four types of activities that affect animals: Implications for animal welfare science and animal ethics philosophy. *Animal Welfare* 20: 581–590.
Gaston, K.J., and R.A. Fuller. 2007. Biodiversity and extinction: Losing the com-mon and the widespread. *Progress in Physical Geography* 31: 213–225.
Gelcich, S., M.J. Kaiser, J.C. Castilla, and G. Edwards-Jones. 2008. Engagement in co-management of marine benthic resources influences environmental perceptions of artisanal fishers. *Environmental Conservation* 35: 36–45.
Goetz, J.L., D. Keltner, and E. Simon-Thomas. 2010. Compassion: An evolutionary analysis and empirical review. *Psychological Bulletin* 136: 351–374.
Harrop, S.R. 1999. Conservation regulation: A backward step for biodiversity? *Biodiversity and Conservation* 8: 679–707.
Hinchliffe, S., M.B. Kearnes, M. Degen, and S. Whatmore. 2005. Urban wild things: A cosmopolitical experiment. *Environment and Planning D* 23: 643–658.
Jepson, P., and S. Canney. 2003. Values-led conservation. *Global Ecology and Biogeography* 12: 271–274.
Jewell, Z. 2013. Effect of monitoring technique on the quality of conservation science. *Conservation Biology* 27: 501–508.
Kareiva, P., and M. Marvier. 2012. What is conservation science? *BioScience* 62: 962–969.
Larson, B. 2007. An alien approach to invasive species: Objectivity and society in invasion biology. *Biological Invasions* 9: 947–956.
Lejano, R.P., and F. Fernandez de Castro. 2013. Norm, network, and commons: The invisible hand of community. *Environmental Science & Policy* 36: 73–85.
Leopold, A. 1949. *A sand county almanac: And sketches here and there*. Oxford/New York: Oxford University Press.
Littin, K.E. 2010. Animal welfare and pest control: Meeting both conservation and animal welfare goals. *Animal Welfare* 19: 171–176.
Lorimer, J. 2007. Nonhuman charisma. *Environment and Planning D* 25: 911–932.
Martin, J., K. French, and R. Major. 2012. Behavioural adaptation of a bird from transient wetland specialist to an urban resident. *PLoS ONE* 7 (art. e50006).
Minteer, B.A., and J.P. Collins. 2005. Why we need an "ecological ethics". *Frontiers in Ecology and the Environment* 3: 332–337.
Naess, A. 1973. The shallow and the deep, long-range ecology movement: A summary. *Inquiry* 16: 95–100.
Nagy, K., and P.D.I., Johnson, (Eds.). 2013. *Trash animals: How we live with nature's filthy, feral, invasive, and unwanted species*. University of Minnesota Press.
O'Sullivan, S. 2011. *Animals, equality and democracy*. Macmillan: Palgrave.
Paquet, P.C., and C.T. Darimont. 2010. Wildlife conservation and animal welfare: Two sides of the same coin? *Animal Welfare* 19: 177–190.
Purcell, B. 2010. *Dingo*. Collingwood, Australia: CSIRO Publishing.
Ramp, D. 2013. Bringing compassion to the ethical dilemma in killing kangaroos for conservation. *Journal of Bioethical Inquiry* 10: 267–272.
Redpath, S.M., et al. 2013. Understanding and managing conservation con-flicts. *Trends in Ecology & Evolution* 28: 100–109.
Regan, T. (1983). *The case for animal rights*. University of California Press.
Routley, R. 1973. *'Is there a need for a new, an environmental, ethic?' 15th World Congress of Philosophy*, 205–210. Sofia, Bulgaria: Sofia Press.
Soulé, M.E. 1985. What is conservation biology? *BioScience* 35: 727–734.

Tisdell, C., and H.S. Nantha. 2007. Comparison of funding and demand for the conservation of the charismatic koala with those for the critically endangered wombat *Lasiorhinus krefftii*. *Biodiversity and Conservation* 16: 1261–1281.

Treves, A., and K.U. Karanth. 2003. Human–carnivore conflict and perspectives on carnivore management worldwide. *Conservation Biology* 17: 1494–1499.

Vucetich, J.A., and M.P. Nelson. 2013. The infirm ethical foundations of conservation. In *Ignoring nature no more: The case for compassionate conservation*, ed. M. Bekoff, 9–25. University of Chicago Press.

Vucetich, J.A., and M.P. Nelson. 2007. What are 60 warblers worth? Killing in the name of conservation. *Oikos* 116: 1267–1278.

Webb, T.J., and D. Raffaelli. 2008. Conversations in conservation: Revealing and dealing with language differences in environmental conflicts. *Journal of Applied Ecology* 45: 1198–1204.

Yeo, J.H., and H. Neo. 2010. Monkey business: Human–animal conflicts in urban Singapore. *Social and Cultural Geography* 11: 681–699.

Comment: We All Live in a Planetary Ark (Planetary Ark, Planetary Ark…)

Hub Zwart

Abstract The Biblical story of the Art (a floating, zoo-like device, constructed to survive climate turmoil and mass extinction) can be regarded as an archetypal image (in the terminology of Gaston Bachelard), capturing structural components of the human-animal relationship. Building on the contributions by Larson and Barr, Keulartz, Bovenkerk and Verweij, and Ramp and Bekoff, I will argue that, in the course of history, the Ark has evolved from a fictional (imaginary) icon into something increasingly real. The agricultural village of the Neolithic era already functioned as a sheltered enclave, a survival machine designed to allow a select number of humans and accompanying species to withstand environmental fluctuations and survival pressure. In the current situation, however, the Ark has developed into a *Gestell* of planetary dimensions. The concept of the anthropocene basically conveys the idea that we have entered a global symbolical Ark, conceptualised by Teilhard de Chardin as the noosphere (the world-wide web of intelligence and policies, technologies and engineering, research and regulations) and emerging against the backdrop of a necro(s)cene: an ambiance of mass extinction. Increasingly, prospects for survival of a disconcertingly large number of species depend on human behaviour and human decision: on our ability or failure to collectively address the daunting challenges of the present.

1 Introduction

The image of the Ark still functions as an archetypal marker in contemporary discourse on human-animal relationships. Perhaps we may even regard it as the archetype of animal conservation research. The Ark is mobilised (that is: endorsed as well as denounced) both in the academic literature (for instance: Norton et al.

H. Zwart (✉)
Radboud University Nijmegen, Nijmegen, The Netherlands
e-mail: h.zwart@science.ru.nl

B. Bovenkerk and J. Keulartz (eds.), *Animal Ethics in the Age of Humans*,
The International Library of Environmental, Agricultural and Food Ethics 23,
DOI 10.1007/978-3-319-44206-8_24

2015) and in the public arena (for instance: Kolbert 2013),[1] but as a basic motif, the Ark is also clearly discernible in the preceding contributions to this section. My aim in this paper is to assess the strengths and weaknesses of the Ark image by commenting on these contributions (by Keulartz, Larson and Barr, Bovenkerk and Verweij, and Ramp and Bekoff), showing how this archetype may help us to understand and assess the present, although it may also become a deceptive obstacle if taken too literally, too stereotypically.

My contribution builds on the work of Gaston Bachelard (1884–1962), a French philosopher of science who developed a 'psychoanalysis of scientific knowledge' to demonstrate how archetypal images may help us to structure the overload of bewildering information we are exposed to, while at the same time pointed out how easily they may become 'epistemological obstacles', preventing us from developing evidence-based and responsive views on the present (Bachelard 1938/1947; cf. Zwart 2008, 39 ff.). But I will also build on the work of Pierre Teilhard de Chardin (1881–1955), whom I consider a philosopher of the Anthropocene *par excellence*, although this is a controversial issue in itself (cf. Hamilton and Grinevald 2015). Insofar as Teilhard may indeed be regarded as such, he was a precursor *avant la lettre*, because he did not actually use the term Anthropocene himself.

I will point out to what extent the concept of the Ark, while serving as a chronic discursive constant in describing human-animal relationships, allows us to point out the discontinuity, the unsettling *newness* of the anthropocenic present compared to Ark-like situations of bygone eras. Notably, in stark contrast to the traditional archetypal image (which evidently evokes the idea of an insulated, floating vessel), the anthropocenic Ark must not only assume planetary propositions, but must also remain emphatically sensitive to what is happening in the outside world;—although, in the topology of the Anthropocene, the very term 'outside' has become more or less outdated, as I will argue.

2 The Archetypal Ark

The biblical Story of the Ark (*Genesis* 6–10) is without doubt one of the most telling and provocative dramatizations of the human-animal relationship. Perhaps we may see it as the founding myth of the Anthropocene itself, projected backwards in time, with animals literally entering *the age of humans*. The story reflects a pastoral framing of human-animal dynamics, which should not come as a surprise, because the story was invented by a culture of livestock farmers. God is casted as an Über-livestock farmer who carefully selects his favoured samples of living beings to sacrifice the rest. The story records what we nowadays would call a bottleneck survival event. During an episode of mass extinction (due to a sudden and cataclysmic period of climate change, brought about by disruptive and unsustainable

[1] http://animals.nationalgeographic.com/animals/photo-ark/.

human misbehaviour), the Ark functioned as a floating, human-made zoo, a conservation device for animals in captivity (ex situ), under the leadership of someone who, unlike his contemporaries, was not only able to foresee the cataclysmic events, but also willing to heed the symptoms and to act accordingly.

From a God's eye view, moreover, even human beings are regarded as cattle in this story. The Book *Genesis* extrapolates the pastoral worldview even to humankind. We see God actively and consciously tending the early human populations He created, as if *Genesis* is actually the record of the process of human domestication, with its eventful ups and downs. The building of the Ark is preceded by a period of exponential population growth: by a massive extension of the human herd. Humankind had begun to increase in numbers, we are told (*Genesis* 6:1), but instead of living up to Divine expectations, most humans had become morally corrupted. The great flood represents a mythological version of a great mass extinction. It rained for 40 days and nights, we are told (*Genesis* 7:4). But it was also an instance of Divine eugenics: only the righteous survived, allowing them to initiate a second wave of population growth. This dynamics of exponential growth, sacrifice and selection, followed by subsequent waves of population growth, clearly reflects the livestock farmers' view on living conditions under domestication. Indeed, the story stages the coming-into-being of the domestication process as such.

The Ark, one could argue, really existed, namely in the form of the Neolithic village, a concept which was developed about ten millennia ago in the very region where the Ark must once have floated: in the Middle-East. During the Neolithic revolution, a new topology, a spatial reorganisation of inside—outside was introduced. Whereas wildlife (animals existing outside the pastoral domains or fences) continued to be exposed to survival challenges, so that many of them eventually became extinct, the agricultural or pastoral villages functioned as Arks or ecological islands, as collective immunisation devices to safeguard their domesticated inhabitants against extinction. The Ark-like village was a small insulated, humanised enclave, surrounded by a sea of wilderness, inhabited by a tiny band of human beings and their accompanying species: domesticated plants and animals which had become fully dependent on humans for their survival; and vice versa (Zwart and Penders 2011).[2]

Thus, the great Deluge, as recorded in *Genesis*, constitutes an act of conscious selection: a human clan with desirable features is singled out, while the rest of humankind, unwilling or unable to convert to the new way of pastoral—agricultural living, finds itself decimated. Those remaining, that is: those found worthy of preservation in the struggle for survival, are allowed to re-colonize the world. Their characteristics will be transferred to future generations. Indeed, the carriers of these favoured dispositions, desirable in the eyes of God, are explicitly encouraged to

[2]We find this reflected in Aesop's fable of the wild and the tame donkey (Perry 183). Whereas the wild donkey initially envies his domesticated cousin, seeing him well-fed and tended, he changes his mind as soon as he sees him being put to work as a pack animal. Domestication comes with a price.

reproduce as exuberantly as possible, and to become as abundant as the stars in heaven (during a night spent in the desert).

In short, the story of the Ark reflects a pastoral view on anthropogenesis, casting God as a super-shepherd in charge of human beings on their way to domestication through a series of selection bottlenecks, transforming them from inhabitants of a forest-like ecosystem called Paradise into producers of an agricultural environment of their own making: a process requiring hard labour, but also a variety of self-made artisanal tools and skills. This view is not exclusively typical for *Genesis*, moreover. A somewhat similar vision can be encountered in Plato's dialogue *The Statesman* (Πολιτικός), where the ideal aristocratic ruler is presented as someone who actively tends the human herd, not only by managing, but also by systematically ameliorating the quality of the human population entrusted to his care, as if humans themselves are to be regarded as cattle; and this includes techniques for selective breeding, all for the benefit of the utopian state (Plato 1925/1995). But one could also think of survival and amelioration as a process governed by the invisible hand of evolution, a view which resonates with the subtitle of Darwin's famous book, which not only addresses *the origin of species*, as is suggested by the title, but also *the preservation of favoured races*, as indicated by the subtitle.

Gradually, the agricultural way of living (bringing together humans and animals in agricultural Arks, to safeguard their preservation), became the Common Human Pattern (Romein and Romein-Verschoor 1954; cf. Zwart 2009), marginalising other ways of struggle for survival. Thus, a large part of human history can be understood as the evolution of artificial, Ark-like ecosystems involving not only human beings themselves, but also the animals and plants they selected, that is: allowed to enter the Ark (Zwart and Penders 2011). The Neolithic Ark-like village was a clearing brought about by human domestication technologies, opening up a particular way of thinking and living. Indeed, in the course of history, human beings 'invited' a broad variety of species into their human-made environment: domesticated, semi-domesticated as well as undomesticated.

But the ancient story of the Ark acquires new relevance against the backdrop of the current era, which is often referred to (also in this volume) as the 'Anthropocene' (Monastersky 2015). We are facing a global environmental crisis, a mass extinction event (Kobler 2014), so that the Ark is on the move again, but this time as a symbolic type of vessel, a structure which has assumed planetary proportions. Yet, as a guiding idea, the archetypal image of the biblical Ark, if taken too literally, may also become deceptive. The Ark of the anthropocenic present clearly has to exceed the size of the agricultural or pastoral villages of the past, has to evolve into a techno-scientific and managerial-regulatory network of global magnitude. Animal species worldwide are now facing an uncanny alternative: they are either allowed to enter the human Ark, the global multiple species herd—for instance by being listed among the 7.368 or so species of vertebrates that are officially threatened with extinction, according to the 2013 IUCN Red List—or they will vanish.

3 Animal Ethics and the Anthropocene

This diagnostics of the present, this endeavour to bring about a wholly new type of Ark, allowing us to address the challenges entailed in living under anthropocenic conditions, is acutely reflected in the contributions to this section of this volume.

The paper by Larson and Barr (this volume) for instance, focuses on the distinction between conservation in situ (in the wild) and ex situ (in captivity). The latter alternative can be regarded as the Ark-like route. Yet, as Larson and Barr argue, this distinction, although it may seem conceptually quite clear, proves increasingly difficult to uphold, especially now that we have entered the Anthropocene, the age of humans, where human impacts are omnipresent. Conservation biologists tend to prefer in situ conservation, Larson and Barr argue, because it maintains the species in a 'natural' state rather than one 'contaminated' by human culture. But as human impacts have grown globally, it increasingly becomes necessary also to manage species under in situ circumstances. Often, endangered species have to be removed from their natural habitats (in situ) so as to be kept in captivity (ex situ), often with the idea of reintroducing them (or their descendants) into the wild at a later date. But increasingly, the intensive management of populations of animals in reserves (in situ) resembles the tending of animals in captivity (ex situ), while captive populations increasingly exist in 'naturalized' conditions (e.g., in zoos or botanical gardens) that are purportedly designed to be as natural as possible. In other words, the 'fingerprints' of human agency are omnipresent, while the topology of the archetypal Ark, distinctively separating inside (ex situ) from outside (in situ), has given way to a completely new situation, where the Ark no longer functions as a protective vessel or floating bubble, but increasingly as a virtual global structure which absorbs and encompasses the biosphere as such. We all live in a planetary Ark as it were, so that the in situ/ex situ dichotomy is inevitably eroding.

Keulartz (this volume) positions this debate against the backdrop of a broader philosophical diagnostics of the present. Building on Jamieson (1995) and others, Keulartz argues that, when it comes to developing an effective survival strategy for endangered animals under anthropocenic conditions, the archetypal image of the Ark may easily deceive us (Keulartz, *Captivity*, this volume; cf. Keulartz 2015). The image of the Ark easily becomes dysfunctional and counterproductive if taken too literally, that is: if we continue to see the Ark as an actual material zoo, as a human-made enclave of limited size: a closed animal ward as it were, where lack of occupation and activity easily gives rise to animal boredom and suffering. As a conservation device, Keulartz argues, a zoo can indeed be regarded as "a kind of Noah's Ark" (2015, p. 337). But the Ark paradigm has "run into rough waters" or even "shipwrecked" and "no longer meets the standard". According to Keulartz, the image of the zoo as an Ark, that is: as a device for ex situ conservation, has lost its credibility and must give way to a new paradigm, which he refers to as the "integrated approach". Still, as I will argue more extensively below, while the metaphor of the Ark may indeed have run aground on the local scale, it remains a relevant concept if we try to transpose it towards the planetary scale. While the idea of the Ark as a closed ward for a limited number of endangered species may be outdated, the discarded image of the Ark emphatically resurges on a global scale, as a concept which captures the planetary situation. Increasingly, all animals (including humans) sooner or later will enter the planetary Ark.

In his contribution to an earlier section of this volume (Keulartz, *Anthropocene*, this volume), Keulartz further elaborates this assessment by arguing that the crisis of the zoo is symptomatic for the crisis affecting human-animal relationships during the Anthropocene as such. As wild animals are currently under increasing pressure from human activity, he argues, we inevitably have become responsible for their habitats. Moreover, living during the planet's "sixth mass extinction", care for the habitat of wild animals cannot rely merely on preservation and protection. The rate and magnitude of "defaunation" now asks for more offensive and interventionist strategies, such as recreation, restoration and rewilding. The unfolding crisis makes it unavoidable to replace the hands-off approach which (until recently) guided mainstream species conservation practices by a more proactive and interventionist strategy. In other words, in order to save the animal world, we have to drastically reorganize it and humanize it. We have to assume full responsibility for it. Indeed, building on the Ark-concept one could argue that we have to *Ark-ize* the global environment as such. In the face of mass extinction and massive ecosystem degradation, our responsibilities now inevitably assume planetary dimensions and proportions, so that the anthropocenic version of the Ark is becoming all-encompassing and omnipresent.

Similar debates can be encountered in the contribution by Bovenkerk and Verweij, who argue that the current situation requires a shift of focus from the individual to the collective dimensions of animal life. But one could even go a step further and claim that, under anthropocenic conditions, the focus must shift towards the supra-collective perspective of the planetary Ark so as to address issues of animal survival *as such*. Moreover, in an increasingly human-dominated world, even the kind of compassion promoted by Ramp and Bekoff (2015) may no longer be sufficient, in the sense that we should rather go for an upscaling of compassion, for something like *pan-passion*: a form of responsibility which aims to take the prospects for animal survival *as such* into account. In other words, the archetypal image of the Ark must be replaced by a symbolic, anthropocenic version, exemplifying a global strategy for coming to terms with the mass extinction threat by building an Ark which encompasses the whole animal planet, more or less. But before explaining this concept in more detail, let me first present a concise summary of the ideas of Pierre Teilhard de Chardin, one of the key preparatory thinkers (quite influential and controversial in the 1950s, 1960s and 1970s, but more or less forgotten nowadays) of the Anthropocene.

4 The Ark, the Anthropocene and the Noosphere

Teilhard's starting point is the claim that evolution has an orientation, an axis, a line of progress, a direction, namely towards increased complexity, increased self-consciousness, increased self-directedness (Teilhard de Chardin 1955). Living entities are increasingly able to consciously co-determine the conditions of their own evolution, and this notably applies to the most recently evolved species, namely humans.[3] Teilhard sees humans beings as "evolution becoming conscious of itself" (idem, 181). Evolution is basically a process of "sublimation" (idem, 106), transposing physiology into culture, culminating in cerebralisation and, ultimately, in self-conscious self-directedness. Teilhard is well aware of the fact that in mainstream biological discourse such claims are encountered with what psychoanalysis refers to as disavowal, but for Teilhard, "sublimation" constitutes an undeniable evolutionary dynamics.

Moreover, according to Teilhard, there is something disconcerting about humans. Scientific portrayals fall short of reality. As seen by science (anatomy, physiology, genetics, molecular biology, etc.), humans are animals just like others animals, but these portraits lack an essential factor, an entire dimension. They fail to reveal how, in humans, evolution becomes an active, self-directed process, so that another world is born, the world of techno-culture. Indeed, due to global human activity, a new layer has emerged, over and above the biosphere, which Teilhard refers to as the *noosphere*, which literally means the "thinking layer", although besides noetic activities it also involves noetic products (technologies, devices, culture, infrastructures, and so on). In other words, the noosphere is evolving into a planetary network of advanced technologies and global communicative circuits. Humans are obviously animals, and yet we represent a discontinuity, a leap, a crisis, a metamorphosis, an awakening, giving rise to the emergence of the noosphere, the thinking layer, relentlessly transforming and absorbing the biosphere. The noosphere represents a conscious reshaping of the world, an epochal transformation affecting the entire planet.

We humans are not in charge, however, and Teilhard emphatically stresses that he does not endorse an anthropocentric view. Rather, we humans are pushed along by this development ourselves, we are subjected to a relentless process of hominization and collectivization, culminating in the emergence of a global 'We': a planetary network of thought and interaction. Indeed, Teilhard has been credited, by Garreau (2005) and others, with predicting the internet: WWW as a global noetic structure, a planetary *We*. For Teilhard, humans are a kind of bridge. We notice the presence of something more powerful than ourselves, he argues, pushing us forward, trawling us along: the inevitable planetisation of the noosphere. A turn of profound importance is taking place throughout the world, and we are only beginning to realise its true dimensions. We have in fact already entered a different

[3]Pierre Teilhard de Chardin was a Jesuit priest as well as a paleo-anthropologist who contributed to the discovery of *Homo erectus* fossils in China in the 1920s.

world. "The future will decide what the best name is to describe the era we are entering", Teilhard tells us (1955, p. 214), but he clearly seems to be pointing at what we nowadays refer to as the Anthropocene. As a consequence, a sense of disquiet comes over us. We seem unable to live up to the daunting challenges and responsibilities emerging directly in front of us. The present situation is without precedent in the history of life and therefore, more than ever, we experience a fundamental existential anguish. An enormous responsibility is looming up in front of us and we seem to fall short in a rather dramatic way. We are waking up to the fact that the planet itself is now becoming thoroughly humanised and technified, but it is doubtful whether we will be able to manage and contain this process. It will require a process of intellectual and technological collectivisation and convergence, resulting in the emergence of a global *We*, empowered to initiate collective action. Somehow, our uneasiness must be transformed into active thinking, a combination of foresight and action. A terrible game is being played, and we are both the players and the cards, but only through global collaboration and scientific convergence can we hope to play a constructive rather than a disruptive role. Passive, natural, Darwinian evolution is being eclipsed by conscious transformation, an active metamorphosis of the planet. The artificial noosphere is carrying on the work of natural selection, and a computer-based, literate culture, an electronic self-consciousness, is increasingly superimposed on genetic heredity (either under domenstication or in the wild).

This calls for a drastic reorganisation of scientific research itself as well, which has to be transformed into a kind of planetary organisation, allowing for global teamwork. And this even involves philosophy, which should no longer be regarded as a solitary calling, but rather as a collective endeavour, as kind of "distributed reflection", as Bachelard once aptly phrased it (1940/1949, p. 2), involving multiple voices working in various places on various case studies, thereby contributing to a diagnostic of the present and prognostics of the future as a joint philosophical practice. Eventually, Teilhard remains an 'optimist', if this is still a viable term under the present circumstances. According to Teilhard, despite our failures and mistakes, we are heading towards a moment of convergence, of science, politics and art, which he refers to as the Omega Point: the end of history as we know it.

What for Teilhard still seems relatively diffuse, has now become painfully discrete, I would argue. We are witnessing a high resolution version of the global tableau drawn out somewhat sketchily by Teilhard, while excavating in the Chinese desert, several decades ago. All animals have entered the human age. And we ourselves have entered the human age as well. We have boarded the planetary Ark. The noosphere is transforming and absorbing the biosphere. Thus we are facing challenges which seem too enormous to live up to, due to lack of vision, sensitivity and coordination. In fact, we seem to have boarded a planetary ship of fools (i.e. the Ark's archetypal photographic negative or reverse image, heading for destruction) rather than an Ark.

5 The Planetary Ark: An Outline

From a Teilhardian perspective, the archetypal Ark image can be helpful to some extent for highlighting specific features of the present, but it can also be misleading in the sense that some of its properties may prove outspokenly counterproductive. In other words, the planetary Ark must build on, but at the same time break away from the archetypal image. The most problematic feature of the archetypal Ark is that it is casted as a closed ward, a dark secluded room, screened off from reality, insensitive to what is happening outside, with no sense of direction, simply adrift. Psychoanalytically speaking, it is an instance of the mother-archetype, a replica of the motherly womb, sheltering us and saving us, protecting us from traumatic experiences, as a guardian against the frightening real, but also holding us captive and keeping us inactive.

In fact, the Ark is the Biblical counterpart of Plato's famous cave, another exemplification of the mother-archetype, involving a group of humans whose legs and necks are fettered from childhood, so that they can only stare at the wall in front of them. A fire is burning higher up, at a distance behind them, and between the fire and the prisoners a low wall has been built, and behind that wall shapes of animals are carried about, as in puppet-shows, whose shadows are cast onto the wall (Plato 1935/2000, pp. 514–515). In short, Plato's story gives us an idea of what the archetypal Ark may have looked like from the inside: a protective womb-like setting, inhabited by foetus-like passengers, chained to the uterus by umbilical cords (fetters), perfectly happy in an environment which, to outsiders, may seem claustrophobic. A facility for human husbandry, as it were, a human hatchery, where humans are kept as cattle. Interestingly, it is a kind of cinema as well, and the prisoners or passengers are watching movies featuring animals: Disney-like movies perhaps, similar to *Finding Nemo*, about imaginary, anthropomorphic animals: phantasmagorias whose main function seems to be an ideological one, namely to obfuscate what is really happening to animals outdoors.

But to see what reality really is like, these prisoners have to be freed from their chains by force and dragged away toward the light, the open air, so that the world of archetypal images projected on the wall (δόξα) is replaced by true knowledge (ἐπιστήμη). But to achieve this, the prisoners have to be *educated* (literally: guided upward): a process of enlightenment and liberation, but also a traumatic experience, similar to the trauma of birth, a painful intellectual awakening.

Humankind currently still has to go through this experience, so it seems. The extinction catastrophe, the massive loss of biodiversity is still met with disavowal and sheer indifference. The emerging crisis is a traumatic experience no doubt, and to really expose ourselves to it, we must leave the archetypal Ark behind and wholeheartedly enter the planetary, anthropocenic version. Science is basically "iconoclastic", Bachelard argues (1938/1947, p. 77), and the archetypal image of the Ark must be demolished and replaced by a completely different type of vessel: open to the world, informed by research; a kind of global observatory, in short. The anthropocenic Ark is not constructed out of wood, but rather consists of symbolical, noetic materials: scientific findings, responsive deliberations, conservation policies. In order to come to terms with the present, the concept of the Ark has to be transferred from the imaginary realm (the archetypal *Gestalt* of the Ark as a secluded ward) into the *symbolical* realm (the planetary, anthropocenic Ark, informed by research-based diagnostics and quantified prognostics). This Ark is no longer steered by one individual, moreover, a Noah-like guide or super-sailor. Rather, the planetary Ark is staffed by a kind of global, collective consciousness, combining multidisciplinary data with a readiness to act.

This also has implications for the role and place of philosophers, who have to leave their traditional Arks, their secluded wards, their introvert libraries, in order to participate in what Bachelard referred to as distributed reflection. Thus, a philosophical diagnostics of the present becomes a joint endeavour to which multiple voices may contribute, via a plethora of philosophical case studies, such as the ones brought together in this section, concerning in situ and ex situ conservation and zoos as conservation devices. Thus philosophers may contribute to a growing convergence of thought and action. A planetary Ark will allow humans to assume responsibility for the present situation, not because their track record of former achievements is so promising, but rather because, in the absence of divine intervention, the global *We* is the only agent available to play this role. Should we fail to achieve this, the planetary fauna will continue to float in the direction of the coming deluge, and the global Ark will definitely revert into a wandering ship of fools.

References

Bachelard, G. 1940/1949. *La Philosophie du non. Essai d'une philosophie du nouvel esprit scientifique*. Paris: Presses Universitaires de France.

Bachelard, G. 1938/1947. *La formation de l'esprit scientifique. Contribution à une psychanalyse de la connaissance objective*. Paris: Vrin.

Bovenkerk, B., and M. Verweij. this volume. Collective dimensions in animal ethics. In *Animal ethics in the age of humans*. Springer.

Garreau, J. 2005. *Radical evolution: The promise and peril of enhancing our minds, our bodies—and what it means to be human*. New York: Broadway Books.

Hamilton, C., and J. Grinevald. 2015. Was the Anthropocene anticipated. *The Anthropocene Review* 2(1): 59–72.

Jamieson, D. 1995. Zoos revisited. In *Ethics on the Ark*, ed. B. Norton, et al., 52–66. Washington: Smithsonian Institution Press.

Keulartz, J. 2015. Captivity for conservation? Zoos at a crossroads. *Journal of Agricultural and Environmental Ethics* 28: 335–351.

Keulartz, J. this volume. Towards an animal ethics for the anthropocene. In *Animal ethics in the age of humans*. Springer.

Kolbert, E. 2013. Building the Ark. *National Geographic*. October 2013.

Kolbert, E. 2014. *The sixth extinction: An unnatural history*. New York: Holt.

Larson, B., and S. Barr. this volume. The flights of the Monarch butterfly: Between in situ and ex situ conservation. In *Animal ethics in the age of humans*. Springer.

Monastersky, R. 2015. The human age. *Nature* 519: 144–147.

Norton, B., M. Hutchins, E. Stevens, and T. Maple, (eds.) 2015. *Ethics on the Ark*. Washington: Smithsonian Institution Press.
Plato. 1925/1995. *The statesman, Philebus, Ion. Plato VIII* [Loeb]. Cambridge: Harvard University Press.
Plato. 1935/2000. *Republic VI-X. Plato VI* [Loeb]. Cambridge: Harvard University Press.
Ramp, D., and M. Bekoff. this volume. Compassion as a practical and evolved ethic for conservation. In *Animal ethics in the age of humans*. Springer.
Romein, J., and A. Romein-Verschoor. 1954. *Aera van Europa*. Leiden: Brill.
Teilhard de Chardin, P. 1955. *The phenomenon of man*. New York: Harper.
Zwart, H. 2008. *Understanding nature: Case studies in comparative epistemology*. Dordrecht: Springer.
Zwart, H. 2009. Biotechnology and naturalness in the genomics era: Plotting a timetable for the biotechnology debate. *Journal of Agricultural and Environmental Ethics* 22: 505–529.
Zwart, H., and B. Penders. 2011. Genomics and the Ark. An ecocentric perspective on human history. *Perspectives in Biology and Medicine* 54(2): 217–231.

Index

B. Bovenkerk and J. Keulartz (eds.), *Animal Ethics in the Age of Humans*,
The International Library of Environmental, Agricultural and Food Ethics 23,
DOI 10.1007/978-3-319-44206-8

FSC
www.fsc.org
MIX
Papier | Fördert
gute Waldnutzung
FSC® C083411

Kolibri 360 GmbH
Ferdinand-Jühlke-Straße 7
99095 Erfurt, Deutschland
produktsicherheit@kolibri360.de